Essential Statistics, Regression, and Econometrics

Essential Statistics, Regression, and Econometrics

Gary Smith
Pomona College

AMSTERDAM • BOSTON • HEIDELBERG • LONDON
NEW YORK • OXFORD • PARIS • SAN DIEGO
SAN FRANCISCO • SINGAPORE • SYDNEY • TOKYO
Academic Press is an imprint of Elsevier

Academic Press is an imprint of Elsevier
225 Wyman Street, Waltham, MA 02451, USA
525 B Street, Suite 1800, San Diego, California 92101-4495, USA
84 Theobald's Road, London WC1X 8RR, UK

Library of Congress Cataloging-in-Publication Data
Smith, Gary, 1945-
Essential statistics, regression, and econometrics / Gary Smith.
p. cm.
Includes bibliographical references and index.
ISBN 978-0-12-382221-5 (hardcover: alk. paper) 1. Regression analysis–Textbooks.
I. Title.
QA278.2.S6127 2012
519.5–dc22 2011006233

British Library Cataloguing-in-Publication Data
A catalogue record for this book is available from the British Library.

ISBN: 978-0-12-382221-5

For information on all Academic Press publications
visit our website: *www.elsevierdirect.com*

Printed in the United States of America
11 12 13 9 8 7 6 5 4 3 2 1

Contents

Introduction

Econometrics is powerful, elegant, and widely used. Many departments of economics, politics, psychology, and sociology require students to take a course in regression analysis or econometrics. So do many business, law, and medical schools. These courses are traditionally preceded by an introductory statistics course that adheres to the fire hose pedagogy: bombard the students with information and hope they do not drown. Encyclopedic statistics courses are a mile wide and an inch deep, and many students remember little after the final exam. This textbook focuses on what students really need to know and remember.

Essential Statistics, Regression, and Econometrics is written for an introductory statistics course that helps students develop the statistical reasoning they need for regression analysis. It can be used either for a statistics class that precedes a regression class or for a one-term course that encompasses statistics and regression analysis.

One reason for this book's focused approach is that there is not enough time in a one-term course to cover the material in more encyclopedic books. Another reason is that an unfocused course overwhelms students with so much nonessential material that they have trouble remembering the essentials.

This book does not cover the binomial distribution and related tests of a population success probability. Also omitted are difference-in-means tests, chi-square tests, and ANOVA tests. These are not crucial for understanding and using regression analysis. Instructors who cover these topics can use the supplementary material at the book's website.

The regression chapters at the end of the book set up the transition to a more advanced regression or econometrics course and are also sufficient for students who take only one statistics class but need to know how to use and understand basic regression analysis.

This textbook is intended to give students a deep understanding of the statistical reasoning they need for regression analysis. It is innovative in its focus on this preparation and in the extended emphasis on statistical reasoning, real data, pitfalls in data analysis, modeling issues, and word problems. Too many students mistakenly believe that statistics courses are too abstract, mathematical, and tedious to be useful or interesting. To demonstrate the power, elegance, and even beauty of statistical reasoning, this book includes a large number of interesting and relevant examples, and discusses not only the uses but also the abuses of statistics. These examples show how statistical reasoning can be used to answer important questions and also expose the errors—accidental or intentional—that people often make. The examples are drawn from many areas to show that statistical reasoning is an important part of everyday life.

The goal is to help students develop the statistical reasoning they need for later courses and for life after college.

I am indebted to the reviewers who helped make this a better book: Woody Studenmund, The Laurence de Rycke Distinguished Professor of Economics, Occidental College; Michael Murray, Bates College; Steffen Habermalz, Northwestern University; and Manfred Keil, Claremont Mckenna College.

Most of all, I am indebted to the thousands of students who have taken statistics courses from me—for their endless goodwill, contagious enthusiasm, and, especially, for teaching me how to be a better teacher.

CHAPTER 1

Data, Data, Data

Chapter Outline

You're right, we did it. We're very sorry. But thanks to you, we won't do it again.
—Ben Bernanke

The Great Depression was a global economic crisis that lasted from 1929 to 1939. Millions of people lost their jobs, their homes, and their life savings. Yet, government officials knew too little about the extent of the suffering, because they had no data measuring output or unemployment.

They instead had anecdotes: "It is a recession when our neighbor loses his job; it is a depression when you lose yours." Herbert Hoover was president of the United States when the Great Depression began. He was very smart and well-intentioned, but he did not know that he was presiding over an economic meltdown because his information came from his equally clueless advisors—none of whom had yet lost their jobs. He had virtually no economic data and no models that predicted the future direction of the economy.

Essential Statistics, Regression, and Econometrics

In his December 3, 1929, State of the Union message, Hoover concluded that "The problems with which we are confronted are the problems of growth and progress" [1]. In March 1930, he predicted that business would be normal by May [2]. In early May, Hoover declared that "we have now passed the worst" [3]. In June, he told a group that had come to Washington to urge action, "Gentlemen, you have come 60 days too late. The depression is over" [4].

A private organization, the National Bureau of Economic Research (NBER), began estimating the nation's output in the 1930s. There were no regular monthly unemployment data until 1940. Before then, the only unemployment data were collected in the census, once every ten years. With hindsight, it is now estimated that between 1929 and 1933, national output fell by one third, and the unemployment rate rose from 3 percent to 25 percent. The unemployment rate averaged 19 percent during the 1930s and never fell below 14 percent. More than a third of the nation's banks failed and household wealth dropped by 30 percent.

Behind these aggregate numbers were millions of private tragedies. One hundred thousand businesses failed and 12 million people lost their jobs, income, and self-respect. Many lost their life savings in the stock market crash and the tidal wave of bank failures. Without income or savings, people could not buy food, clothing, or proper medical care. Those who could not pay their rent lost their shelter; those who could not make mortgage payments lost their homes. Farm income fell by two-thirds and many farms were lost to foreclosure. Desperate people moved into shanty settlements (called Hoovervilles), slept under newspapers (Hoover blankets), and scavenged for food where they could. Edmund Wilson [5] reported that

> *There is not a garbage-dump in Chicago which is not haunted by the hungry. Last summer in the hot weather when the smell was sickening and the flies were thick, there were a hundred people a day coming to one of the dumps.*

1.1 Measurements

Today, we have a vast array of statistical data that can help individuals, businesses, and governments make informed decisions. Statistics can help us decide which foods are healthy, which careers are lucrative, and which investments are risky. Businesses use statistics to monitor production, estimate demand, and design marketing strategies. Government statisticians measure corn production, air pollution, unemployment, and inflation.

The problem today is not a scarcity of data, but rather the sensible interpretation and use of data. This is why statistics courses are taught in high schools, colleges, business schools, law schools, medical schools, and Ph.D. programs. Used correctly, statistical reasoning can help us distinguish between informative data and useless noise, and help us make informed decisions.

Flying Blind and Clueless

U.S. government officials had so little understanding of economics during the Great Depression that even when they finally realized the seriousness of the problem, their policies were often counterproductive. In 1930, Congress raised taxes on imported goods to record levels. Other countries retaliated by raising their taxes on goods imported from the United States. Worldwide trade collapsed with U.S. exports and imports falling by more than 50 percent.

In 1931, Treasury Secretary Andrew Mellon advised Hoover to "liquidate labor, liquidate stocks, liquidate the farmers, liquidate real estate" [6]. When Franklin Roosevelt campaigned for president in 1932, he called Hoover's federal budget "the most reckless and extravagant that I have been able to discover in the statistical record of any peacetime government anywhere, anytime" [7]. Roosevelt promised to balance the budget by reducing government spending by 25 percent. One of the most respected financial leaders, Bernard Baruch, advised Roosevelt to "Stop spending money we haven't got. Sacrifice for frugality and revenue. Cut government spending—cut it as rations are cut in a siege. Tax—tax everybody for everything" [8]. Today—because we have models and data—we know that cutting spending and raising taxes are exactly the wrong policies for fighting an economic recession. The Great Depression did not end until World War II caused a massive increase in government spending and millions of people enlisted in the military.

The Federal Reserve (the "Fed") is the government agency in charge of monetary policy in the United States. During the Great Depression, a seemingly clueless Federal Reserve allowed the money supply to fall by a third. In their monumental work, *A Monetary History of the United States*, Milton Friedman and Anna Schwartz argued that the Great Depression was largely due to monetary forces, and they sharply criticized the Fed's perverse policies. In a 2002 speech honoring Milton Friedman's 90th birthday, Ben Bernanke, who became Fed chairman in 2006, concluded his speech: "I would like to say to Milton and Anna: Regarding the Great Depression. You're right, we did it. We're very sorry. But thanks to you, we won't do it again" [9].

During the economic crisis that began in the United States in 2007, the president, Congress, and Federal Reserve did not repeat the errors of the 1930s. Faced with a credit crisis that threatened to pull the economy into a second Great Depression, the government did the right thing by pumping billions of dollars into a deflating economy.

Why do we now know that cutting spending, raising taxes, and reducing the money supply are the wrong policies during economic recessions? Because we now have reasonable economic models that have been tested with data.

1.2 Testing Models

The great British economist John Maynard Keynes observed that the master economist "must understand symbols and speak in words" [10]. We need words to explain our reasoning, but we also need models so that our theories can be tested with data.

In the 1930s, Keynes hypothesized that household spending depends on income. This "consumption function" was the lynchpin of his explanation of business cycles. If people spend less, others will earn less and then spend less, too. This fundamental interrelationship between spending and income explains how recessions can persist and grow like a snowball rolling downhill.

If, on the other hand, people buy more coal from a depressed coal-mining area, the owners and miners will then buy more and better food, the farmers will buy new clothes, and the tailors will start going to the movies again. Not only the coal miners gain; the region's entire economy prospers.

At the time, Keynes had no data to test his theory. It just seemed reasonable that households spend more when their income increases and spend less when their income falls. Eventually, a variety of data were assembled that confirmed his intuition. Table 1.1 shows estimates of U.S. aggregate disposable income (income after taxes) and spending for the years 1929 through 1940. When income fell, so did spending; and when income rose, so did spending.

Table 1.2 shows a very different type of data based on a household survey during the years 1935–1936. As shown, families with more income tended to spend more.

Today, economists agree that Keynes' hypothesis is correct—that spending does depend on income—but that other factors also influence spending. These more complex models can be tested with data, and we do so in later chapters.

Table 1.1: U.S. Disposable Personal Income and Consumer Spending, Billions of Dollars [11]

	Income	Spending
1929	83.4	77.4
1930	74.7	70.1
1931	64.3	60.7
1932	49.2	48.7
1933	46.1	45.9
1934	52.8	51.5
1935	59.3	55.9
1936	67.4	62.2
1937	72.2	66.8
1938	66.6	64.3
1939	71.4	67.2
1940	76.8	71.3

Table 1.2: Family Income and Spending, 1935–1936 [12]

Income Range ($)	Average Income ($)	Average Spending ($)
<500	292	493
500–999	730	802
1,000–1,499	1,176	1,196
1,500–1,999	1,636	1,598
2,000–2,999	2,292	2,124
3,000–3,999	3,243	2,814
4,000–4,999	4,207	3,467
5,000–10,000	6,598	4,950

The Political Business Cycle

There seems to be a political business cycle in the United States, in that the unemployment rate typically increases after a presidential election and falls before the next presidential election. The unemployment rate has increased in only three presidential election years since the Great Depression. This is no doubt due to the efforts of incumbent presidents to avoid the wrath of voters suffering through an economic recession. Two exceptions were the reelection bids of Jimmy Carter in 1980 (the unemployment rate went up 1.3 percentage points) and George H. W. Bush in 1992 (the unemployment rate rose 0.7 percentage points). In each case, the incumbent was soon unemployed, too. The third exception was in 2008, when George W. Bush was president; the unemployment rate rose 1 percent and the Republicans lost the White House. In later chapters, we test the political business cycle model.

1.3 Making Predictions

Models help us understand the world and are often used to make predictions; for example, a consumption function can be used to predict household spending, and the political business cycle model can be used to predict the outcome of a presidential election. Here is another example.

Okun's Law

The U.S. unemployment rate was 6.6 percent when John F. Kennedy became president of the United States in January 1961 and reached 7.1 percent in May 1961. Reducing the unemployment rate was a top priority because of the economic and psychological distress felt by the unemployed and because the nation's aggregate output would be higher if these people were working. Not only would the unemployed have more income if they were working, but also they would create more food, clothing, and homes for others to eat, wear, and live in.

One of Kennedy's advisers, Arthur Okun, estimated the relationship between gross domestic product (GDP) and the unemployment rate. His estimate, known as *Okun's law*, was that output would be about 3 percent higher if the unemployment rate were 1 percentage point lower. Specifically, if the unemployment rate had been 6.1 percent, instead of 7.1 percent, output would have been about 3 percent higher.

This prediction was used to help sell the idea to Congress and the public that there are both private and public benefits from reducing the unemployment rate. Later in this book, we estimate Okun's law using more recent data.

1.4 Numerical and Categorical Data

Unemployment, inflation, and other data that have natural numerical values—5.1 percent unemployment, 3.2 percent inflation—are called *numerical* or *quantitative* data. The income and spending in Tables 1.1 and 1.2 are quantitative data.

Some data, for example, whether a person is male or female, do not have natural numerical values (a person cannot be 5.1 percent male). Such data are said to be *categorical* or *qualitative* data. With categorical data, we count the number of observations in each category. The data can be described by frequencies (the number of observations) or relative frequencies (the fraction of the total observations); for example, out of 1,000 people surveyed, 510, or 51 percent, were female.

The Dow Jones Industrial Average, the most widely reported stock market index, is based on the stock prices of 30 of the most prominent U.S. companies. If we record whether the Dow went up or down each day, these would be categorical data. If we record the percentage change in the Dow each day, these would be numerical data.

From 1901 through 2007, the Dow went up on 13,862 days and went down on 12,727 days. The relative frequency of up days is 52.1 percent:

$$\frac{13{,}862}{13{,}862 + 12{,}727} = 0.521$$

For the numerical data on daily percentage changes, we might calculate a summary statistic, such as the average percentage change (0.021 percent), or we might separate the percentage changes into categories, such as the number of days when the percentage change in the Dow was between 1 and 2 percent, between 2 and 3 percent, and so on.

1.5 Cross-Sectional Data

Cross-sectional data are observations made at the same point in time. These could be on a single day, as in Table 1.3, which shows the percentage changes in each of the Dow Jones stocks on January 29, 2008. Cross-sectional data can also be for a single week, month, or

Table 1.3: Percentage changes in the prices of Dow stocks, January 29, 2008

Company	Percent Price Change	Company	Percent Price Change
Alcoa	3.78	Johnson & Johnson	−0.10
AIG	1.98	JPMorgan Chase	1.88
American Express	0.40	CocaCola	−0.86
Boeing	3.36	McDonald's	−0.32
Citigroup	0.26	3M Company	0.58
Caterpillar	0.78	Altria	1.11
Du Pont	0.31	Merck	−1.24
Walt Disney	−0.57	Microsoft	−0.12
General Electric	0.04	Pfizer	0.19
General Motors	0.68	Procter & Gamble	−0.61
Home Depot	0.57	AT&T	1.49
Honeywell	−0.48	United Technologies	−0.55
Hewlett-Packard	−0.42	Verizon	0.60
IBM	1.12	Wal-Mart	0.30
Intel	0.21	Exxon	−0.05

year; for example, the survey data on annual household income and spending in Table 1.2 are cross-sectional data.

The Hamburger Standard

The *law of one price* says that, in an efficient market, identical goods should have the same price. Applied internationally, it implies that essentially identical goods should cost about the same anywhere in the world, once we convert prices to the same currency. Suppose the exchange rate between U.S. dollars and British pounds (£) is 2 dollars/pound. If a sweater sells for £20 in Britain, essentially identical sweaters should sell for $40 in the United States. If the price of the American sweaters were more than $40, Americans would import British sweaters instead of buying overpriced American sweaters. If the price were less than $40, the English would import American sweaters.

The law of one price works best for products (like wheat) that are relatively homogeneous and can be imported relatively inexpensively. The law of one price does not work very well when consumers do not believe that products are similar (for example, Hondas and Fords) or when there are high transportation costs, taxes, and other trade barriers. Wine can be relatively expensive in France if the French prohibit wine imports or tax imports heavily. A haircut and round of golf in Japan can cost more than in the United States, because it is impractical for the Japanese to have their hair cut in Iowa and play golf in Georgia.

Since 1986, *The Economist*, an influential British magazine, has surveyed the prices of Big Mac hamburgers around the world. Table 1.4 shows cross-sectional data on the prices of Big Mac hamburgers in 20 countries. The law of one price does not apply to Big Macs, because an American will not travel to Argentina to buy a hamburger for lunch.

Table 1.4: The Hamburger Standard, July 2007 [13]

	Big Mac Price	
	In Local Currency	**In U.S. Dollars**
United States	$3.41	$3.41
Argentina	Peso 8.25	$2.63
Australia	A$3.45	$3.09
Brazil	Real 6.9	$3.95
Britain	£1.99	$3.90
China	Tuan 11	$1.51
Egypt	Pound 9.54	$1.73
Euro area	€3.06	$4.54
Hong Kong	HK$12	$1.54
Indonesia	Rupiah 15900	$1.70
Japan	¥280	$2.58
Mexico	Peso 29	$2.65
Norway	Kroner 40	$7.58
Pakistan	Rupee 140	$2.23
Philippines	Peso 85	$2.09
Russia	Ruble 52	$2.14
Saudi Arabia	Riyal 9	$2.40
South Africa	Rand 15.5	$2.29
South Korea	Won 2900	$3.09
Taiwan	NT$75	$2.32

1.6 Time Series Data

Time series data are a sequence of observations at different points of time. The aggregate income and spending in Table 1.1 are time series data.

Silencing Buzz Saws

During the Great Depression, the Federal Reserve was ineffectual because it had little reliable data and did not understand how monetary policies affect the economy. Today, the Fed has plenty of data and uses models tested with these data to understand how monetary policies affect the economy. As a result, the Fed does not act perversely when the economy threatens to sink into an unwanted recession. On the other hand, the Fed sometimes uses tight-money policies to cool the economy and resist inflationary pressures. As a cynic (I) once wrote, the Fed raises interest rates enough to cause a recession when it feels it is in our best interest to lose our jobs.

The Federal Reserve's tight-credit policies during the years 1979–1982 are a striking example. In 1979, the rate of inflation was above 13 percent, and in October of that year, the Fed decided that its top priority was to reduce the rate of inflation substantially. Over the next three years, the Fed tightened credit severely and interest rates rose to unprecedented levels.

When Paul Volcker, the Fed chairman, was asked in 1980 if the Fed's stringent monetary policies would cause an economic recession, he replied, "Yes, and the sooner the better" [14]. In another 1980 conversation, Volcker remarked that he would not be satisfied "until the last buzz saw is silenced" [15]. In 1981, interest rates reached 18 percent on home mortgages and were even higher for most other bank loans. As interest rates rose, households and businesses cut back on their borrowing and on their purchases of automobiles, homes, and office buildings. Construction workers who lost their jobs were soon spending less on food, clothing, and entertainment, sending ripples through the economy. The unemployment rate rose from 5.8 percent in 1979 to above 10 percent in September 1982, the highest level since the Great Depression. Table 1.5 shows that the Fed achieved its single-minded objective, as the annual rate of inflation fell from 13.3 percent in 1979 to 3.8 percent in 1982.

In the fall of 1982, the Fed decided that the war on inflation had been won and that there were ominous signs of a possible financial and economic collapse. The Fed switched to easy money policies, supplying funds as needed to bring down interest rates, encourage borrowing and spending, and fuel the economic expansion that lasted the remainder of the decade. In later chapters, we see how interest rates, inflation, and unemployment are interrelated.

Table 1.5: The Fed's 1979–1982 War on Inflation

	Inflation	10-Year Interest Rate	Unemployment
1970	5.6	7.3	4.9
1971	3.3	6.2	5.9
1972	3.4	6.2	5.6
1973	8.7	6.8	4.9
1974	12.3	7.6	5.6
1975	6.9	8.0	8.5
1976	4.9	7.6	7.7
1977	6.7	7.4	7.1
1978	9.0	8.4	6.1
1979	13.3	9.4	5.8
1980	12.5	10.8	7.1
1981	8.9	13.9	7.6
1982	3.8	13.0	9.7
1983	3.8	11.1	9.6
1984	3.9	12.4	7.5
1985	3.8	10.6	7.2
1986	1.1	7.7	7.0
1987	4.4	8.4	6.2
1988	4.4	8.8	5.5
1989	4.6	8.5	5.3
1990	6.1	8.6	5.6

1.7 Longitudinal (or Panel) Data

Longitudinal data (or *panel* data) involve repeated observations of the same things at different points in time. Table 1.6 shows data on the prices between 2003 and 2007 of computer hard drives of various sizes. If we look at the prices of different hard drives in a given year, such as 2004, these are cross-sectional data. If we instead look at the price of a 200 GB hard drive in 2003, 2004, 2005, 2006, and 2007, these are time series data. If we look at the prices of hard drives of four different sizes in those five years, these are longitudinal data.

1.8 Index Numbers (Optional)

Some data, such as home prices, have natural values. *Index numbers*, on the other hand, measure values relative to a base value, often set equal to 100. Suppose, for example, that a house sold for $200,000 in 1980, $300,000 in 1990, and $400,000 in 2005. If we want to express these data as index numbers, we can set the base value equal to 100 in 1980. Because this house's price was 50 percent higher in 1990 than in 1980, the 1990 index value equals 150 (Table 1.7). Similarly, because the house price was 100 percent higher in 2005 than in 1980, the 2005 index value equals 200.

As in this example, the index values (100, 150, and 200) have no natural interpretation. It is not $100, 100 square feet, or $100 per square foot. Instead, a comparison of index values is used to show the percentage differences. A comparison of the 1980 and 1990 index values shows that the price was 50 percent higher in 1990 than in 1980.

In practice, index numbers are not used for individual homes, but for averages or aggregated data, such as average home prices in the United States, where the underlying data are unwieldy and we are mostly interested in percentage changes.

Table 1.6: Prices of Hard Drives of Various Sizes [16]

Size (Gigabytes)	2003	2004	2005	2006	2007
80	$155	$115	$85	$80	$70
120	$245	$150	$115	$115	$80
160	$500	$200	$145	$120	$85
200	$749	$260	$159	$140	$100

Table 1.7: Index Numbers of House Prices

	House Price	Index
1980	$200,000	100
1990	$300,000	150
2005	$400,000	200

The Consumer Price Index

The *Consumer Price Index* (*CPI*) measures changes in the cost of living by tracking changes in the prices of goods and services that households buy. The CPI was created during World War I to determine whether incomes were keeping up with prices, and it is still used for this purpose. Employers and employees both look at the CPI during wage negotiations. Social Security benefits are adjusted automatically for changes in the CPI. The U.S. Treasury issues Treasury Inflation-Protected Securities (TIPS) whose payments rise or fall depending on changes in the CPI.

Because it is intended to measure changes in the cost of living, the CPI includes the cost of food, clothing, housing, utilities, and transportation. Every 10 years or so, the Bureau of Labor Statistics (BLS) surveys thousands of households to learn the details of their buying habits. Based on this survey, the BLS constructs a market basket of thousands of goods and services and tracks their prices every month. These prices are used to compute price indexes that measure the current cost of the market basket relative to the cost in the base period:

$$P = 100\left(\frac{\text{current cost of market basket}}{\text{cost of market basket in base period}}\right) \tag{1.1}$$

The logic can be illustrated by the hypothetical data in Table 1.8 for a market basket of three items. This basket cost $50.00 in 1990 and $64.00 in 2000, representing a 28 percent increase: ($64.00 – $50.00)/$50.00 = 0.28. If we use 1990 as the base year, Equation 1.1 gives our price index values:

$$P_{1990} = 100\left(\frac{\$50.00}{\$50.00}\right) = 100$$

$$P_{2000} = 100\left(\frac{\$64.00}{\$50.00}\right) = 128$$

As intended, the price index shows a 28 percent increase in prices.

The Dow Jones Index

In 1880, Charles Dow and Edward Jones started a financial news service that they called Dow-Jones. Today, the most visible offspring are the *Wall Street Journal*, one of the most

Table 1.8: A Price Index Calculation

		1990		2000	
	Quantity	Price of One	Cost of Basket	Price of One	Cost of Basket
Loaf of bread	10	$2.00	$20.00	$2.50	$25.00
Pound of meat	6	$3.00	$18.00	$4.00	$24.00
Gallon of milk	3	$4.00	$12.00	$5.00	$15.00
Total			$50.00		$64.00

widely read newspapers in the world, and the Dow Jones Industrial Average (the *Dow*), the most widely reported stock market index.

Since 1928, the Dow has been based on 30 stocks that are intended to be "substantial companies—renowned for the quality and wide acceptance of their products or services—with strong histories of successful growth" [17]. The editors of the *Wall Street Journal* periodically alter the composition of the Dow either to meet the index's objectives or to accommodate mergers or reorganizations.

The Dow Jones Industrial Average is calculated by adding together the prices of the 30 Dow stocks and dividing by a divisor k, which is modified whenever one stock is substituted for another or a stock splits (increasing the number of shares outstanding and reducing the price of each share proportionately). Suppose, for instance, that the price of each of the 30 stocks is \$100 a share and the divisor is 30, giving a Dow average of 100:

$$\text{DJIA} = \frac{100 + 100 + \cdots + 100}{30} = \frac{30(100)}{30} = 100$$

Now one of these stocks is replaced by another stock, which has a price of \$50. If the divisor stays at 30, the value of the Dow drops by nearly 2 percent:

$$\text{DJIA} = \frac{100 + 100 + \cdots + 100 + 50}{30} = \frac{2950}{30} = 98.33$$

indicating that the average stock price dropped by 2 percent, when, in fact, all that happened was a higher-priced stock was replaced by a lower-priced stock.

The Dow allows for these cosmetic changes by adjusting the divisor. In our example, we want a divisor k that keeps the average at 100:

$$\text{DJIA} = \frac{100 + 100 + \cdots + 100 + 50}{k} = \frac{2950}{k} = 100$$

We can solve this equation for $k = 2950/100 = 29.5$, rather than 30. Thus the Dow average would now be calculated by dividing the sum of the 30 prices by 29.5 rather than 30.

The divisor is also adjusted every time a stock splits. The cumulative effect of these adjustments has been to reduce the Dow divisor to 0.132129493 in March 2011.

1.9 Deflated Data

A nation's population generally increases over time and so do many of the things that people do: marry, work, eat, and play. If we look at time series data for these human activities without taking into account changes in the size of the population, we will not be able to distinguish changes that are due merely to population growth from those that reflect changes in people's behavior. To help make this distinction, we can use *per capita* data, which have been adjusted for the size of the population.

For example, the number of cigarettes sold in the United States totaled 484.4 billion in 1960 and 525 billion in 1990, an increase of 8 percent. To put these numbers into perspective, we need to take into account the fact that the population increased by 39 percent during this period, from 179.3 million to 248.7 million people. We can do so by dividing each year's cigarette sales by the population to obtain per capita data:

$$1960: \frac{484.4\text{ billion}}{179.3\text{ million}} = 2{,}702$$

$$1990: \frac{525\text{ billion}}{248.7\text{ million}} = 2{,}111$$

Total sales increased by 8 percent, but per capita consumption fell by 22 percent.

Nominal and Real Magnitudes

Economic and financial data that are expressed in a national currency (for example, U.S. dollars) are called *nominal data*. If you are paid $10 an hour, that is your nominal wage. If you earn $20,000 a year, that is your nominal income.

However, we do not work solely for the pleasure of counting and recounting our dollars. We earn dollars so that we can buy things. We therefore care about how much our dollars buy. Data that are denominated in dollars (such as wages and income) need to be adjusted for the price level so that we can measure their purchasing power. Data that have been adjusted in this way are called *real data*.

For instance, if you are working to buy loaves of bread, your real wage would be determined by dividing your nominal wage by the price of bread. If your nominal wage is $10 an hour and bread is $2 a loaf, your real wage is 5 loaves an hour:

$$\text{real wage} = \frac{\text{nominal wage}}{\text{price}} = \frac{\$10/\text{hour}}{\$2/\text{loaf}} = 5\text{ loaves/hour}$$

Similarly, if you earn $20,000 a year, your real income is 10,000 loaves a year:

$$\text{real income} = \frac{\text{nominal income}}{\text{price}} = \frac{\$20{,}000/\text{year}}{\$2/\text{loaf}} = 10{,}000\text{ loaves/year}$$

The underlying economic principle behind the calculation of real data is that real, rather than nominal, magnitudes are what matter. When choosing between working and playing this summer, you will not just think about the nominal wage but also about how much your wages will buy. Fifty years ago, when a dollar would buy a lot, most people would have jumped at a chance to earn $10 an hour. Now, when a dollar buys little, many people would rather go to the beach than work for $10 an hour.

People who think about nominal income, rather than real income, suffer from what economists call *money illusion*. Someone who feels better off when his or her income goes up by 5 percent although prices have gone up by 10 percent is showing definite signs of money illusion.

Our illustrative calculation uses a single price, the price of a loaf of bread. However, we do not live by bread alone, but also by meat, clothing, shelter, medical care, computers, and on and on. The purchasing power of our dollars depends on the prices of a vast array of goods and services. To get a representative measure of purchasing power, we need to use a price index such as the CPI. Table 1.9 shows the CPI values for five selected years using a base of 100 in 1967.

The CPI is only meaningful in comparison to its value in another period. Thus a comparison of the 116.3 value in 1970 with the 246.8 value in 1980 shows that prices more than doubled during this 10-year period.

We calculate real values by dividing the nominal value by the CPI. For instance, if we divide a wage of 10 dollars an hour by the 2007 value of the CPI, the real wage is

$$\text{real wage} = \frac{\text{nominal wage}}{\text{price}} = \frac{10}{617.7} = 0.01619$$

Because the price level is an index that has been arbitrarily scaled to equal 100 in the base period, we cannot interpret this real wage as 0.01619 loaves of bread or anything else. Like price indexes, real wages are meaningful only in comparison to other real wages.

Table 1.10 shows the average hourly earnings of production and nonsupervisory workers for the same years that are shown in Table 1.9.

Table 1.9: Consumer Price Index (CPI = 100 in 1967)

1970	116.3
1980	246.8
1990	391.4
2000	515.8
2010	653.2

Table 1.10: Average Hourly Earnings

1970	$3.40
1980	$6.85
1990	$10.20
2000	$14.02
2010	$19.07

To see whether real wages increased between 1970 and 1980, we can use the data in Tables 1.9 and 1.10 to compute real wages in each year:

$$1970\text{: real wage} = \frac{\text{nominal wage}}{\text{price}} = \frac{\$3.40}{116.3} = 0.02923$$

$$1980\text{: real wage} = \frac{\text{nominal wage}}{\text{price}} = \frac{\$6.85}{246.8} = 0.02776$$

Real wages dropped by about 5 percent:

$$\frac{\text{1980 real wage} - \text{1970 real wage}}{\text{1970 real wage}} = \frac{0.02776 - 0.02923}{0.02923} = -0.0506$$

There is another way to get this answer. We can convert the \$3.40 1970 wage into 1980 dollars by multiplying \$3.40 by the 1980 price level relative to the 1970 price level:

$$\text{1970 wage in 1980 dollars} = (\text{1970 wage})\left(\frac{\text{1980 CPI}}{\text{1970 CPI}}\right) = \$3.40\left(\frac{246.8}{116.3}\right) = \$7.215$$

This value, \$7.215, can be interpreted as the amount of money needed in 1980 to buy what \$3.40 bought in 1970. Again, we see that real wages dropped by about 5 percent:

$$\frac{\text{1980 wage} - \text{1970 wage in 1980 dollars}}{\text{1970 wage in 1980 dollars}} = \frac{6.85 - 7.215}{7.215} = -0.0506$$

There are multiple paths to the same conclusion. In each case, we use the CPI to adjust wages for the change in prices—and find that wages did not keep up with prices between 1970 and 1980.

The Real Cost of Mailing a Letter

In May 2008, the cost of mailing a first-class letter in the United States was increased to 42¢. Table 1.11 shows that in 1885 the cost was 2¢. Was there an increase in the real cost of mailing a first-class letter between 1885 and 2008, that is, an increase relative to the prices of other goods and services? Using a base of 100 in 2008, the value of the CPI was 4.3 in 1885 and 100 in 2008. Therefore, 2¢ bought as much in 1885 as 47¢ bought in 2008:

$$\text{1885 postage in 2008 dollars} = (\text{1885 postage})\left(\frac{\text{2008 CPI}}{\text{1885 CPI}}\right) = \$0.02\left(\frac{100}{4.3}\right) = \$0.47$$

If the cost of mailing a first-class letter had increased as much as the CPI over this 123-year period, the cost in 2008 would have been 47¢ instead of 42¢. Over this 123-year period, the real cost of mailing a first-class letter fell by about 11 percent:

$$\frac{\text{2008 postage} - \text{1885 postage in 2008 dollars}}{\text{1885 postage in 2008 dollars}} = \frac{0.42 - 0.47}{0.47} = -0.11$$

Table 1.11: Cost of Mailing a First-Class Letter

Date Introduced	First-Class Postage Rates ($)	CPI (2008 = 100)
7/1/1885	0.02	4.3
11/3/1917	0.03	6.0
7/1/1919	0.02	8.2
7/6/1932	0.03	6.4
8/1/1958	0.04	13.6
1/7/1963	0.05	14.4
1/7/1968	0.06	16.4
5/16/1971	0.08	19.0
3/2/1974	0.10	23.2
12/31/1975	0.13	25.3
5/29/1978	0.15	30.7
3/22/1981	0.18	42.8
11/1/1981	0.20	42.8
2/17/1985	0.22	50.7
4/3/1988	0.25	55.7
2/3/1991	0.29	64.3
1/1/1995	0.32	71.7
1/10/1999	0.33	78.4
1/7/2001	0.34	83.3
6/30/2002	0.37	84.7
1/8/2006	0.39	94.9
5/14/2007	0.41	97.1
5/12/2008	0.42	100.0

Real Per Capita

Some data should be adjusted for both population growth and inflation, thereby giving *real per-capita* data. The dollar value of a nation's aggregate income, for example, is affected by the population and the price level, and unless we adjust for both, we will not be able to tell whether there has been a change in living standards or just changes in the population and price level.

Other kinds of data may require other adjustments. An analysis of the success of domestic automakers against foreign competitors should not use just time series data on sales of domestically produced automobiles, even if these data are adjusted for population growth and inflation. More telling data might show the ratio of domestic sales to total automobile sales.

Similarly, a study of motor vehicle accidents over time needs to take into account changes in the number of people traveling in motor vehicles and how far they travel. In 1990, for example, there were 44,500 motor vehicle deaths in the United States and it was estimated that 2,147 billion miles were driven that year. The number of deaths per mile driven was (44,500)/(2,147 billion) = 0.000000021. Instead of working with so many zeros, we can multiply this number by 100 million to obtain a simpler (and more intelligible) figure of 2.1 deaths per 100 million miles driven.

In every case, our intention is to let the data inform us by working with meaningful numbers. In the next chapter, we begin seeing how to draw useful statistical inferences from data.

Exercises

1.1 Which of these data are quantitative and which are qualitative?
   a. The number of years a professor has been teaching.
   b. The number of books a professor has written.
   c. Whether a professor has a Ph.D.
   d. The number of courses a professor is teaching.

1.2 Which of these data are categorical and which are numerical?
   a. Whether a family owns or rents their home.
   b. A house's square footage.
   c. The number of bedrooms in a house.
   d. Whether a house has central air conditioning.

1.3 Which of these data are categorical and which are numerical?
   a. Whether a computer is made by Apple or by another company.
   b. The size of a computer monitor.
   c. The speed of a computer processor.
   d. Whether a computer has a firewire port.

1.4 Which of these data are quantitative and which are qualitative?
   a. A country's unemployment rate.
   b. A country's population.
   c. A country's gross domestic product (GDP).
   d. Whether a country belongs to the European Union.

1.5 Identify the following data as cross-section, time series, or panel data:
   a. Unemployment rates in Germany, Japan, and the United States in 2010.
   b. Unemployment and inflation rates in Germany, Japan, and the United States in 2009.
   c. Unemployment rates in Germany in 2007, 2008, 2009, and 2010.
   d. Unemployment rates in Germany and the United States in 2007, 2008, 2009, and 2010.

1.6 The Framingham Heart Study began in 1948 with 5,209 adult subjects, who have returned every two years for physical examinations and laboratory tests. The study now includes the children and grandchildren of the original participants. Identify the following data as cross-section, time series, or panel data:
   a. The number of people who died each year from heart disease.
   b. The number of men who smoked cigarettes in 1948, 1958, 1968, and 1978.

c. The ages of the women in 1948.
d. Changes in HDL cholesterol levels between 1948 and 1958 for each of the females.

1.7 (continuation) Identify the following data as cross-section, time series, or panel data:
a. Blood pressure of one woman every two years.
b. The average HDL cholesterol level of the men in 1948, 1958, and 1968.
c. The number of children each woman had in 1958.
d. The age at death of the 5,209 subjects.

1.8 Table 1.12 lists the reading comprehension scores for nine students at a small private school known for its academic excellence. Twenty students were admitted to the kindergarten class, and the nine students in the table stayed at the school through eighth grade. The scores are percentiles relative to students at suburban public schools; for example, Student 1 scored in the 98th percentile in first grade and in the 53rd percentile in second grade. Identify the following data as cross-section, time series, or panel data:
a. Student 4's scores in grades 1 through 8.
b. The nine students' eighth grade scores.
c. The nine students' scores in first grade and eighth grade.
d. The nine students' scores in first grade through eighth grade.

1.9 Table 1.13 shows index data on the overall CPI and three items included in the CPI. Explain why you either agree or disagree with these statements:
a. Food cost more than housing in 2010.
b. Housing cost more than food in 2000 but cost less than food in 2010.
c. The cost of food went up more than the cost of housing between 2000 and 2010.

Table 1.12: Exercise 1.8

	Student								
Grade	**1**	**2**	**3**	**4**	**5**	**6**	**7**	**8**	**9**
1	98	87	87	92	80	80	72	98	87
2	53	72	89	72	45	35	62	62	81
3	98	52	94	60	73	26	26	94	88
4	72	55	96	51	68	34	68	94	59
5	62	62	68	35	47	18	99	95	95
6	63	53	80	38	33	53	69	69	85
7	97	61	79	8	47	32	65	65	53
8	74	65	31	15	31	74	46	55	31

Table 1.13: Exercise 1.9

	CPI	Food	Apparel	Housing
2000	172.2	168.4	129.6	169.6
2010	218.1	219.6	119.5	216.3

1.10 (continuation) Explain why you either agree or disagree with these statements:
  a. Apparel cost less than food in 2000.
  b. The cost of apparel went down between 2000 and 2010.
  c. The cost of food went up more than the overall CPI between 2000 and 2010.

1.11 (continuation) Calculate the percentage change in the cost of food, apparel, and housing between 2000 and 2010.

1.12 Table 1.14 shows data on four apparel price indexes that are used to compute the CPI. Explain why you either agree or disagree with these statements:
  a. Men's apparel cost more than women's apparel in 2000.
  b. Men's apparel cost more than women's apparel in 2010.
  c. Men's apparel cost more in 2000 than in 2010.

1.13 (continuation) Explain why you either agree or disagree with these statements:
  a. Men's apparel is more expensive than women's apparel, but boy's apparel is less expensive than girl's apparel.
  b. The cost of boy's apparel went down between 2000 and 2010.
  c. The cost of boy's apparel went down more than did the cost of men's apparel between 2000 and 2010.

1.14 (continuation) Calculate the percentage change in the cost of each of these four types of apparel between 2000 and 2010.

1.15 Why does it make no sense to deflate a country's population by its price level in order to obtain the "real population"?

1.16 It has been proposed that real wealth should be calculated by deflating a person's wealth by hourly wages rather than prices. Suppose that wealth is $100,000, the price of a hamburger is $2, and the hourly wage is $10. What conceptual difference is there between these two measures of real wealth?

1.17 Table 1.15 shows the prices of three items in 2000 and 2010.
  a. Construct a price index that is equal to 100 in 2000. What is the value of your index in 2010? What is the percentage change in your index between 2000 and 2010?
  b. Construct a price index that is equal to 100 in 2010. What is the value of your index in 2000? What is the percentage change in your index between 2000 and 2010?

Table 1.14: Exercise 1.12

	Men	Women	Boys	Girls
2000	133.1	121.9	116.2	119.7
2010	117.5	109.5	91.5	95.4

Table 1.15: Exercise 1.17

	Quantity	2000 Price	2010 Price
Loaf of bread	10	$2.50	$3.00
Pound of meat	6	$4.00	$5.00
Gallon of milk	3	$5.00	$7.00

1.18 Answer the questions in Exercise 1.17, this time assuming that the quantities purchased are 12 loaves of bread, 6 pounds of meat, and 2 gallons of milk.

1.19 The Dow Jones Industrial Average was 240.01 on October 1, 1928 (when the Dow expanded to 30 stocks), and 14,087.55 on October 1, 2007. The CPI was 51.3 in 1928 and 617.7 in 2007. Which had a larger percentage increase over this period, the Dow or the CPI?

1.20 On January 11, 2008, the value of the Dow Jones Industrial Average was 12,606.30, and the divisor was 0.123017848. Explain why you either agree or disagree with these statements:

a. The average price of the 30 stocks in the Dow was 12,606.30.
b. The average price of the 30 stocks in the Dow was 12,606.30 divided by the divisor.
c. The average price of the 30 stocks in the Dow was 12,606.30 multiplied by the divisor.

1.21 Look up the values of the CPI in December 1969 and December 1989 and the Dow Jones Industrial Average on December 31, 1969, and December 31, 1989. Did consumer prices or stock prices increase more over this 20-year period? Did the real value of stocks increase or decline?

1.22 Look up the values of the CPI in December 1989 and in December 2009 and the Dow Jones Industrial Average on December 31, 1989, and on December 31, 2009. Did consumer prices or stock prices increase more over this 20-year period? Did the real value of stocks increase or decline?

1.23 Two professors calculated real stock prices (adjusted for changes in the CPI) back to 1857 and concluded that [18]

inflation-adjusted stock prices were approximately equal to nominal stock prices in 1864–1865, were greater than nominal prices through 1918, tracked them rather closely ... following World War I and were nearly equal to them throughout the World War II period. Since the end of World War II, however, the nominal and real stock price series have begun to diverge, with the real price moving further below the nominal price.

What does this conclusion, by itself, tell us about stock prices and consumer prices during these years?

1.24 The authors of the study cited in the preceding exercise go on to consider what would have to happen for nominal stock prices to again equal real stock prices. What is the correct answer?

1.25 Use the data in Tables 1.9 and 1.10 to determine the
   a. 1970 wage in 2010 dollars.
   b. Percentage change in real wages between 1970 and 2010.

1.26 Use the data in Tables 1.9 and 1.10 to determine the
   a. 1980 wage in 2000 dollars.
   b. Percentage change in real wages between 1980 and 2000.

1.27 Table 1.16 shows the official U.S. poverty thresholds for families of four people. Use the CPI data in Table 1.9 to calculate the real poverty thresholds in 1960 dollars in each of the years shown.

1.28 Babe Ruth was paid $80,000 in 1931. When it was pointed out that this was more than President Hoover made, he replied, "I had a better year than he did." Using a base 100 in 1967, the CPI was 45.6 in 1931 and 653.2 in 2010. What was the value of Babe Ruth's 1931 salary in 2010 dollars? Is this value higher or lower than $3,297,828, the average Major League Baseball player's salary in 2010?

1.29 The first professional football player was W. W. Pudge Heffelfinger, a former All-American at Yale, who was working on a railroad in 1892 when he was paid $500 to play one game for the Allegheny Athletic Association against the Pittsburgh Athletic Club. (Pudge scored the game's only touchdown, on a 25-yard run.) Using a scale of 100 in 1967, the CPI was 27 in 1892 and 653.2 in 2010. What was the value in 2010 dollars of Pudge's one-game salary in 1892? Is this value higher or lower than $68,750, the average National Football League player's salary per game in 2010?

1.30 A researcher wants to see if *U.S. News & World Report* rankings influence the number of applications received by colleges and universities. He obtains the unadjusted data shown in Table 1.17 on applications to one small liberal arts college. To take into account the nationwide growth of college applications, this researcher uses data from the Consortium on Financing Higher Education (COFE) showing that, during this period, college applications increased by an average of 4 percent per year. He consequently

Table 1.16: Exercise 1.27

1960	$3,022
1970	$3,968
1980	$8,414
1990	$13,359
2000	$17,604
2010	$22,050

Table 1.17: Exercise 1.30

	Number of Applications	
	Unadjusted	Adjusted
1987	3,192	3,069.2
1988	2,945	2,831.7
1989	3,176	3,053.8
1990	2,869	2,758.7
1991	2,852	2,742.3
1992	2,883	2,772.1
1993	3,037	2,920.2
1994	3,293	3,166.3
1995	3,586	3,448.1

adjusted the number of applications to this college by dividing each value by 1.04. Identify his mistake and explain how you would calculate the adjusted values correctly. Do not do the actual calculations; just explain how you would do them.

1.31 When a child loses a baby tooth, an old U.S. tradition is for the tooth to be put under the child's pillow, so that the tooth fairy will leave money for it. A survey by a Northwestern University professor [19] estimated that the tooth fairy paid an average of 12 cents for a tooth in 1900 and 1 dollar for a tooth in 1987. Using a base of 100 in 1967, the CPI was 25 in 1900, 340.4 in 1987, and 653.2 in 2010. Did the real value of tooth fairy payments rise or fall between 1900 and 1987? If tooth fairy payments had kept up with inflation between 1900 and 2010, how large should the 2010 payment have been?

1.32 The U.S. Social Security Administration estimated that someone 22 years old who earned $40,000 a year in 2008, works steadily for 45 years, and retires at age 67 in 2053, would begin receiving annual benefits of about $90,516 in 2054. This calculation assumes a 3.5 percent annual rate of inflation between 2008 and 2054. If so, how much will this $90,516 buy in terms of 2008 dollars, that is, how much money would be needed in 2008 to buy as much as $90,516 will buy in 2054?

1.33 The U.S. government reported that a total of 38.6 million pounds of toxic chemicals had been released into the air above New Jersey in 1982. A spokesman for the New Jersey Department of Environmental Protection boasted that "It speaks well of our enforcement," since this was less chemicals than had been released in 21 other states [20]. What problem do you see with his argument?

1.34 The Dow Jones Industrial Average was 838.92 on December 31, 1970, and 11,577.51 on December 31, 2010. The Consumer Price Index was at 39.8 in December 1970 and 219.2 in December 2010 (using a base of 100 in 1982–1984). Did the Dow increase by more or less than consumer prices over this 40-year period? If the Dow had increased by just as much as consumer prices, what would the value of the Dow have been on December 31, 2010?

1.35 During the 1922–1923 German hyperinflation, people reported receiving more money for returning their empty beer bottles than they had originally paid to buy full bottles. Could they have made a living buying beer, emptying the bottles, and returning them?

1.36 Between August 1945 and July 1946, the number of Hungarian pengos (the unit of Hungarian currency) in circulation increased by a factor of 12,000,000,000,000,000,000,000,000, while the price level increased by a factor of 4,000,000,000,000,000,000,000,000,000. Did the real value of the money supply increase or decrease?

1.37 Critically evaluate this economic commentary [21]:

When it comes to measuring inflation, the average consumer can do a far better job than the economics experts.... Over the years I have been using a system which is infallible.... The Phindex [short for the Phillips index] merely requires you to divide the total dollar cost of a biweekly shopping trip by the number of brown paper bags into which the purchases are crammed. You thus arrive at the average cost per bagful.

When I started this system some 10 years ago, we would walk out of the store with about six bags of groceries costing approximately $30—or an average of $5 per bag.... On our most recent shopping trip, we emerged with nine bagsful of stuff and nonsense, totaling the staggering sum of $114.... [T]he Phindex shows a rise from the initial $5 to almost $13, a whopping 153 percent.

1.38 Use the data in Table 1.11 to determine the 1971 cost in 2008 dollars of mailing a first-class letter. Is this value higher or lower than the actual $0.42 cost of mailing a letter in 2008?

1.39 Use the data in Table 1.11 to determine the 1991 cost in 2008 dollars of mailing a first-class letter. Is this value higher or lower than the actual $0.42 cost of mailing a letter in 2008?

1.40 Use the data in Table 1.11 to determine the percentage increase in the CPI between 2007 and 2008.

1.41 Use the data in Table 1.11 to determine the percentage increase in the CPI between 1978 and 1981. (Do not calculate the annual rate of increase, just the overall percentage increase.)

1.42 Table 1.11 shows the CPI with the index set equal to 100 in 2008. Suppose that we redo the CPI so that the index equals 100 in 1958; if so, what is the value of the CPI in 2008? What is the percentage increase in the CPI between 1958 and 2008 if we use the CPI with the index equal to 100 in 2008? If we use the CPI with the index equal to 100 in 1958?

1.43 Table 1.11 shows the CPI with the index set equal to 100 in 2008. Suppose that we redo the CPI so that the index equals 100 in 1985; if so, what is the value of the CPI in 2008? What is the percentage increase in the CPI between 1985 and 2008 if we use the CPI with the index equal to 100 in 2008? If we use the CPI with the index equal to 100 in 1985?

1.44 Instead of using the total population, we might deflate cigarette sales by the number of people over the age of 18, since a person must be at least 18 years old to buy cigarettes legally. There were 116 million people over the age of 18 in the United States in 1960 and 186 million in 1990. Use the data in the text to calculate the percentage change between 1960 and 1990 in cigarette consumption per person over the age of 18.

1.45 Mr. Bunker lives on beer and pretzels. In 2000, he bought 1,000 six-packs of beer for \$2.00 per six-pack and 500 bags of pretzels for \$1.00 per bag. In 2010, Bunker's beer cost \$3.00 per six-pack and his pretzels cost \$2.00 per bag. One way to calculate Bunker's rate of inflation is

$$100\frac{C_{2010} - C_{2000}}{C_{2000}}$$

where C_{2010} = cost of 1,000 six-packs and 500 bags of pretzels in 2010 and C_{2000} = cost of 1,000 six-packs and 500 bags of pretzels in 2000. What was the percentage increase in Bunker's cost of living between 2000 and 2010? If we calculate a Bunker Price Index (BPI) and scale this BPI to equal 100 in 2000, what is the 2010 value of the BPI?

1.46 (continuation) Another way to calculate Bunker's rate of inflation is

$$\begin{pmatrix}\text{\% increase in}\\ \text{price of beer}\end{pmatrix}\begin{pmatrix}\text{fraction of 2000 budget}\\ \text{spent on beer}\end{pmatrix} + \begin{pmatrix}\text{\% increase in}\\ \text{price of pretzels}\end{pmatrix}\begin{pmatrix}\text{fraction of 2000 budget}\\ \text{spent on pretzels}\end{pmatrix}$$

What is Bunker's rate of inflation using this approach?

1.47 Ms. Vigor lives on carrot juice and tofu. In 2000, she bought 1,000 bottles of carrot juice for \$2.00 per bottle and 400 pounds of tofu for \$3.00 per pound. In 2010, Vigor's carrot juice cost \$2.50 per bottle and her tofu cost \$5.00 per pound. One way to calculate Vigor's rate of inflation is

$$100\frac{C_{2010} - C_{2000}}{C_{2000}}$$

where C_{2000} = cost of 1,000 bottles of carrot juice and 400 pounds of tofu in 2000 and C_{2010} = cost of 1,000 bottles of carrot juice and 400 pounds of tofu in 2010. What was the percentage increase in Vigor's cost of living between 2000 and 2010? If we calculate a Vigor Price Index (VPI) and scale this VPI to equal 100 in 2000, what is the 2010 value of the VPI?

1.48 (continuation) Because of the large increase in the price of tofu, Ms. Vigor changed her diet to 1,200 bottles of carrot juice and 250 pounds of tofu in 2010. Calculate Vigor's rate of inflation using C_{2000} = cost of 1,200 bottles of carrot juice and 250 pounds of tofu in 2000 and C_{2010} = cost of 1,200 bottles of carrot juice and 250 pounds of tofu in 2010. What was the percentage increase in Vigor's cost of living between 2000 and 2010? Is the percentage increase in her cost of living higher, lower, or the same when we use her 2010 purchases in place of her 2000 purchases? Explain why your answer makes sense intuitively.

1.49 Explain the error in this interpretation of inflation data [22]:

In the 12-month period ending in December of 1980, consumer prices rose by 12.4 percent—after a 13.3 percent increase the year before. Similar measures of inflation over the next three years were 8.9 percent, 3.9 percent, and 3.8 percent.... We are certainly paying less for what we buy now than we were at the end of [1980].

1.50 Answer this letter to Ann Landers [23]:

I've read your column for ages and almost always agree with you. One subject on which we do not see eye-to-eye, however, is senior citizens driving. According to the Memphis Commercial Appeal, recent statistics by the National Highway Traffic Safety Administration indicate that drivers over 70 were involved in 4,431 fatal crashes in 1993. That is far fewer than any other age group. The 16- to 20-year-olds were involved in 7,711 fatal crashes. Elderly drivers also had the lowest incidence of drunk driving accidents. Now, will you give seniors the praise they deserve?

CHAPTER 2

Displaying Data

Chapter Outline

A picture is worth a thousand words.
—***Anonymous***

Data are the building blocks of statistics. In later chapters, we summarize data and draw statistical inferences from data. In this chapter, we use visual displays to interpret data and draw inferences. Graphs can help us see tendencies, patterns, trends, and relationships. A picture can be worth not only a thousand words but a thousand numbers.

2.1 Bar Charts

Table 2.1 shows the percentage of families in 1960 and 2000 with no children under the age of 18, one child under 18, and so on. In 1960, 43 percent of U.S. families had no children and 9 percent had four or more children; in 2000, 52 percent had no children and only 3 percent had four or more children.

Table 2.1: Percentage Distribution of U.S. Families by Number of Children under Age 18 [1]

	1960	2000
No children	43	52
One child	19	20
Two children	18	18
Three children	11	7
Four or more children	9	3

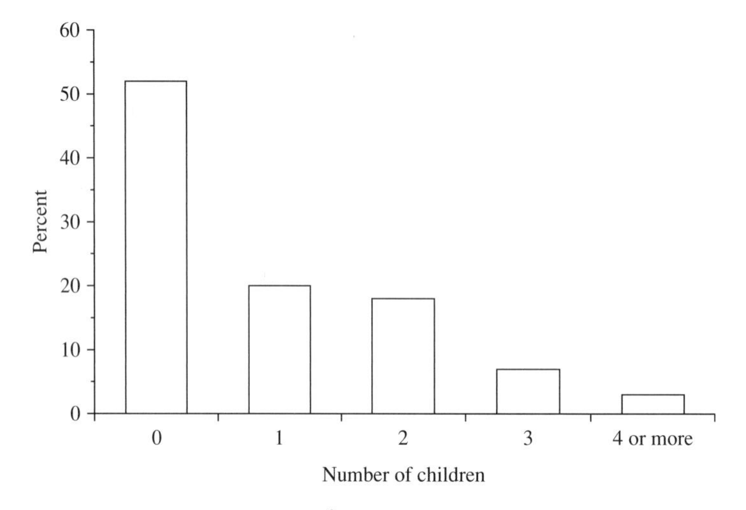

Figure 2.1
Percentage of families with no, one, two, three, or four or more children, 2000.

These data can be displayed in a *bar chart*, where the heights of the bars are used to show the number of observations (or the percentage of observations) in each category. For example, Figure 2.1 shows the 2000 data with the bar heights showing the percentage of families. If we instead used the number of families or the fraction of families, the bar chart would look the same, but the units on the vertical axis would be different.

Because we instinctively gauge size by area rather than by height, each bar should be the same width; otherwise the wider bars look disproportionately large. Suppose, for example, that we have only two categories, 0–1 children or more than one child, and that 75 percent of the families have 0–1 children and 25 percent have more than one child. The bar for 0–1 children should therefore be three times the height of the bar for more than one child. Now suppose that we want to liven up our bar chart by replacing the boring bars with cute pictures of a child. If we simply increase the height of the child's picture without changing the width, as in Figure 2.2, the stretched picture is distorted and unprofessional.

To keep the pictures from looking like fun house mirrors, we must triple both the height and width of the bar for 0–1 children, as in Figure 2.3. But now the area of the first bar is

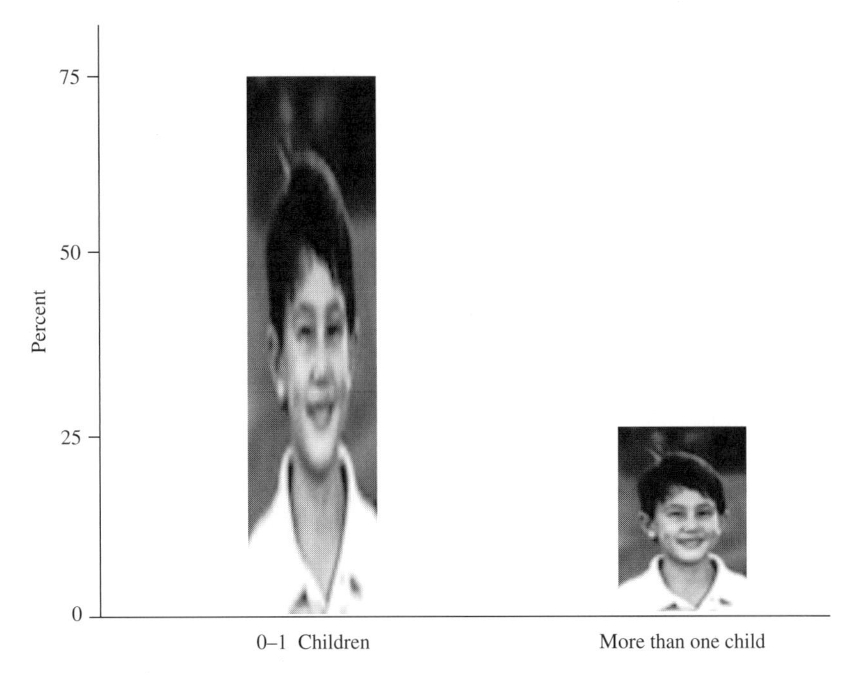

Figure 2.2
Bar pictures are bizarre if the height is changed but the width is not.

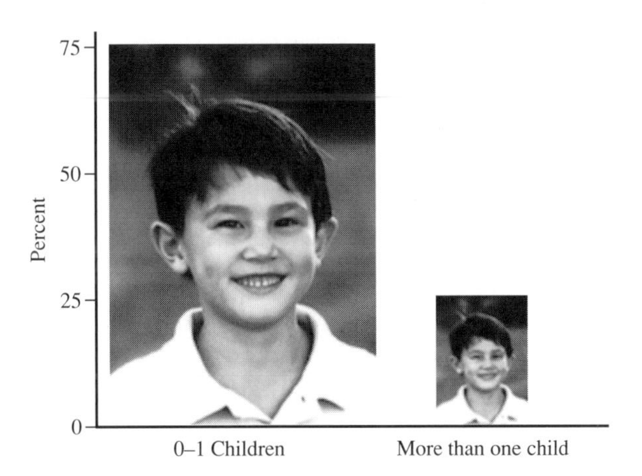

Figure 2.3
A bar chart with unequal widths.

nine times the area of the second bar, which makes it seem that there are nine times as many families with 0–1 children, when there are really only three times as many families. The solution? Keep the bar widths the same—even if it means using simple bars instead of eye-catching pictures. Remember, our objective is to convey information fairly and accurately.

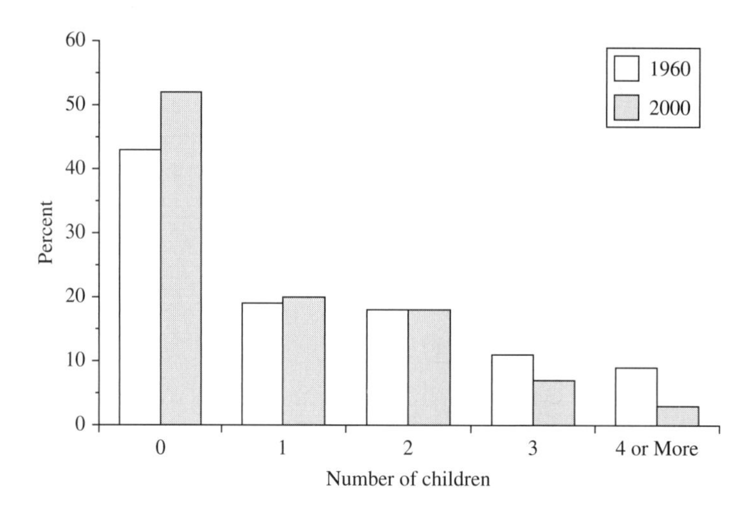

Figure 2.4
Percentage of families with no, one, two, three, or four or more children, 1960 and 2000.

Bar charts can also be drawn with multiple data sets, as in Figure 2.4, which compares family sizes in 1960 and 2000. This chart shows that, over this 40-year period, there was an increase in the percentage of families with fewer than two children and a decrease in the percentage with more than two children.

These simple bar charts are little more than a visual depiction of the data in Table 2.1. Bar charts often do not reveal any insights beyond what could be seen by looking at a table, though the visual images may be easier to understand and remember than numbers in a table.

Bowlers' Hot Hands

It is commonly thought that athletic performances are streaky, in that athletes sometimes "get hot" and perform exceptionally well before falling back to their normal performance level. A basketball player might make five shots in a row. A football quarterback might complete five passes in a row. A baseball player might get hits in five consecutive games. However, a careful statistical investigation [2] concluded that the common belief that basketball players have hot streaks is erroneous, in that memorable performances happen no more often than would be predicted by chance. For example, basketball players making five shots in a row happens no more often than a coin lands heads five times in a row. In addition, the players were no more likely (indeed, some were *less* likely) to make a basket after having made a basket than after missing a basket.

Unfortunately, these basketball data have serious flaws. A player's two successive shots might be taken 30 seconds apart, 5 minutes apart, in different halves of a game, or even in different games. Another problem is that the shots a player chooses to try may be affected by his recent experience. A player who makes several shots may take more difficult shots than he would try after missing several shots. In addition, the opposing team may guard a player differently when he is perceived to be hot or cold.

Bowling is a sport that does not have these complications. Every roll is made from the same distance at regular intervals with intense concentration and no strategic considerations. An analysis of Professional Bowler Association (PBA) data indicates that, for many bowlers, the probability of rolling a strike is not independent of previous outcomes.

Table 2.2 summarizes data based on the frequencies with which bowlers rolled strikes immediately after having rolled one to four consecutive strikes or one to four consecutive nonstrikes. All these data are from within games; we do not consider whether the first roll of a game is influenced by the last roll of the previous game. For each category, we consider only bowlers who had at least ten observations in both situations. For example, 43 bowlers had at least ten opportunities to bowl after rolling four consecutive strikes and also had at least ten opportunities to bowl after rolling four consecutive nonstrikes. Of these 43 bowlers, 34 (79 percent) were more likely to roll a strike after four consecutive strikes than after four consecutive nonstrikes.

Figure 2.5 shows these bowling data in a bar chart. Again, the main advantage of a bar chart is that it paints a visual picture of the data.

Table 2.2: Bowlers Who Rolled Strikes More Often after Consecutive Strikes or Nonstrikes [3]

Bowlers Who Rolled Strikes More Often	Number of Consecutive Strikes/Nonstrikes			
	1	2	3	4
After strikes	80	77	59	34
After nonstrikes	54	33	22	9
Total number of bowlers	134	110	81	43

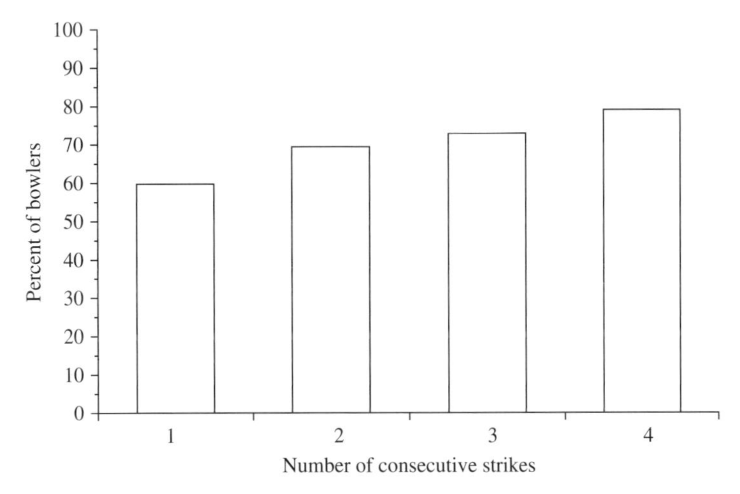

Figure 2.5
Percentage of bowlers who rolled a strike more often after one, two, three, or four consecutive strikes than after an equal number of consecutive nonstrikes.

Bar Charts with Interval Data

The data on family size and bowling involve discrete categories—two children, four consecutive strikes. The bars are drawn with spaces between them to show that you cannot have 2.5 children or 3.5 strikes. We can also use bar charts when numerical data are divided into intervals.

For example, Table 1.3 in Chapter 1 shows the percentage price changes in each of the 30 Dow stocks on January 29, 2008. Table 2.3 puts these percentage changes in order from largest to smallest. Prices went up for 19 stocks and down for 11 stocks, ranging from a 3.78 percent increase for Alcoa to a 1.24 percent decline for Merck.

We can separate the percentage price changes into intervals. Because the price changes range from –1.24 percent to 3.78 percent, we can specify six equally spaced intervals

Table 2.3: Percentage Changes in the Prices of Dow Stocks, January 29, 2008

Company	Percent Price Change
Alcoa	3.78
Boeing	3.36
AIG	1.98
JPMorgan Chase	1.88
AT&T	1.49
IBM	1.12
Altria	1.11
Caterpillar	0.78
General Motors	0.68
Verizon	0.60
3M Company	0.58
Home Depot	0.57
American Express	0.40
DuPont	0.31
Wal-Mart	0.30
Citigroup	0.26
Intel	0.21
Pfizer	0.19
General Electric	0.04
Exxon	–0.05
Johnson & Johnson	–0.10
Microsoft	–0.12
McDonald's	–0.32
Hewlett-Packard	–0.42
Honeywell	–0.48
United Technologies	–0.55
Walt Disney	–0.57
Procter & Gamble	–0.61
Coca-Cola	–0.86
Merck	–1.24

ranging from –2 percent to +4 percent. Now we count the number of price changes between –2 percent and –1 percent. Between –1 percent and 0 percent. Between 0 percent and 1 percent. And so on.

Table 2.4 shows the results. Two stocks had price changes in the interval 3–4 percent, none had price changes in the interval 2–3 percent, and so on. There are no price changes that are exactly on the border between intervals, for example, a price change of exactly 3 percent. When there are borderline observations, we need to adopt a rule; for example, having the interval "2–3" mean "greater than 2 to less than or equal to 3," in which case a price change of exactly 2 percent goes into the next lower interval. If there are a large number of observations, the specific rule does not matter much; what is important is that our rule is consistent.

Figure 2.6 is a bar chart showing the number of stocks in each of the six intervals. This is a simple visual representation of the information in Table 2.4 but perhaps in a way that is more easily processed. Unlike in a bar chart for categorical data, the bars touch because they span numerical intervals.

Table 2.4 also shows the relative frequencies of observations in each interval:

$$\text{relative frequency} = \frac{\text{observations in interval}}{\text{total observations}}$$

When there are lots of data, it may be hard to interpret the fact that an interval has, say, 736 observations, but easy to interpret the fact that 15 percent of the observations are in this interval.

Figure 2.7 is a bar chart of the data in Table 2.4, this time showing the relative frequencies. Figures 2.6 and 2.7 look identical; the only difference is the different scales on the vertical axis.

Table 2.4: Percentage Changes in the Prices of Dow Stocks

Percent Price Change	Number of Stocks	Relative Frequency
3 to 4	2	$\frac{2}{30} = 0.067$
2 to 3	0	$\frac{0}{30} = 0.000$
1 to 2	5	$\frac{5}{30} = 0.167$
0 to 1	12	$\frac{12}{30} = 0.400$
–1 to 0	10	$\frac{10}{30} = 0.333$
–2 to –1	1	$\frac{1}{30} = 0.033$
Total	30	1.000

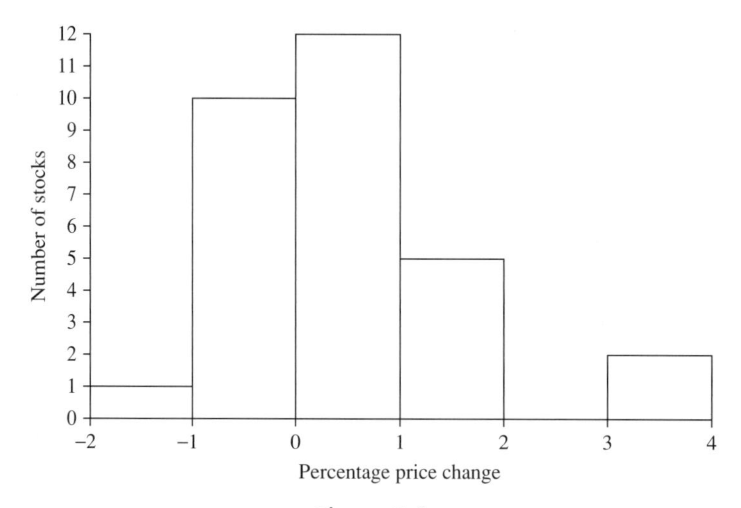

Figure 2.6
A bar chart showing the number of observations.

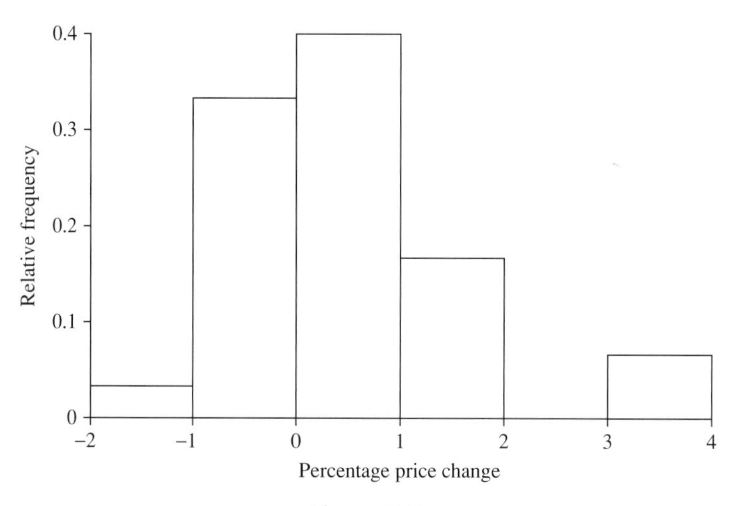

Figure 2.7
A bar chart showing relative frequencies.

2.2 Histograms

Figures 2.6 and 2.7 use intervals that are equally wide, which is generally a good idea. But sometimes there are persuasive reasons for using unequal intervals. If we are looking at years of education, one interval might be 0 to 8 years and another interval might be 9 to 12 years. If we are looking at income, we might group the data by tax bracket. The 2008 federal income tax brackets for a married couple filing jointly are in Table 2.5. For example, a married couple with a taxable income of $40,000 paid a federal income tax of

\$1,605.00 + 0.15(\$40,000 – \$16,050) = \$1,510.00 + \$3,592.50 = \$5,179.50. Their average tax rate was 13 percent:

$$\frac{\$5{,}179.50}{\$40{,}000.00} = 0.130$$

Their marginal tax rate is 15 percent in that they would pay an extra 15¢ in tax on every extra dollar of taxable income (until their income exceeds \$65,101 and they move into the 25 percent tax bracket). Because different income tax brackets have different marginal tax rates, it makes sense to use intervals that match the income tax brackets.

Bar graphs can be misleading when the intervals are not equally wide. Table 2.6 reproduces the data in Table 2.4, with the intervals –2 to –1 and –1 to 0 combined into a single interval, –2 to 0.

Figure 2.8 shows the corresponding bar chart. Now, look back at Figure 2.7. Figures 2.7 and 2.8 use the same data with the relative frequencies calculated correctly and drawn correctly using the same scale on the vertical axis, yet the figures appear to be quite different. The reason for this visual disparity is that our eyes gauge size by the areas of

Table 2.5: Federal Income Tax Brackets for a Married Couple Filing Jointly in 2008

Taxable Income Bracket	Tax
\$0–\$16,050	10% of the amount over \$0
\$16,051–\$65,100	\$1,605.00 plus 15% of the amount over \$16,050
\$65,101–\$131,450	\$8,962,50 plus 25% of the amount over \$65,100
\$131,451–\$200,300	\$25,550.00 plus 28% of the amount over \$131,450
\$200,301–\$357,700	\$44,828.00 plus 33% of the amount over \$200,300
\$357,701–No limit	\$96,770.00 plus 35% of the amount over \$357,700

Table 2.6: Percentage Changes in the Prices of Dow Stocks, Unequal Intervals

Percent Price Change	Number of Stocks	Relative Frequency
3 to 4	2	$\frac{2}{30} = 0.067$
2 to 3	0	$\frac{0}{30} = 0.000$
1 to 2	5	$\frac{5}{30} = 0.167$
0 to 1	12	$\frac{12}{30} = 0.400$
–2 to 0	11	$\frac{11}{30} = 0.367$
Total	30	1.000

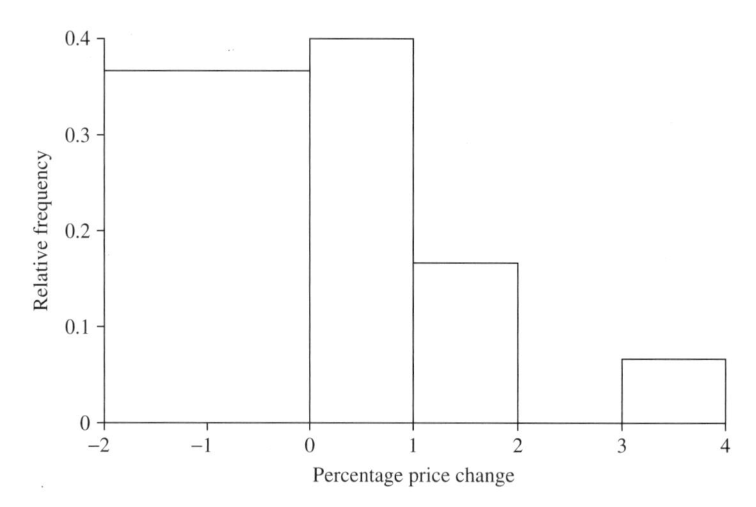

Figure 2.8
A bar chart with unequal intervals.

the bars rather than their heights. The heights of the bars are correct, but the areas are misleading. The interval –2 to 0 percent contains 36.7 percent of the observations and the height of the bar is 0.367, but the double-wide bar in Figure 2.8 gives this interval 54 percent of the total area encompassed by the four bars. It seems that more than half of the stocks had a price decline, when only about a third did.

If the problem is that we look at areas instead of heights, the solution is to make the areas equal to the relative frequencies. Remembering that the area of a bar is equal to its width times its height, we want

$$\text{relative frequency} = \text{bar area} = (\text{bar width})(\text{bar height})$$

Therefore, the bar height must be set equal to the relative frequency divided by the bar width:

$$\text{bar height} = \frac{\text{relative frequency}}{\text{bar width}}$$

When the bar heights are adjusted in this way, so that the bar areas are equal to the relative frequencies, the graph is called a *histogram*. Table 2.7 shows these calculations for our Dow price change data. For the first four intervals, we simply divide by 1 because these interval widths are all equal to 1. For the fifth interval (the double-wide interval –2 to 0), we divide by 2 to get the histogram height.

Figure 2.9 shows a histogram using the calculations in Table 2.7. The area of the first bar is equal to its relative frequency

$$\text{bar area} = (\text{bar width})(\text{bar height}) = 2(0.183) = 0.367$$

Table 2.7: Histogram Heights for the Percentage Changes in the Prices of Dow Stocks

Percent Price Change	Number of Stocks	Relative Frequency	Histogram Bar Height
3 to 4	2	$\frac{2}{30} = 0.067$	$\frac{0.067}{1} = 0.067$
2 to 3	0	$\frac{0}{30} = 0.000$	$\frac{0.000}{1} = 0.000$
1 to 2	5	$\frac{5}{30} = 0.167$	$\frac{0.167}{1} = 0.167$
0 to 1	12	$\frac{12}{30} = 0.400$	$\frac{0.400}{1} = 0.400$
−2 to 0	11	$\frac{11}{30} = 0.367$	$\frac{0.367}{2} = 0.183$
Total	30	1.000	

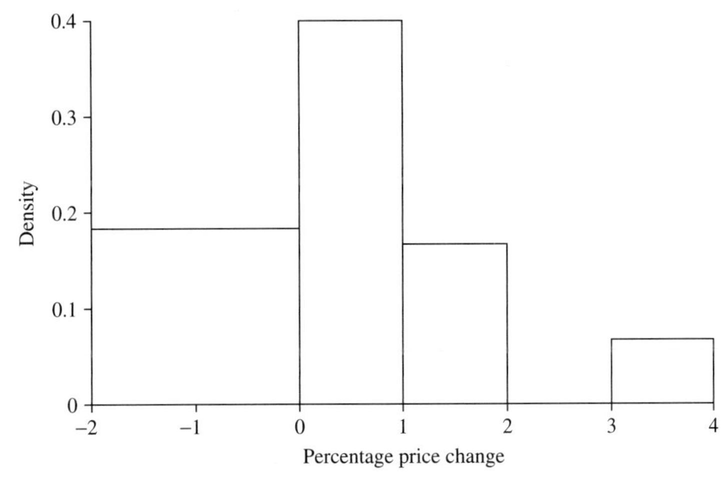

Figure 2.9
A histogram for the Dow price changes.

The fact that the area of this first bar is equal to 36.7 percent of the total area correctly reflects the fact that 36.7 percent of the observations are in this interval. Because the area of each bar is equal to a relative frequency, the total area of the bars is equal to 1.

The height of a bar in a histogram is called its *density*, because it measures the relative frequency per unit of the variable we are analyzing. In Figure 2.9, the density is frequency per percentage point. When the data are measured in dollars, the density is frequency per dollar. When the data are measured in years, the density is frequency per year. The general term *density* covers all these cases. Because the term is unfamiliar to nonstatisticians and histograms are standardized to have a total area equal to 1, histograms are often drawn with no label on the vertical axis or no vertical axis at all!

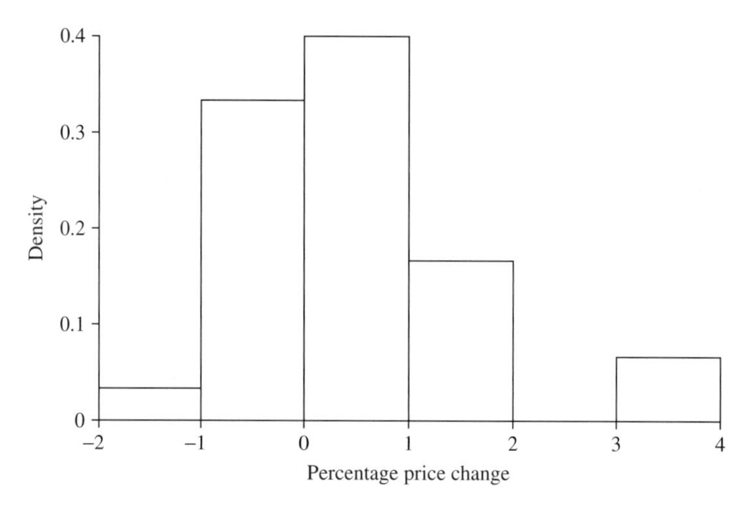

Figure 2.10
A histogram with equal intervals.

Figure 2.10 shows a histogram of the price change data, using equal intervals. If we compare the two bars in Figure 2.10 for the intervals –2 to –1 and –1 to 0 with the single bar in Figure 2.9 for the combined interval –2 to 0, we see that the single bar is just an average of the two separate bars. The single bar loses some of the detail provided by two separate bars but does not misrepresent how often the percentage price changes were negative.

Letting the Data Speak

Within reasonable bounds, the number of intervals used in a histogram is a matter of taste. Here are a few guidelines. Histograms with fewer than four intervals usually convey too little information to justify drawing a figure. If all we want to say is that 63 percent of the prices went up, we can say this in words more easily than in graphs. At the other extreme, if the number of intervals is too large, a histogram can turn into a jumble of narrow spikes, with each spike representing only a few observations. Histograms usually have five to ten intervals, using more intervals when we have more data.

In choosing the width and location of the intervals, the simplest approach is equally wide intervals that begin with round numbers, like 10 and 20, and cover all of the data. Some intervals may be intrinsically interesting; for example, 0 is often an informative place to end one interval and begin another.

If our data have an open-ended interval, such as "more than $100,000 income," we can omit this interval entirely. Alternatively, we can close the interval with an assumed value and hope that it represents the data fairly; for instance, if we are working with education data that have an interval "more than 16 years," we might make this interval 16–20 years, and assume that this encompasses almost all of the data.

Understanding the Data

A histogram is intended to help us understand the data. One of the first things we notice in a histogram is the center of the data, which is between 0 and 1 percent in Figure 2.10. Figure 2.11 shows how a histogram might have multiple peaks.

Another consideration, as in Figure 2.12, is whether the data are bunched closely around the center of the data or widely dispersed.

We also notice if the histogram is roughly symmetrical about the center. The two histograms in Figures 2.12 are perfectly symmetrical, in that the bars to the right of center are an exact mirror image of the bars to the left of center. Data are seldom perfectly symmetrical but are often roughly so. On the other hand, histograms sometimes have an asymmetrical shape that resembles a playground slide: steep on the side with the ladder then declining more gradually on the side with the slide. The asymmetrical data in the top half of Figure 2.13 are *positively skewed*, or skewed to the right. The data in the bottom half of Figure 2.13 are negatively skewed and run off to the left.

Histograms may also reveal *outliers*—values that are very different from the other observations. If one value is very different from the rest, as in Figure 2.14, we should check our data to see why. Sometimes an outlier is just a clerical error, the misplacement of a decimal point. Sometimes, it reflects a unique situation—perhaps the year that a natural disaster occurred—and can be discarded if we are not interested in such unusual events. In other cases, the outlier may be a very informative observation that should be investigated further. It might be very interesting to see how the stock market reacts to a natural disaster, a presidential assassination, or the end of a war.

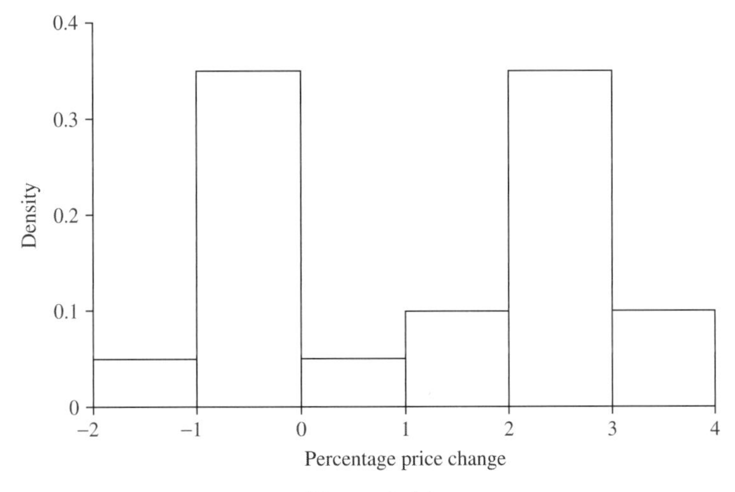

Figure 2.11
A histogram with multiple peaks.

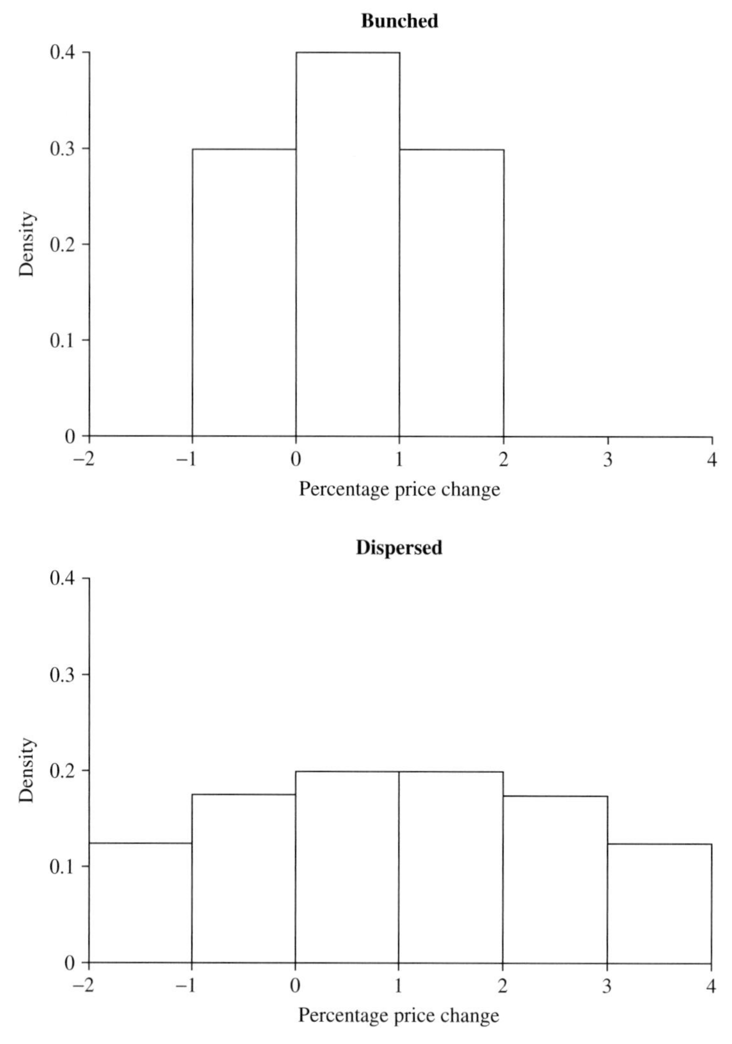

Figure 2.12
Data may be bunched or dispersed.

A dramatic example occurred in the 1980s, when scientists discovered that the software analyzing satellite ozone readings over the South Pole had been automatically omitting a substantial number of very low readings because these were outliers in comparison to readings made in the 1970s. The computer program assumed that readings this far from what had been normal in the 1970s must be mistakes. When scientists reanalyzed the data, including the previously ignored outliers, they found that an ominous hole in the ozone layer had been developing for several years, with Antarctic ozone declining by 40 percent between 1979 and 1985. Susan Solomon, of the National Oceanic and Atmospheric Administration's Aeronomy Laboratory, said that, "this is a change in the ozone that's of

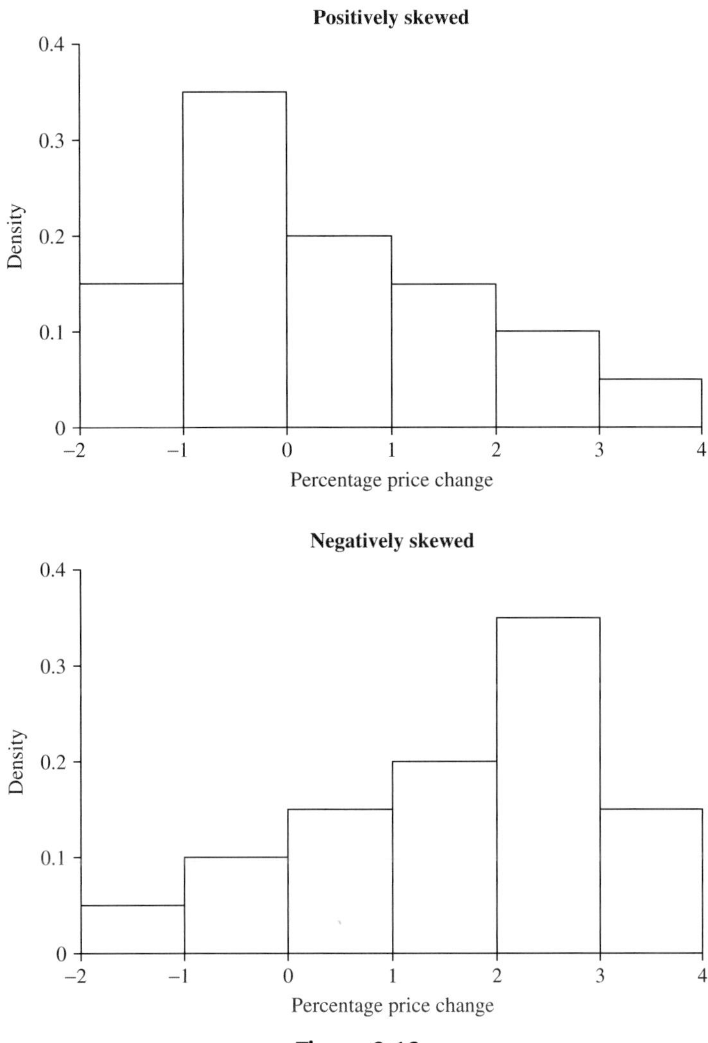

Figure 2.13
Positively and negatively skewed data.

absolutely unprecedented proportions. We've just never seen anything like what we're experiencing in the Antarctic" [4]. Outliers are sometimes clerical errors, measurement errors, or flukes that, if not corrected or omitted, will distort our analysis of the data; other times, as with the ozone readings, they are the most interesting observations.

Histograms are often roughly bell-shaped, with a single peak and the bars declining symmetrically on either side of the peak—at first gradually, then more quickly, then gradually again. The histogram in Figure 2.15 of annual stock returns going back to 1926 (the earliest year that these data are available) has a very rough bell shape. This bell-shaped curve, known as the *normal distribution*, is encountered frequently in later chapters.

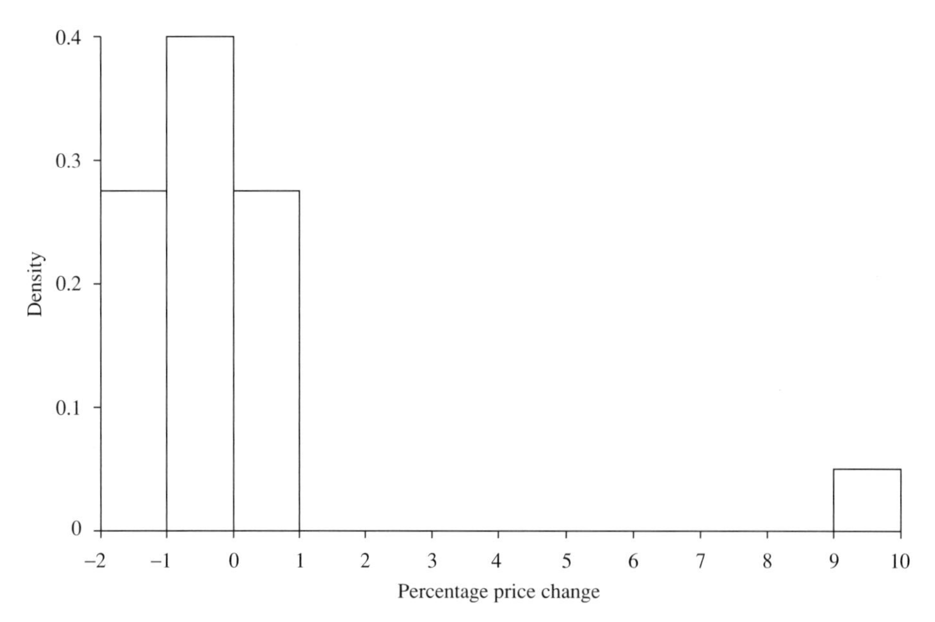

Figure 2.14
An outlier.

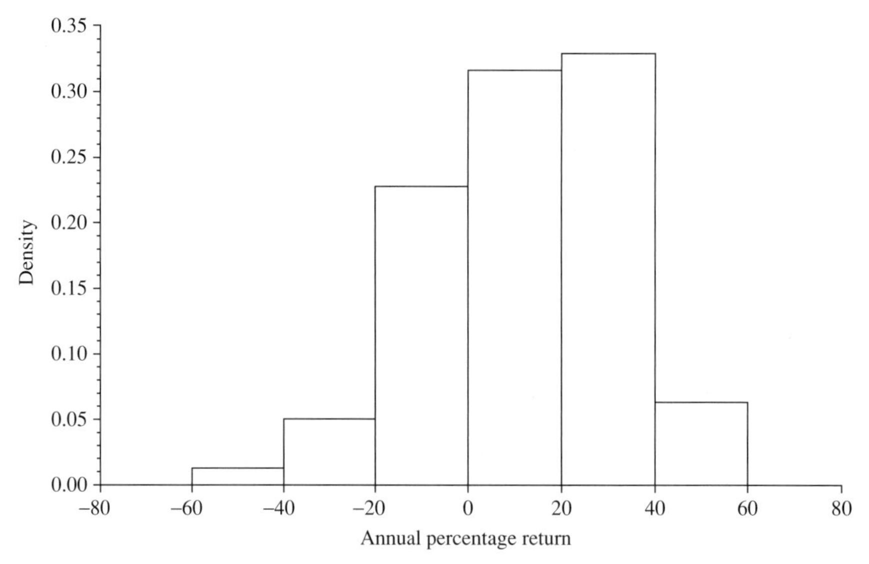

Figure 2.15
Annual percentage stock returns (1926–2009).

2.3 Time Series Graphs

Time series graphs show data over time, with time on the horizontal axis and the data of interest on the vertical axis. Time series graphs can be used to reveal several characteristics of data:

1. Whether there is a long-run trend upward or downward.
2. Whether there are any unusual deviations from the long-run trend.
3. Whether there has been a change in the trend.
4. Whether there are seasonal or other patterns in the data.

Figure 2.16 shows daily values of the Dow Jones Industrial Average between 1920 and 1940. Very clearly, we see a run up in the late 1920s, followed by a spectacular collapse. After the 1929 crash, the Dow recovered to nearly 300 in the spring of 1930 but then began a long, tortuous slide, punctuated by brief but inadequate rallies before finally touching bottom at 41.22 on July 8, 1932—down 89 percent from the September 1929 peak of 381.17. It was not until 1956, 27 years later, that the stock market regained its 1929 peak.

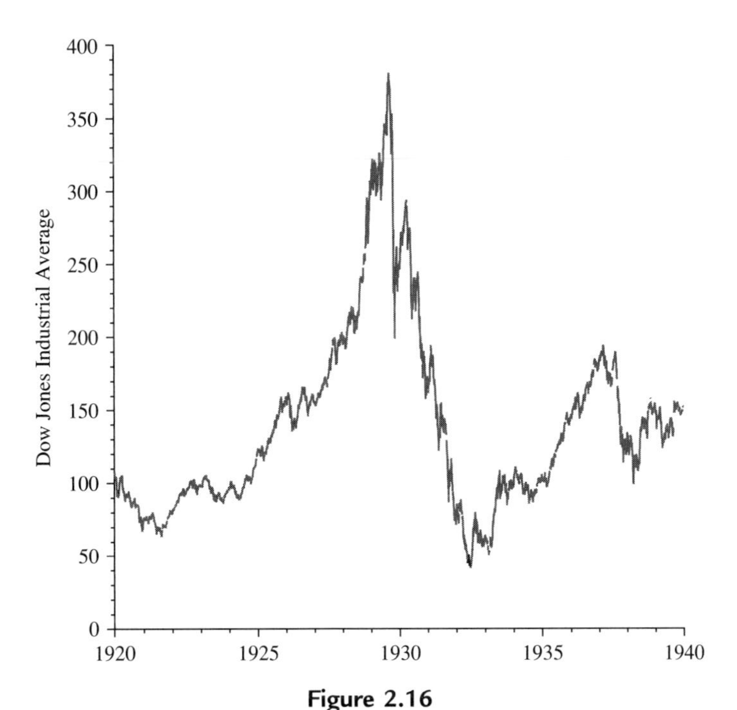

Figure 2.16
The stock market crash during the Great Depression.

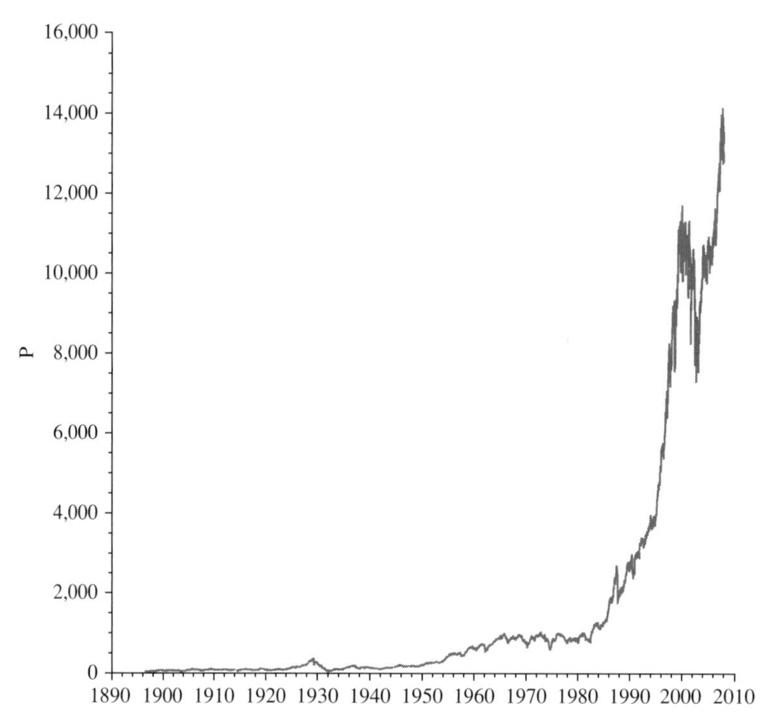

Figure 2.17
The Dow Jones Industrial Average, 1896–2007.

Figure 2.17 shows daily values of the Dow from when it began in 1896 until December 31, 2007. Clearly, there has been a long-run upward trend, punctuated by sharp increases and declines.

The 89 percent drop in 1929–1932 barely registers in Figure 2.17, but the 35 percent drop between April 11, 2000, and October 9, 2002, looks enormous. This is because the 89 percent drop in 1929–1932 was a 340-point fall, which was huge relative to the 381 value of the Dow in 1929 but small relative to the 13,000 value of the Dow in 2007. In contrast, the 38 percent drop in 2000–2002 was a 4,437-point decline.

What matters more to investors, the number of points the market drops or the percentage by which it drops? If it is the percentage ups and downs that matter, then we can graph the logarithm of prices, as in Figure 2.18. The slope of a logarithmic graph shows the percentage change. Figure 2.18 puts the data in proper perspective if it is percentage changes that matter. Figure 2.18 shows that, in percentage terms, the 1929–1932 stock market crash was much worse than the 2000–2002 crash.

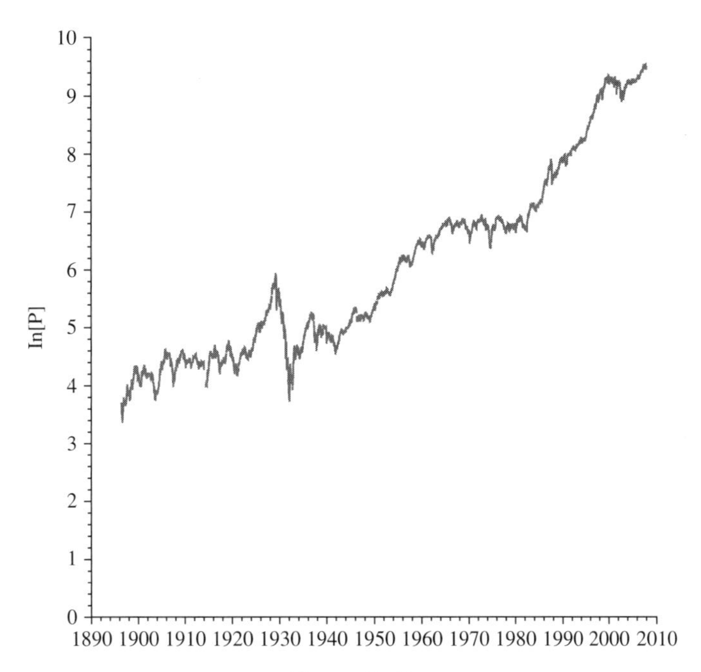

Figure 2.18
The logarithm of the Dow Jones Industrial Average, 1896–2007.

Unemployment during the Great Depression

In Chapter 1, we saw that government officials had very little economic data during the Great Depression. Government analysts mostly drew inferences from what they heard from friends or witnessed themselves. Researchers have subsequently pieced together estimates of the unemployment rate for several years before, during, and after the Great Depression. The federal government's Bureau of Labor Statistics has compiled official monthly estimates of the unemployment rate from household surveys since 1948.

Figure 2.19 is a time series graph of these after-the-fact estimates of the unemployment rate back to 1920—what government officials might have seen if unemployment data had been collected at that time. Figure 2.19 shows the spike in unemployment during the 1920–1921 recession, at the time the sharpest recession the United States had ever experienced, and also shows the devastating increase in unemployment between 1929 and 1932. The unemployment rate declined after 1932 but still was at levels higher than during the 1920–1921 crisis. The unemployment rate was above 14 percent for a full decade, and plunged to tolerable levels only when the United States became involved in World War II.

Cigarette Consumption

Time series data are often adjusted for size of the population or the price level to help us distinguish between changes that happened because the population grew or prices increased, and changes that happened because people changed their behavior.

In Chapter 1, we saw that total cigarette sales in the United States increased by 8 percent between 1960 and 1990, but that per capita sales fell by 22 percent. A time series graph can give a more complete picture. Figure 2.20 shows total sales each year between 1900 and

Figure 2.19
Unemployment during the Great Depression [5].

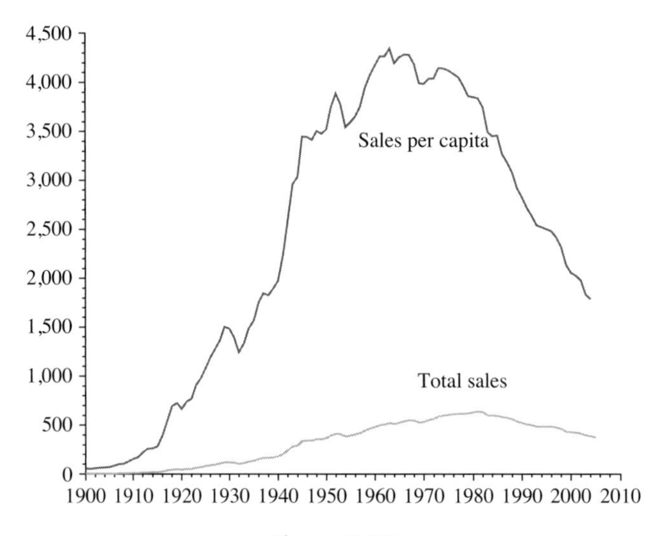

Figure 2.20
Total cigarette sales (billions) and sales per capita.

2007 and total sales per adult. Total sales generally increased up until 1981, but sales per capita peaked in the 1960s and declined steadily after the release of the 1964 *Surgeon General's Report* warning of the health risks associated with smoking. Figure 2.20 convincingly demonstrates a long-term upward trend in sales per capita before the *Surgeon General's Report* and a downward trend afterward. By looking at sales per capita, we can see the change in people's behavior.

2.4 Scatterplots

Scatterplots depict the relationship between two variables. If there is thought to be a causal relationship between these variables, with changes in the independent variable causing changes in the dependent variable, then the independent variable is put on the horizontal axis and the dependent variable is put on the vertical axis.

For example, Table 1.1, reproduced here as Table 2.8, shows the values of aggregate U.S. income and spending each year from 1929 through 1940. Spending seems to go up and down with income, but it is not clear whether this is a close relationship and how we would quantify this relationship. When income goes up by a dollar, how much does consumption tend to increase?

A scatter diagram can help us answer these questions. Keynes' theory was that spending is positively related to income—so income should be on the horizontal axis and spending should be on the vertical axis. Figure 2.21 shows a scatterplot of these data. There does, indeed, appear to be a pretty tight relationship between income and spending. In Chapter 8, we see how to fit a line to these data so that we can quantify the relationship between income and spending. It turns out that, when income increases by a dollar, spending tends to increase by about 84 cents.

Table 2.8: U.S. Disposable Personal Income and Consumer Spending, Billions of Dollars

Year	Income	Spending
1929	83.4	77.4
1930	74.7	70.1
1931	64.3	60.7
1932	49.2	48.7
1933	46.1	45.9
1934	52.8	51.5
1935	59.3	55.9
1936	67.4	62.2
1937	72.2	66.8
1938	66.6	64.3
1939	71.4	67.2
1940	76.8	71.3

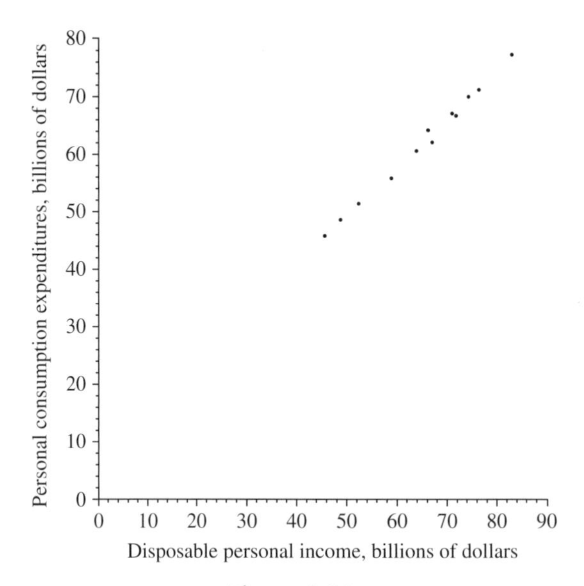

Figure 2.21
Time series relationship between income and spending, 1929–1940.

Okun's Law

Chapter 1 discussed Okun's law, the idea that there is an inverse relationship between changes in output and the unemployment rate. The causal direction is unclear here. It could be argued that an increase in the unemployment rate reduces the number of people working and the amount they produce. Or it could be argued that, when consumers, businesses, and governments buy less, firms hire fewer workers and the unemployment rate increases. We follow this latter line of reasoning and put the percentage change in real GDP on the horizontal axis and the percentage-point change in the unemployment rate on the vertical axis. A scatterplot of annual data for the years 1947–2007 (Figure 2.22) shows a clear inverse relationship. When GDP goes up, unemployment goes down; when GDP goes down, unemployment goes up. In Chapter 8, we see how to quantify this inverse relationship by estimating Okun's law.

The Fed's Stock Valuation Model

In a speech on December 5, 1996, Federal Reserve Board Chairman Alan Greenspan [6] worried aloud about "irrational exuberance" in the stock market. The Fed subsequently developed a stock valuation model in which the "fair-value price" P for the S&P 500 stock index is equal to the ratio of the earnings E of the companies in the S&P 500 to the interest rate R on 10-year Treasury bonds:

$$P = \frac{E}{R}$$

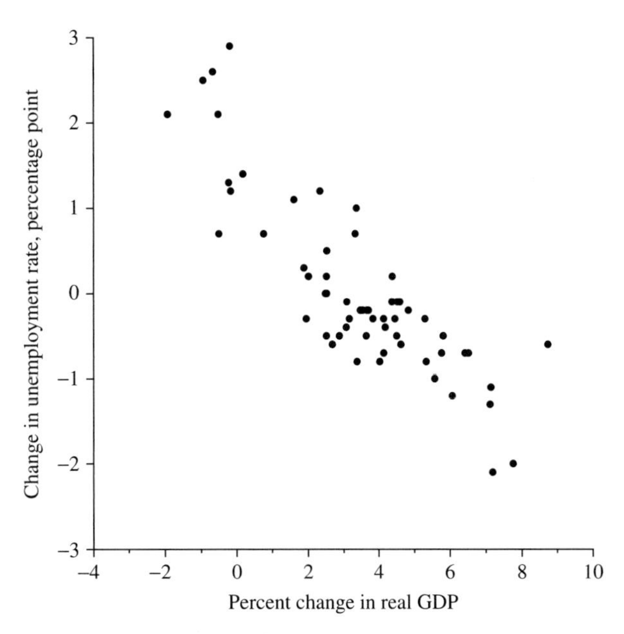

Figure 2.22
Output and unemployment, 1947–2007.

Another way of expressing this model is that the fair-value earnings/price ratio is equal to the interest rate on 10-year Treasury bonds:

$$\frac{E}{P} = R \tag{2.1}$$

We can think of the earnings/price ratio, the left-hand side of Equation 2.1, as a rough estimate of the anticipated return from buying stock. If you buy stock for \$100 a share and the company earns \$10 a share, then your rate of return is evidently 10 percent. This is not quite right because firms typically do not pay all of their earnings out to shareholders. They pay out some of their earnings as dividends and reinvest the rest in the firm. The shareholders' returns depend on how profitable this reinvestment is. Still, the earnings/price ratio is a rough estimate of the shareholders' returns.

The right-hand side of Equation 2.1 is the interest rate on 10-year Treasury bonds, which is a reasonable estimate of an investor's rate of return from buying these bonds. So, Equation 2.1 equates the anticipated rate of return from stocks to the rate of return from bonds. It is unlikely that these are exactly equal to each other, but it is plausible that they are positively related. If the interest rate on Treasury bonds goes up, the return on stocks has to increase too; otherwise, investors flee from stocks to bonds.

Figure 2.23 shows that, overall, there is a remarkably close correspondence between R and E/P. One big divergence was in the 1973–1974 stock market crash, when stock prices fell

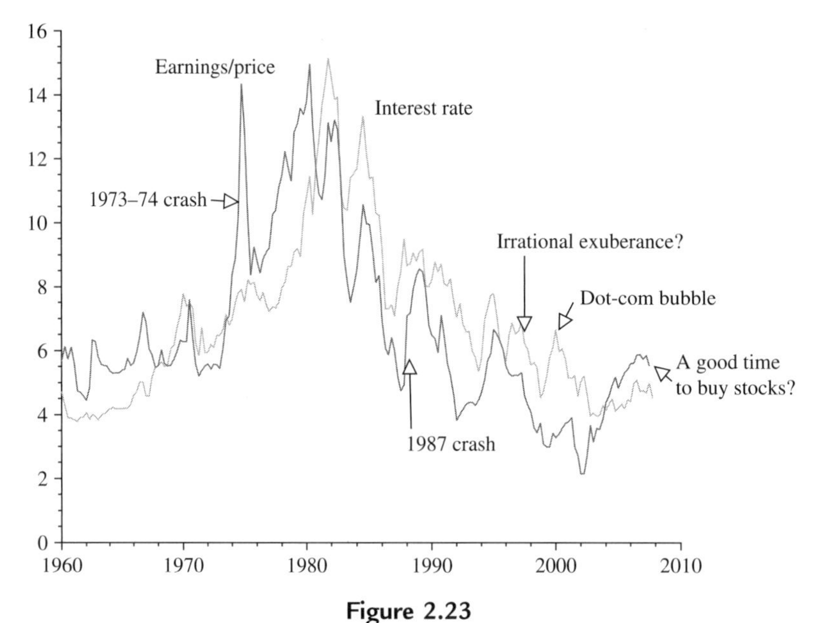

Figure 2.23
Stock earnings/price ratio and the interest rate on 10-year Treasury bonds.

dramatically and increased the earnings/price ratio substantially. Another divergence was in 1987, shortly before the biggest one-day crash in U.S. stock market history. Ironically, there seems to be nothing unusual in December 1996, when Fed Chair Alan Greenspan worried aloud about "irrational exuberance" in the stock market. After his pronouncement, however, prices raced far ahead of earnings, driving the earnings/price ratio to a record low, before the dot-com bubble burst.

In 2009, the earnings/price ratio seemed high—suggesting that stocks prices were, if anything, low. Time will tell if this suggestion is correct.

Using Categorical Data in Scatter Diagrams

Sometimes, we want to separate the data in a scatter diagram into two or more categories. We can do this by using different symbols or colors to identify the data in the different categories.

Table 1.6 shows the prices of hard drives of various sizes in the years 2003 through 2007. Figure 2.24 shows the data for 2006 and 2007, separated into these two categories by using asterisks for the 2006 data and dots for the 2007 data. We see that (a) in each year, there is a positive relationship between size and price; and (b) prices fell between 2006 and 2007.

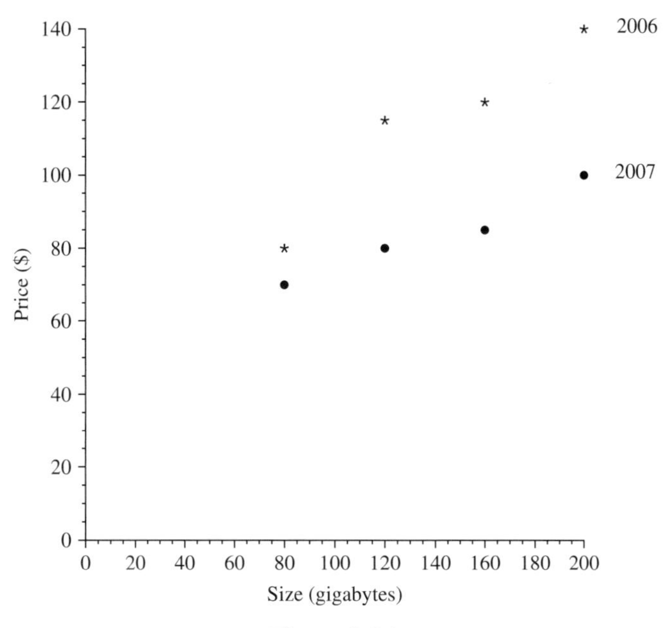

Figure 2.24
Price and storage capacity.

2.5 Graphs: Good, Bad, and Ugly

Graphs are supposed to help us understand data. They should display data accurately and encourage the reader to think about the data rather than admire the artwork. Unfortunately, graphs can be misleading if not done carefully and honestly.

Omitting the Origin

Sometimes, it is helpful to omit 0 from the vertical axis so that we can see the fine detail. But, the resulting magnification of the ups and downs in the data can give a misleading impression of how big the ups and downs are.

For example, Figure 2.25 shows a graph of weekly data on the assets of money market mutual funds similar to one that was used to illustrate a newspaper writer's claim that, "After peaking at $392 billion in March, money market fund assets have fallen sharply" [7]. This figure omits 0 from the vertical axis so that we can see clearly that money market assets peaked on March 25 then declined. This decline is not only easier to see, but also appears much bigger than it really was. The height of the line drops 30 percent, which suggests that fund assets fell sharply; however, the actual numbers show that there was only a 2 percent decline, from $392 billion to $384 billion.

Figure 2.26 shows the same data, but with 0 included on the vertical axis. Now, the values (ranging from $373 billion to $392 billion) are so far from zero that any patterns in the data have been ironed out of the graph, and we cannot tell when fund assets peaked. On the other hand, the graph does tell us correctly that the dip was minor.

In general, the omission of 0 magnifies variations in the heights of bars in bar charts and histograms and in the heights of lines in a time series graph, allowing us to detect differences that might otherwise be ambiguous. However, once 0 has been omitted, we can

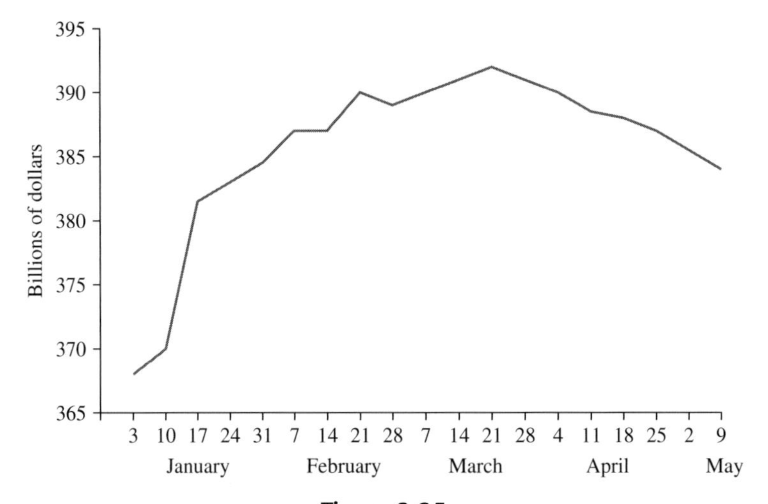

Figure 2.25
Money market fund assets fall sharply.

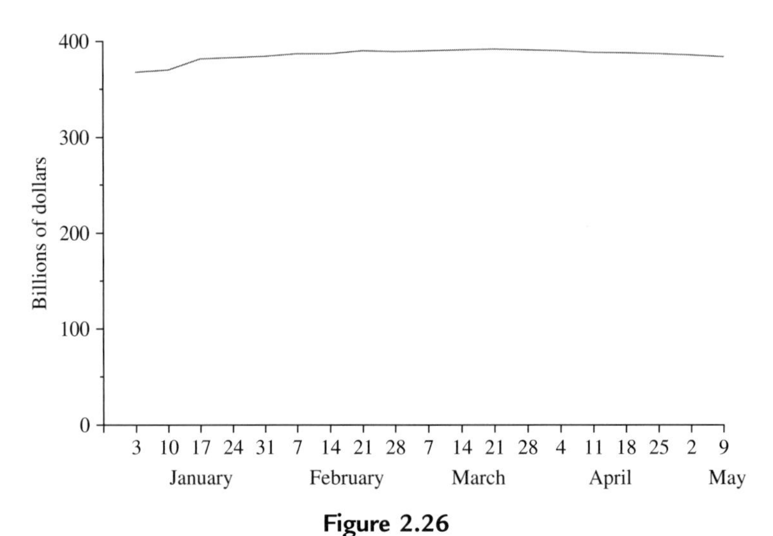

Figure 2.26
Not so sharply when the origin is shown.

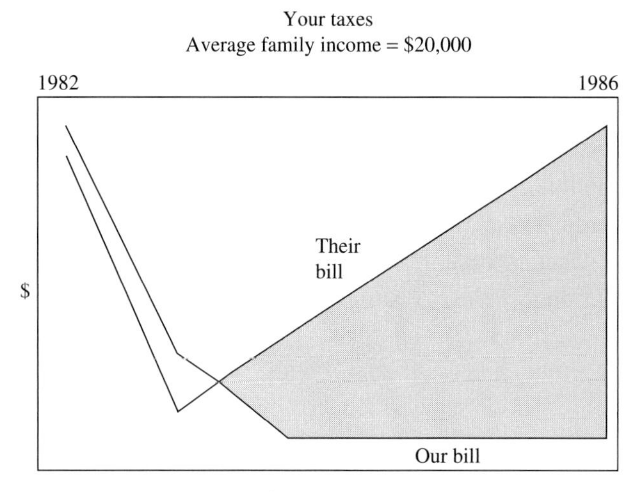

Figure 2.27
How much would taxes be?

no longer gauge relative magnitudes by comparing the heights of the bars or lines. Instead, we need to look at the actual numbers.

Figure 2.27 gives a famous example in which President Ronald Reagan showed the American people how much a family with $20,000 income would pay in taxes under the administration's tax proposal, in comparison to a plan drafted by the House Ways and Means Committee [8]. Because the vertical axis shows no numbers (just a dollar sign), we have no basis for gauging the magnitude of the gap between the lines. The 1986 gap in fact represented about a 9 percent tax reduction, from $2,385 to $2,168. The omission of 0 magnifies this 9 percent difference to 90 percent of the height of the line labeled "Their bill," and the omission of numbers from the vertical axis prevents the reader from detecting this magnification. Afterward, in response to criticism that this figure distorted the data, a White House spokesman told reporters, "We tried it with numbers and found that it was very hard to read on television, so we took them off. We were trying to get a point across" [9].

Changing the Units in Mid-Graph

For a graph to represent the data accurately, the units on each axis must be consistent. One inch should not represent $1,000 for part of the axis and $10,000 for the rest of the axis. If the units change in mid-graph, the figure surely distorts the information contained in the data. For example, a *Washington Post* graph like Figure 2.28 showed a steady increase in doctor income between 1939 and 1976. Or does it? Look at the horizontal axis. The equally spaced marks initially represent a difference of 8 years between 1939 and 1947, then a difference of 4 years between 1947 and 1951 and 4 years between 1951 and 1955. Later differences range from 1 year to 8 years. These puzzling intervals may have been chosen so that the line is approximately straight. A straight line may be visually appealing but misrepresents the data.

Figure 2.29 uses consistent intervals on the horizontal axis to show what really happened to doctor income during this time period. Doctor income did not increase in a straight line, and there may have been a significant change in 1964, when Medicare began. To explore this more fully we should adjust doctor income for inflation; we could also graph the logarithm of doctor income, as explained earlier, so that the slope of the line shows the percentage change. No matter what data we use, each axis should use consistent units.

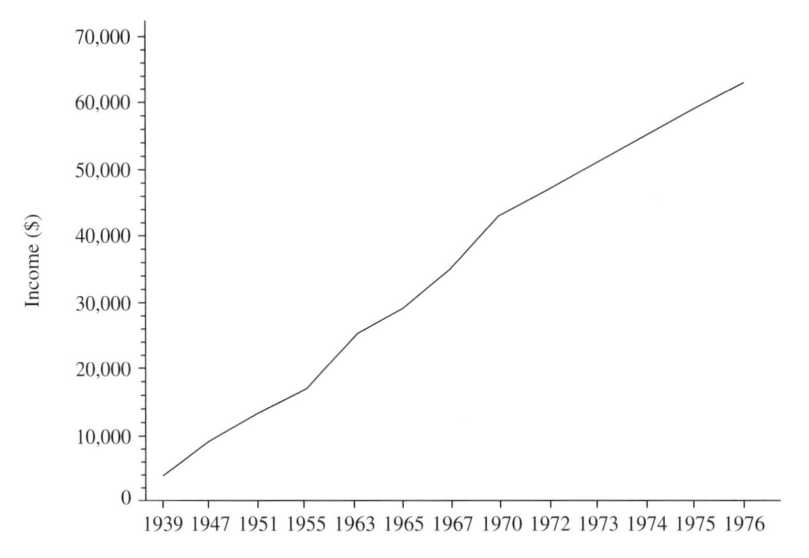

Figure 2.28
Doctor income marches steadily upward [10].

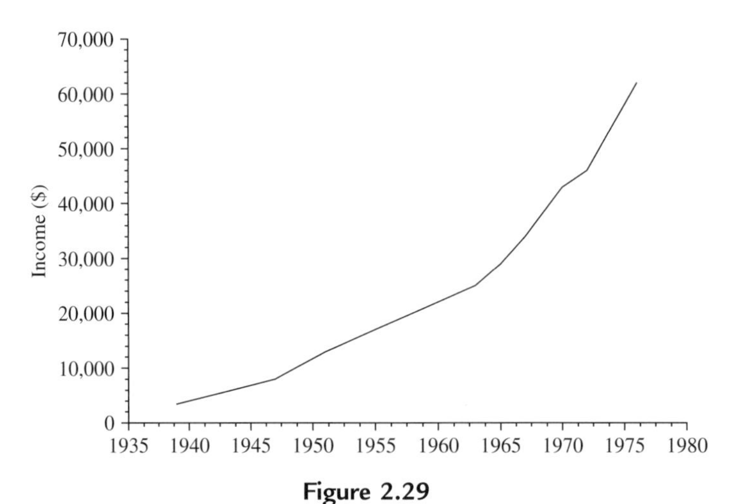

Figure 2.29
A better picture of doctor income.

In 1976, the National Science Foundation (NSF) used a graph like Figure 2.30 to show a decline in the number of Nobel Prizes in science (chemistry, physics, and medicine) awarded to U.S. citizens. Again, look at the time axis. Each of the first seven observations are for decades, but the eighth is for a 4-year period, 1971–1974. The NSF, which certainly knows better, altered the time axis in mid-graph!

Because the number of prizes awarded during a 4-year period is fewer than during a 10-year period, the NSF was able to create an illusion of an alarming drop in the number of science prizes awarded to Americans. Using data for the full decade, Figure 2.31 shows that U.S. citizens ended up winning even more prizes in the 1970s than in the 1960s. (Of course, this trend could not continue forever, unless the Nobel Committee increased the total number of prizes; in the 1970s, the U.S. won more than half of the science Nobel Prizes.)

Choosing the Time Period

One of the most bitter criticisms of statisticians is that, "Figures don't lie, but liars figure." The complaint is that an unscrupulous statistician can prove anything by carefully choosing favorable data and ignoring conflicting evidence. For example, a clever selection of when to begin and end a time series graph can be used to show a trend that would be absent in a more complete graph. Figure 2.32 illustrates this point with daily data on the Dow Jones Industrial Average from January 2004 through December 2008. If we were to show data for only October 2004 through October 2007, it would appear that stock prices increase

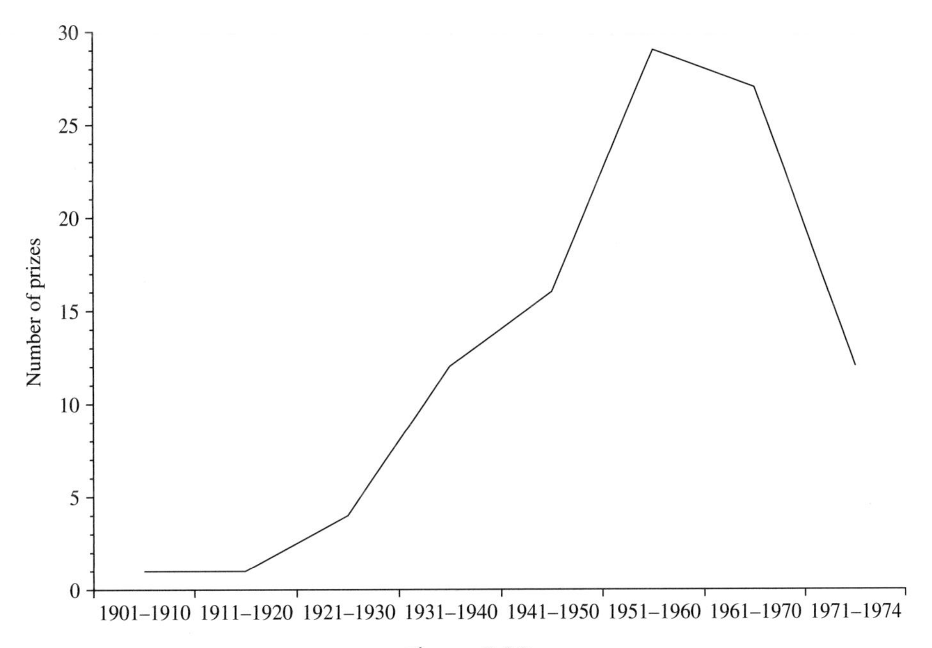

Figure 2.30
A precipitous decline in U.S. Nobel Prizes in science? [11].

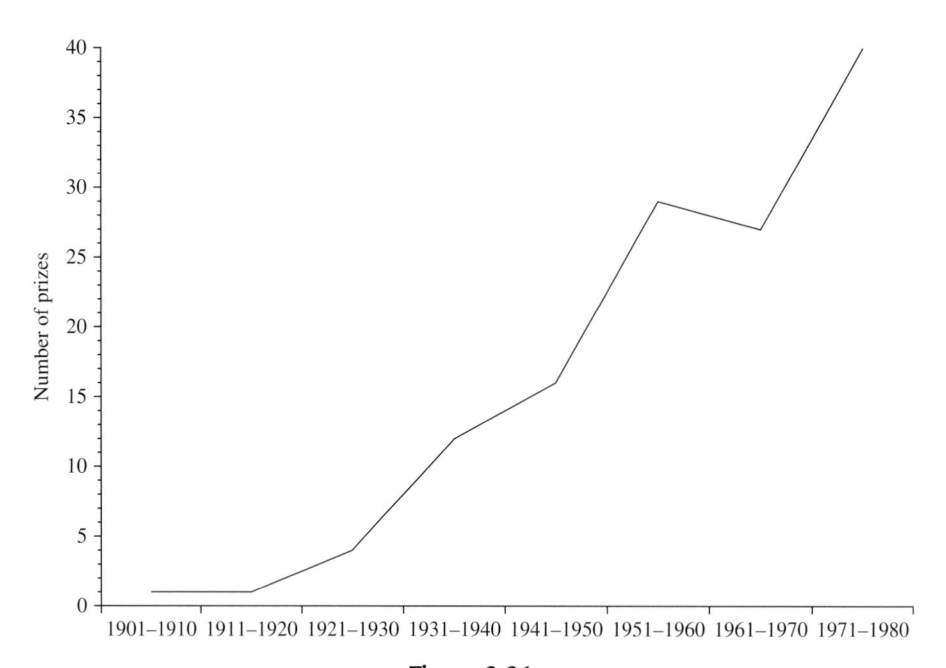

Figure 2.31
U.S. Nobel Prizes in science keep increasing.

Figure 2.32
Stocks are a great/terrible investment.

by 39 percent (more than 11 percent a year); if, instead, we were to show data for only November 2007 through November 2008, it would appear that stock prices fell by 46 percent a year. Which is correct? Both time periods are too brief to describe meaningful trends in something so volatile. A much longer perspective is needed for a more balanced view. Over the past 100 years, stock prices have increased, on average, by about 4 percent a year.

Would anyone be so naive as to extrapolate stock prices on the basis of only a few years of data? Investors do it all the time, over even shorter periods, hoping that the latest zig is the prelude to many happy profits or fearing that a recent zag is the beginning of a crash. In October 2007, a financial planner's clients told her that they wanted to get a home equity loan so that they could borrow more money and earn double-digit returns in the stock market. A year later, these clients said they wanted to sell all their stocks because they could not afford to lose another 50 percent in the stock market.

We should be suspicious whenever the data begin or end at peculiar points in time. If someone publishes a graph in 2009 using data for October 2002 to January 2007, we should be puzzled why this person did not use data for before October 2002 or after January 2007, and why the data begin in October and end in January. A general rule is that, if the beginning and ending points seem to be peculiar choices that one would make only after scrutinizing the data, then they were probably chosen to distort the historical record. There may be a perfectly logical explanation, but we should insist on hearing this explanation.

A graph is essentially descriptive—a picture meant to tell a story. As with any story, bumblers may mangle the punch line and dishonest people may tell lies. Those who are conscientious and honest can use graphs to describe data in a simple and revealing way. There is room for mistakes and chicanery, but there is also room for enlightenment.

The Dangers of Incautious Extrapolation

While time series graphs provide a simple and useful way of describing the past, we must not forget that they are merely descriptive. Graphs may reveal patterns in the data, but they do not explain why these patterns occur. We, the users, must provide these logical explanations. Sometimes, explanations are easily found and provide good reasons for predicting that patterns in past data will continue in the future. Because of the Thanksgiving holiday, the consumption of turkey has shown a bulge in November for decades and will probably continue to do so for years to come. Because of the cold weather, construction activity declines in winter in the northern part of the United States and will most likely continue doing so.

If we have no logical explanation for the patterns in the data and nonetheless assume they will continue, then we have made an *incautious extrapolation* that may well turn out to be embarrassingly incorrect. Extrapolations are riskier the further into the future we look, because the underlying causes have more time to weaken or evaporate. We can be more certain of turkey sales next November than in the year 5000.

In his second State of the Union address, Abraham Lincoln predicted, based on an extrapolation of data for the years 1790 to 1860, that the U.S. population would be 251,689,914 in 1930 [12]. (In addition to the incautious extrapolation, note the unjustified precision in this estimate of the population 70 years hence.) The actual U.S. population in 1930 turned out to be 123 million, less than half the size that Lincoln had predicted. A 1938 presidential commission [13] erred in the other direction, predicting that the U.S. population would never exceed 140 million. Just 12 years later, in 1950, the U.S. population was 152 million. Fifty years after that, in 2000, the U.S. population was 281 million.

Many ludicrous examples have been concocted to dramatize the error of incautious extrapolation. In 1940, there were an average of 3.2 people in each car on the highway in the United States. By 1960 this average had dropped to 1.4. At this rate, by the year 2080, seven out of eight cars on the highway will be empty!

Unnecessary Decoration

Some well-meaning people add pictures, lines, blots, and splotches to graphs, apparently intending to enliven the pictures, but too often creating what Edward Tufte [14] calls *chartjunk*, as in Figure 2.33, unattractive graphs that distract the reader and strain the eyes. Another way to create chartjunk is to use multiple colors, preferably bright and clashing.

Computers also make possible *textjunk*, printed documents that look like a ransom note made by pasting together characters in mismatched sizes, styles, and fonts. I once received a two-page newsletter that used 32 fonts, not counting variations caused by letters that were bold, italicized, or in different sizes. It was painful to read.

The purpose of a graph is to reveal information that would not be as apparent or as easily remembered if presented in a table. Figure 2.33 is an example that does neither. The nine

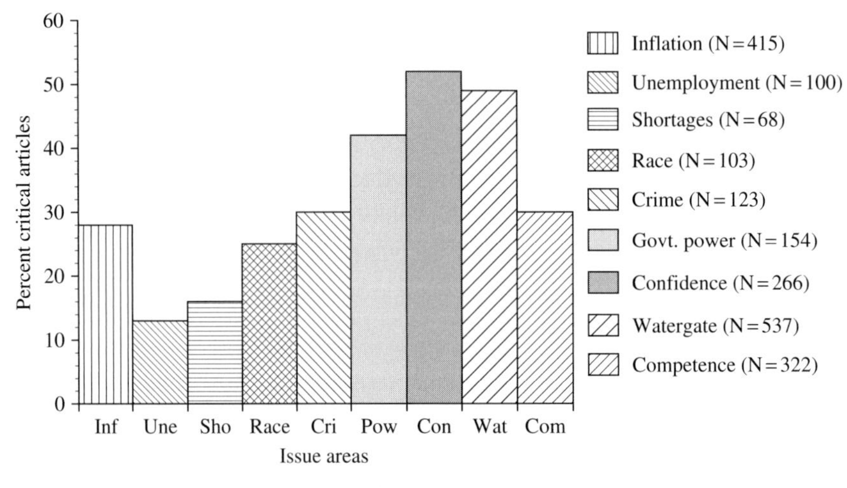

Figure 2.33
An unhelpful bar graph [15].

Table 2.9: Sometimes, a Simple Table Is Better Than a Cluttered Graph [16]

Content of Front-Page Articles	Number of Articles	Percent of Articles That Criticized a Specific Person or Policy
Watergate	537	49
Inflation	415	28
Government competence	322	30
Confidence in government	266	52
Government power	154	42
Crime	123	30
Race	103	25
Unemployment	100	13
Shortages of energy or food	68	16

bars refer to nine types of articles from the front pages of U.S. newspapers, with the heights of the bars showing the percentage of the articles in each category that were critical of a specific person or policy. The order in which the categories are presented is arbitrary; there is no reason, for example, why the inflation bar should precede rather than follow the unemployment bar. The pattern in the bars—dipping, rising, then dipping again—reveals no useful information. It would be more sensible to arrange the bars from tallest to smallest.

The bar labels along the horizontal axis are abbreviated and jumbled together, making it difficult to tell which label goes with which bar and what some labels mean. Longer versions of the labels appear in a legend to the right of the bar chart, but the similarity in the bar patterns makes it tedious to look back and forth, and difficult to remember which label goes with which pattern. The bar patterns themselves are distracting and make it hard to focus on the heights of the bars. Finally, if the reader does manage to concentrate on the heights of the bars, it is not easy to line up these heights with the numerical values on the vertical axis. This bar chart is not an improvement over a simple table, such as Table 2.9.

The first full-time graph specialist at *Time* magazine was an art school graduate who asserted that "the challenge is to present statistics as a visual idea rather than a tedious parade of numbers." Too often, graphs are drawn to be artistic, rather than informative. Graphs are not mere decoration, to enliven numbers for the easily bored. They are intended to summarize and display data to help our analysis and interpretation. A useful graph displays data accurately and coherently and encourages the reader to understand the data's substance.

Exercises

2.1 For each of the following studies, identify the type of graph (histogram, time series graph, or scatter diagram) that would be the *most* appropriate. (You can use more than one graph of each type, for example, two histograms.)

a. Are stock returns more volatile than bond returns?

b. Do interest rates affect home construction?'

c. Is the rate of inflation higher or lower now than in the past?
d. Do cigarette sales depend on cigarette prices?
e. Is there more variation in the unemployment rate or the rate of inflation?

2.2 For each of the following studies, identify the type of graph (histogram, time series graph, or scatter diagram) that would be the *most* appropriate. (You can use more than one graph of each type, for example, two histograms.)
a. Has air pollution in Los Angeles generally risen or fallen during the past 20 years?
b. Do colleges that accept a large percentage of their students in early-decision programs have higher yields (percentage of accepted students who enroll)?
c. Can college grade point averages be predicted from high school grade point averages?
d. Do students who get more sleep get higher grades?
e. Is there more dispersion in the starting salaries of economics majors or English majors?

2.3 For each of the following studies, identify the type of graph (histogram, time series graph, or scatter diagram) that would be the *most* appropriate. (You can use more than one graph of each type, for example, two histograms.)
a. Do countries with lots of smokers have lots of lung-cancer deaths?
b. Does the time between eruptions of the Old Faithful geyser depend on the duration of the preceding eruption?
c. Is there more variation in annual rainfall in Los Angeles or in New York?
d. Have temperatures in Los Angeles gone up or down over the past 100 years?
e. Could the states that John McCain won and lost in 2008 have been predicted from how well George W. Bush did in each state in 2004?

2.4 For each of the following studies, identify the type of graph (histogram, time series graph, or scatter diagram) that would be the *most* appropriate. (You can use more than one graph of each type, for example, two histograms.)
a. Has the average grade in statistics classes risen or fallen in the past 20 years?
b. Can final exam scores in statistics classes be predicted from homework scores?
c. Can grades in statistics classes be predicted from overall GPAs?
d. Is there more dispersion in homework scores or final exam scores?
e. Is there more dispersion in grades in statistics classes or history classes?

2.5 Use the data in Table 1.4 to make a histogram of the U.S. dollar prices of a Big Mac in these 20 countries. Use these intervals for the prices: 1–2, 2–3, 3–4, 4–5, 5–8.

2.6 Use the data in Table 1.4 to make a histogram of the U.S. dollar prices of a Big Mac in these 20 countries. Use these intervals for the prices: 1–3, 3–4, 4–5, 5–8.

2.7 Use the data in Table 1.4 for the 19 countries other than the United States to make a histogram of the U.S. dollar prices of a Big Mac. Use these intervals for the prices: 1.41–2.41, 2.41–3.41, 3.41–4.41, 4.41–5.41, 5.41–8.41.

2.8 Use these household taxable income data (in thousands of dollars) to make a histogram, using intervals equal to the federal income tax brackets shown in Table 2.5: 10, 20, 30, 40, 50, 60, 70, 80, 90, 100.

2.9 Use these household taxable income data (in thousands of dollars) to make a histogram, using intervals equal to the federal income tax brackets shown in Table 2.5: 15, 23, 32, 37, 48, 48, 75, 80, 110, 130.

2.10 Use these household taxable income data (in thousands of dollars) to make a histogram, using intervals equal to the federal income tax brackets shown in Table 2.5: 10, 15, 20, 20, 30, 35, 40, 40, 50, 55, 60, 60, 70, 75, 80, 80, 90, 95, 100, 100.

2.11 Use these household taxable income data (in thousands of dollars) to make a histogram, using intervals equal to the federal income tax brackets shown in Table 2.5: 12, 15, 23, 24, 32, 37, 37, 37, 48, 48, 52, 59, 75, 80, 80, 85, 110, 114, 130, 150.

2.12 Use the tax brackets in Table 2.5 to calculate the taxes and average tax rates for each of these incomes (in thousands of dollars) for these ten married couples filing a joint tax return: 10, 20, 30, 40, 50, 60, 70, 80, 90, 100. Now display these average tax rates in a histogram with these intervals: 0%–5%, 5%–10%, 10%–15%, 15%–20%, 20%–25%.

2.13 Use the tax brackets in Table 2.5 to calculate the taxes and average tax rates for each of these incomes (in thousands of dollars) for these ten married couples filing a joint tax return: 15, 23, 32, 37, 48, 48, 75, 80, 110, 130. Now display these average tax rates in a histogram with these intervals: 0%–5%, 5%–10%, 10%–15%, 15%–20%, 20%–25%.

2.14 Do not collect any data, but use your general knowledge to draw a rough sketch of a histogram of the ages of the students at your school. Be sure to label the horizontal and vertical axes.

2.15 Do not collect any data, but use your general knowledge to draw a rough sketch of a histogram of the first-year salaries of last year's graduates from your school. Be sure to label the horizontal and vertical axes.

2.16 Use the data in Tables 1.9 and 1.10 to make a time series graph of real average hourly earnings.

2.17 Use the data in Table 1.11 to make a time series graph of the real cost of mailing a letter.

2.18 Use the data in Table 1.2 to make a scatterplot of income and spending. Does there seem to be a close positive relationship?

2.19 Explain why you think the following pairs of data are positively related, negatively related, or essentially unrelated. If you think there is a causal relationship, which is the explanatory variable?

a. The height of a father and the height of his oldest son.

b. The age of a mother and the age of her oldest child.

c. The number of scalp hairs and the age of a man between the ages of 30 and 40.
d. A woman's pre-pregnancy weight and the weight of her baby.
e. A woman's age and her cost of a 1-year life insurance policy.

2.20 Explain why you think the following pairs of data are positively related, negatively related, or essentially unrelated. If you think that there is a relationship, which is the explanatory variable?
a. The age (in days) and weight of male children less than 1 year old.
b. The age and chances of having breast cancer, for women between the ages of 40 and 80.
c. The age and height of people between the ages of 30 and 40.
d. The weight and gasoline mileage (miles per gallon) of 2010 cars.
e. The income and age of male lawyers aged 21 to 54.

2.21 Table 2.10 shows the percentage of the Jewish population of various ages in Germany in 1928 and in 1959 and of a theoretical stationary population that is neither growing nor contracting. Make three histograms, one for each population. (Assume that the interval 60 and over is 60–79.) What conclusions can you draw from these graphs?

2.22 A survey of the prices of houses recently sold in Indianapolis obtained these data. Display these data in a histogram.

Sale Price ($)	Number of Houses
75,000–100,000	10
100,000–150,000	30
150,000–200,000	20

2.23 A survey of the prices of houses recently sold in Boston obtained these data. Display these data in a histogram.

Sale Price ($)	Number of Houses
100,000–200,000	10
200,000–500,000	60
500,000–1,000,000	30

Table 2.10: Exercise 2.21 [17]

Age	1928	1959	Stationary Model
0–19	25.9	14.2	33
20–39	31.3	19.8	30
40–59	27.7	28.2	25
60 and over	14.3	28.2	12

2.24 Use the test scores in Exercise 1.8 two make two histograms, one using the first-grade scores and the other the eighth-grade scores. In each case, use the intervals 0–9, 10–19,..., 90–99. Summarize the main differences between these two histograms.

2.25 Use a histogram with intervals 0–9.99, 10–14.99, 15–19.99, and 20–34.99 to summarize the data in Table 2.11 on annual inches of precipitation from 1961 to 1990 at the Los Angeles Civic Center. Be sure to show your work.

2.26 From the 30 stocks in the Dow Jones Industrial Average, students identified the ten stocks with the highest dividend/price (*D*/*P*) ratio and the ten stocks with the lowest *D*/*P* ratio. The percentage returns on each stock were then recorded over the next year (Table 2.12). Display these returns in two histograms using the same units on the horizontal axis. Do the data appear to have similar distributions?

2.27 Plutonium has been produced in Hanford, Washington, since the 1940s, and some radioactive waste has leaked into the Columbia River. A 1965 study of cancer incidence in nearby communities [18] compared an exposure index and the cancer mortality rate per 100,000 residents for nine Oregon counties:

Exposure Index	8.34	6.41	3.41	3.83	2.57	11.64	1.25	2.49	1.62
Cancer Mortality	210.3	177.9	129.9	162.3	130.1	207.5	113.5	147.1	137.5

Which variable is the explanatory variable and which is the dependent variable? Make a scatter diagram and describe the relationship as either positive, negative, or nonexistent.

Table 2.11: Exercise 2.25

5.83	15.37	12.31	7.98	26.81	12.91	23.66	7.58	26.32	16.54
9.26	6.54	17.45	16.69	10.70	11.01	14.97	30.57	17.00	26.33
10.92	14.41	34.04	8.90	8.92	18.00	9.11	9.98	4.56	6.49

Table 2.12: Exercise 2.26

High *D*/*P*	Low *D*/*P*
39.77	−37.74
32.01	46.83
28.93	−5.81
32.30	48.04
31.28	20.91
12.17	20.24
40.68	1.05
14.33	−28.10
40.39	17.03
7.74	−1.55

2.28 The Soviet Antarctica Expeditions at Vostok analyzed an ice core that was 2,083 meters long to estimate the atmospheric concentrations of carbon dioxide during the past 160,000 years [19]. Table 2.13 shows some of their data, where D is the depth of the ice core, in meters; A is the age of the trapped air, in years before present; and C is the carbon dioxide concentration, in parts per million by volume. Draw a time series graph of atmospheric carbon dioxide concentration over the past 160,000 years. (Show the most recent period on the right-hand side of the horizontal axis.) What patterns do you see?

2.29 Table 2.14 shows U.S. federal government saving and GDP (in billions of dollars). Government saving is government income minus government expenditures; a positive value represents a budget surplus, a negative value a budget deficit. Calculate the government savings as a percentage of GDP each year and display your data in a time series graph. Write a brief paragraph summarizing your graph.

Table 2.13: Exercise 2.28

Depth D	Age A	Carbon C	Depth D	Age A	Carbon C
126.4	1,700	274.5	1,225.7	80,900	222.5
302.6	9,140	259.0	1,349.0	90,630	226.0
375.6	12,930	245.0	1,451.5	98,950	225.0
474.2	20,090	194.5	1,575.2	110,510	233.5
602.3	30,910	223.0	1,676.4	119,500	280.0
748.3	42,310	178.5	1,825.7	130,460	275.0
852.5	50,150	201.0	1,948.7	140,430	231.0
975.7	59,770	201.0	2,025.7	150,700	200.5
1,101.4	70,770	243.0	2,077.5	159,690	195.5

Table 2.14: Exercise 2.29

Year	Saving	GDP	Year	Saving	GDP	Year	Saving	GDP
1960	−1.0	526.4	1977	−46.0	2,030.1	1994	−262.3	7,085.2
1961	−8.0	544.8	1978	−31.9	2,293.8	1995	−244.5	7,414.7
1962	−8.3	585.7	1979	−26.7	2,562.2	1996	−179.7	7,838.5
1963	−4.1	617.8	1980	−75.7	2,788.1	1997	−73.8	8,332.4
1964	−7.9	663.6	1981	−73.5	3,126.8	1998	27.0	8,793.5
1965	−5.2	719.1	1982	−161.3	3,253.2	1999	64.4	9,353.5
1966	−8.0	787.7	1983	−201.8	3,534.6	2000	146.6	9,951.5
1967	−21.2	832.4	1984	−190.4	3,930.9	2001	−65.1	10,286.2
1968	−12.1	909.8	1985	−215.5	4,217.5	2002	−422.4	10,642.3
1969	1.4	984.4	1986	−238.5	4,460.1	2003	−553.3	11,142.1
1970	−21.5	1,038.3	1987	−208.4	4,736.4	2004	−531.1	11,867.8
1971	−31.3	1,126.8	1988	−186.7	5,100.4	2005	−418.3	12,638.4
1972	−16.8	1,237.9	1989	−181.7	5,482.1	2006	−291.6	13,398.9
1973	−3.8	1,382.3	1990	−251.8	5,800.5	2007	−408.1	14,061.8
1974	−15.7	1,499.5	1991	−301.3	5,992.1	2008	−912.3	14,369.1
1975	−86.7	1,637.7	1992	−373.1	6,342.4	2009	−1,592.7	14,119.0
1976	−62.5	1,824.6	1993	−338.3	6,667.4	2010	−1,559.1	14,660.4

2.30 Use the data in the preceding exercise to calculate the percentage change in government saving and the percentage change in GDP each year from 1990 through 2010. When you calculate the percentage change from a negative base, divide by the absolute value of the base. For example, the percentage change in saving from 1990 to 1991 is 100*(–301.3 – (–251.8))/abs(–251.8) = –19.6%. Make a scatterplot of these percentage changes and write a brief paragraph summarizing this graph.

2.31 A study of the average sentence length of British public speakers produced the data in Table 2.15. Plot these data using a horizontal axis from 1500 to 2200 and a vertical axis from 0 to 80. Draw a freehand straight line that fits these data and extend this line to obtain a rough prediction of the average number of words per sentence in the year 2200.

2.32 In 1974, Congress tried to reduce U.S. fuel usage by imposing a nationwide 55 miles per hour speed limit. As time passed, motorists increasingly disregarded this speed limit; and in 1987, the speed limit on rural interstate highways was increased to 65 miles per hour. Plot the data on U.S. motor vehicle deaths per 100 million miles driven, shown in Table 2.16, and see if anything unusual seems to have happened in 1974 and during the next several years.

2.33 In 1992, two UCLA professors compared the trends over time in male and female world records in several running events. Their data (in minutes) for the marathon,

Table 2.15: Exercise 2.31 [20]

Speaker	Year	Words per Sentence
Francis Bacon	1598	72.2
Oliver Cromwell	1654	48.6
John Tillotson	1694	57.2
William Pitt	1777	30.0
Benjamin Disraeli	1846	42.8
David Lloyd George	1909	22.6
Winston Churchill	1940	24.2

Table 2.16: Exercise 2.32

Year	Deaths	Year	Deaths	Year	Deaths	Year	Deaths
1960	5.1	1969	4.9	1978	3.3	1987	2.4
1961	5.1	1970	4.7	1979	3.3	1988	2.3
1962	5.1	1971	4.5	1980	3.3	1989	2.2
1963	5.3	1972	4.3	1981	3.2	1990	2.1
1964	5.4	1973	4.1	1982	2.8	1991	1.9
1965	5.3	1974	3.5	1983	2.6	1992	1.8
1966	5.5	1975	3.4	1984	2.6	1993	1.8
1967	5.4	1976	3.3	1985	2.5	1994	1.7
1968	5.2	1977	3.3	1986	2.5		

a race covering 42,195 meters, are in Table 2.17. Convert each of these times to velocity, in meters per minute, by dividing 42,195 meters by the time. Then plot the velocity data on a single time series graph, with the velocity axis running from 0 to 400 and the time axis running from 1900 to 2000. (Plot each point at the midpoint of the decade; for example, plot the 1910s data at 1915.) The professors wrote that, "Despite the potential pitfalls, we could not resist extrapolating these record progressions into the future." Draw one freehand straight line that fits the male data and another that fits the female data. Do these lines cross before 2000? Extrapolating the female line backward, what was the implied female world record in 1900?

2.34 What is wrong with Figure 2.34, showing the distribution of family income? How would you define the middle class?

2.35 What is wrong with the histogram in Figure 2.35?

Table 2.17: Exercise 2.33 [21]

	1910s	1920s	1930s	1940s	1950s	1960s	1970s	1980s
Males	156.12	149.03	146.70	145.65	135.28	128.57	128.57	126.83
Females					217.12	187.43	147.55	141.10

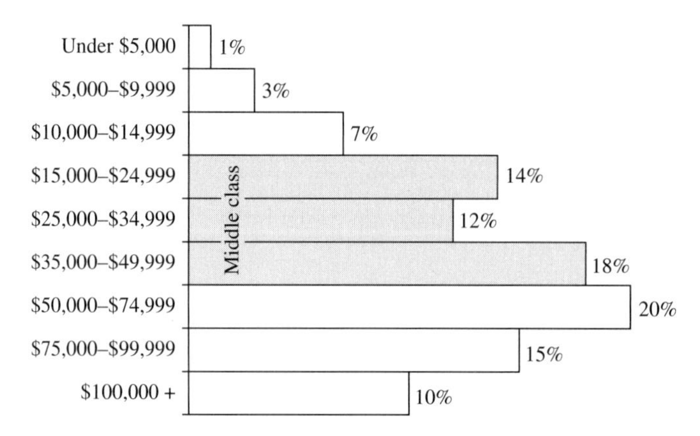

Figure 2.34
Exercise 2.34.

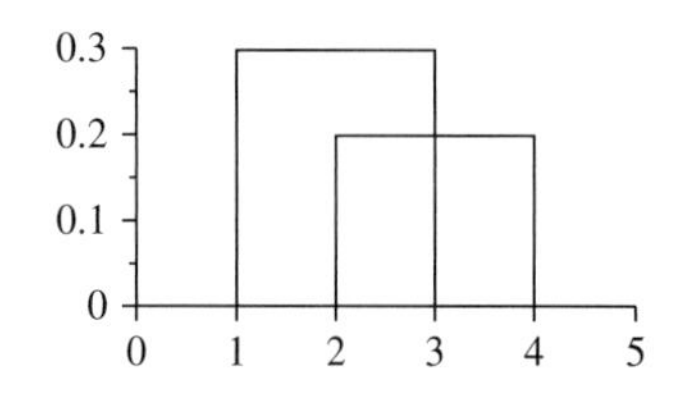

Figure 2.35
Exercise 2.35.

2.36 What is wrong with the histogram in Figure 2.36?

2.37 What is wrong with the histogram in Figure 2.37?

2.38 What is wrong with the histogram in Figure 2.38?

2.39 What is wrong with the histogram in Figure 2.39 that was constructed from annual rainfall data for the years 1901 to 2000? (Density is the rainfall for each year divided by the total rainfall for all 100 years.)

2.40 Explain why Figure 2.40, a graph of the Dow Jones Industrial Average of stock prices on January 27, 1997, gives a misleading visual impression of the volatility of stock prices that day.

2.41 What is misleading about Figure 2.41, showing the increase in the cost of mailing a letter in the U.S. between 1971 and 2009?

2.42 Modern-day adult British women experience a pronounced loss of bone density as they grow older, contributing to an increasing incidence of osteoporotic hip fractures. During the restoration of a London church in the 1990s, a crypt was opened that contained the

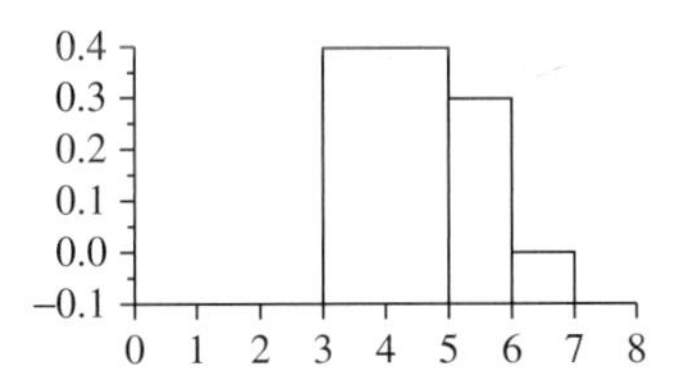

Figure 2.36
Exercise 2.36.

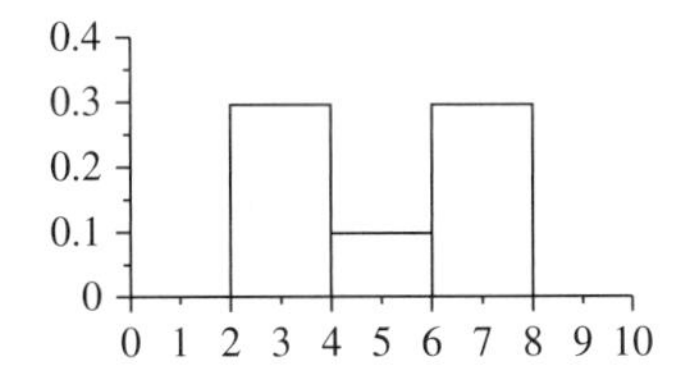

Figure 2.37
Exercise 2.37.

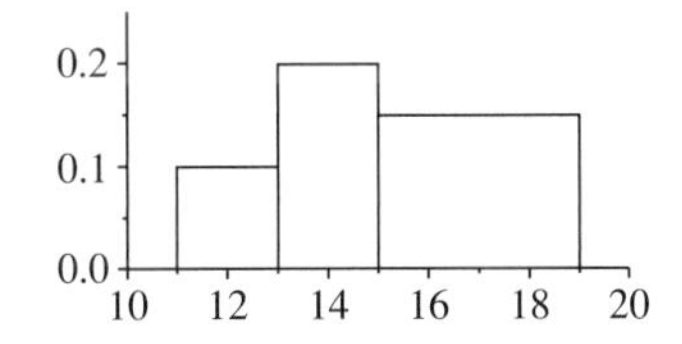

Figure 2.38
Exercise 2.38.

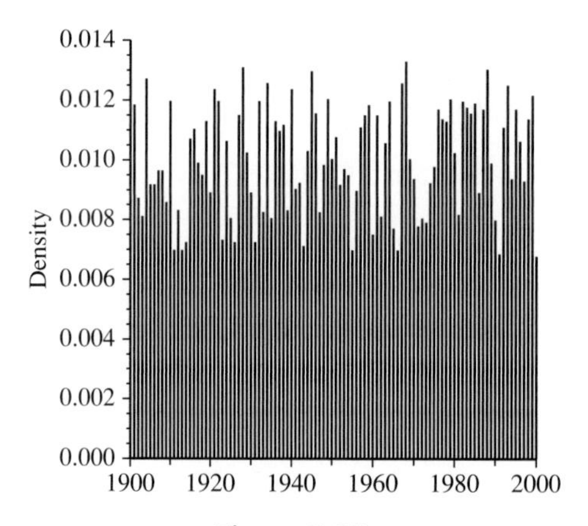

Figure 2.39
Exercise 2.39.

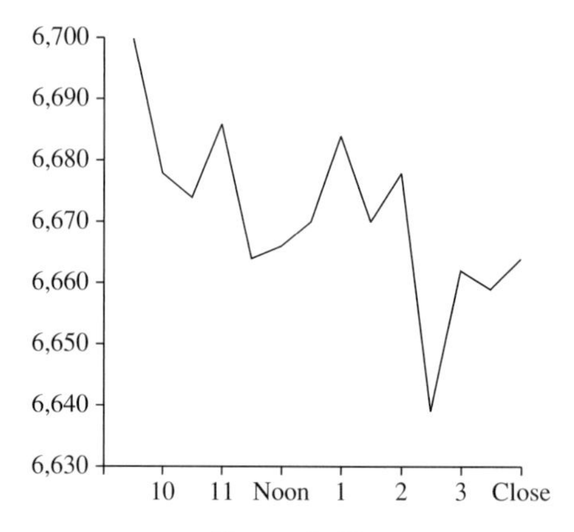

Figure 2.40
Exercise 2.40.

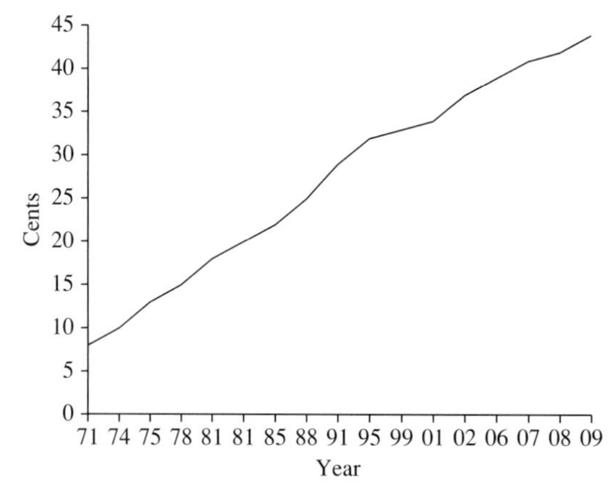

Figure 2.41
Exercise 2.41.

skeletons of more than 100 persons buried between 1729 and 1852. Figure 2.42 is a graph of bone density in the femoral neck region of the skeletons of 25 females who died between the ages of 15 and 45. These bone densities were calculated relative to the average bone density of women in the sample of age 18–30 years. Thus, a figure of 120 indicates that this person's bone density was 20 percent larger than the average 18–30-year-old woman in the sample. Does there seem to be a clear positive relationship, clear negative relationship, or essentially no relationship? What is the implication?

2.43 What is wrong with Figure 2.43 of Table 2.18's quarterly earnings per share data for Gordon Jewelry, a leading retail jewelry shop? Redraw the graph correctly, then describe any patterns you see.

2.44 Identify what is misleading about Figure 2.44, showing a precipitous decline in 2011 in the number of passengers carried by three major airlines [23].

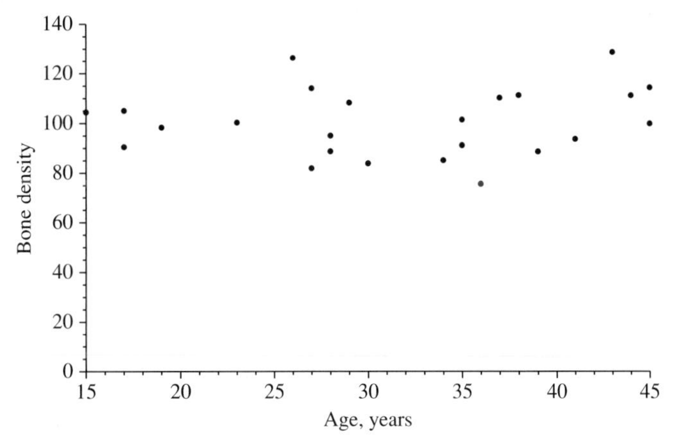

Figure 2.42
Exercise 2.42 [22].

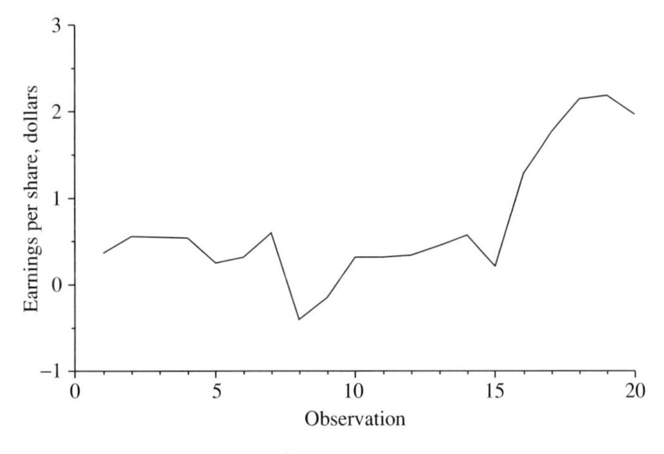

Figure 2.43
Exercise 2.43.

2.45 What is misleading about Figure 2.45, illustrating the fact the purchasing power of the U.S. dollar fell by 50 percent between 1986 and 2010?

2.46 What is misleading about Figure 2.46, showing that the number of U.S. families with annual incomes of $100,000 or more exploded between 1980 and 1993?

2.47 What is misleading about Figure 2.47, comparing the populations of three cities?

2.48 What is misleading about Figure 2.48, showing a leveling off of U.S. federal tax revenue?

Table 2.18: Exercise 2.43

	1978	1979	1980	1981	1982
March 1–May 31	0.37	0.56	0.55	0.54	0.25
June 1–August 31	0.32	0.60	−0.40	−0.15	0.32
September 1–November 30	0.32	0.34	0.45	0.57	0.21
December 1–February 28	1.29	1.77	2.15	2.19	1.97

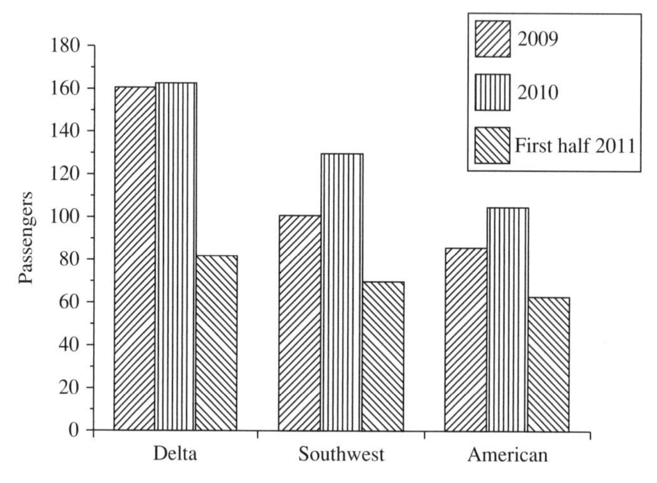

Figure 2.44
Exercise 2.44.

Figure 2.45
Exercise 2.45 [24].

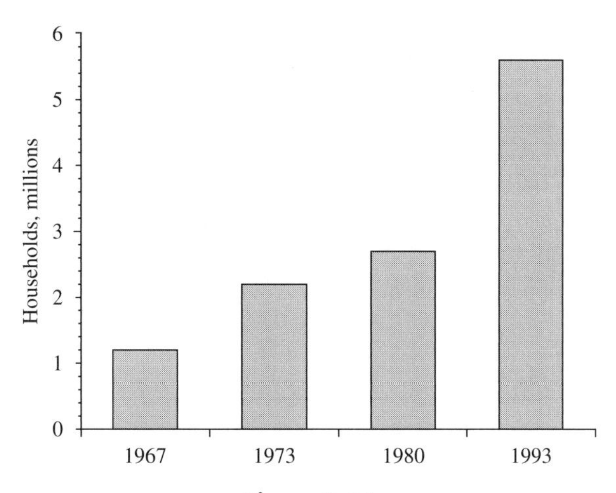

Figure 2.46
Exercise 2.46 [25].

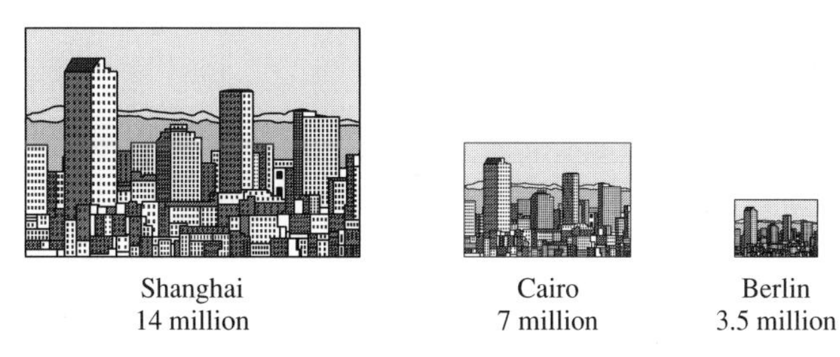

Figure 2.47
Exercise 2.47.

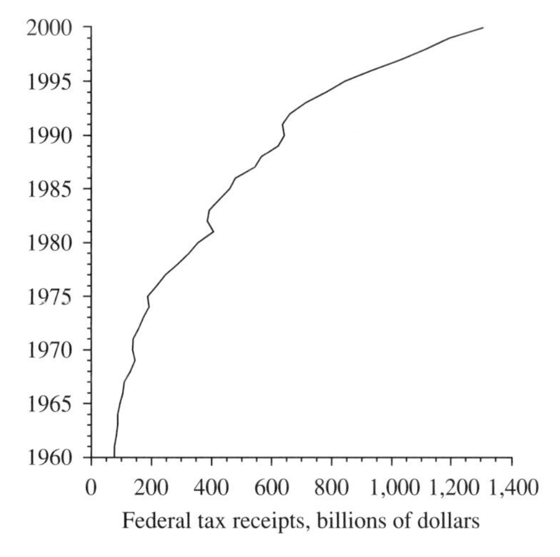

Figure 2.48
Exercise 2.48.

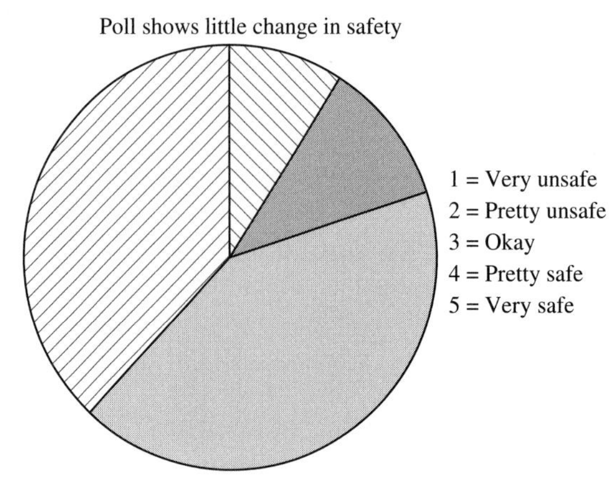

Figure 2.49
Exercise 2.49.

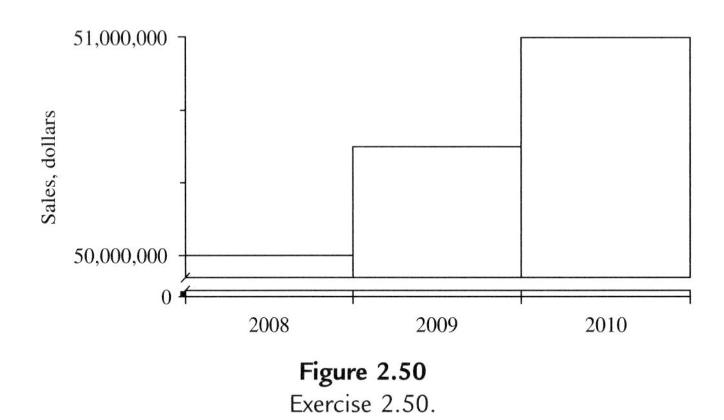

Figure 2.50
Exercise 2.50.

2.49 Identify several problems that prevent Figure 2.49 from conveying useful information.

2.50 What is misleading about Figure 2.50, which shows a large increase in a company's annual sales?

CHAPTER 3

Descriptive Statistics

Chapter Outline

And that's the news from Lake Wobegon, where all the women are strong, all the men are good-looking, and all the children are above-average.

—***Garrison Keillor***

Chapter 2 explained how graphs help us understand data by identifying typical values, outliers, variations, trends, and correlations. For formal statistical inference, we need to work with precise numerical measures. It is not enough to say "It looks like output and unemployment are inversely related." We want to measure the closeness of the relationship and the size of the relationship by making statements like this: "There is a −0.85 correlation between the change in real GDP and the change in the unemployment rate; when real GDP increases by 1 percent, the unemployment rate tends to fall by about 0.4 percentage points." We would also like to have some numerical measure of how much confidence we have in such statements.

In this chapter, we look at several statistical measures used to describe data and draw statistical inferences.

3.1 Mean

The most well-known descriptive statistic is the *mean*, or average value, which is obtained by adding up the values of the data and dividing by the number of observations:

$$\begin{aligned}\overline{X} &= \frac{X_1 + X_2 + \cdots + X_n}{n} \\ &= \frac{1}{n}\sum_{i=1}^{n} X_i\end{aligned} \tag{3.1}$$

Essential Statistics, Regression, and Econometrics

where the n observations are denoted as X_i and the upper case Greek letter Σ (pronounced "sigma") indicates that the values should be summed up. The standard symbol for the mean, $\overline{X}$, can be pronounced "X-bar."

For example, Table 2.3 (in Chapter 2) gives the percentage changes in the prices of the 30 Dow stocks on January 29, 2008. The average is obtained by adding up the 30 values and dividing by 30:

$$\begin{aligned}\overline{X} &= \frac{3.78 + 3.36 + \cdots - 1.24}{30} \\ &= \frac{14.32}{30} \\ &= 0.48\end{aligned}$$

The average percentage change was 0.48 percent.

One interesting property of the mean is that, if we calculate how far each value is from the mean, the average of these deviations is 0:

$$\frac{1}{n}\sum_{i=1}^{n}(X_i - \overline{X}) = 0$$

Table 3.1 gives three simple numerical examples. In the first example, the data are symmetrical about the mean of 20. The first observation is 10 below the mean, the second is equal to the mean, and the third is 10 above the mean. In the second example, the increase in the value of the third observation from 30 to 90 pulls the mean up to 40. Now, the first observation is 30 below the mean, the second is 20 below the mean, and the third is 50 above the mean. In the third example, the increase in the third observation to 270 pulls the mean up to 100. Now, the first observation is 90 below the mean, the second is 80 below the mean, and the third is 170 above the mean. In every case, the average deviation is 0.

Outliers are values that are very different from the other observations and pull the mean toward them. A national magazine once reported [1] that a group of Colorado teachers had failed a history test, with an average score of 67. It turned out that only four teachers had taken the test and one received a score of 20. The other three averaged 83. The one very low score pulled the mean down to 67 and misled a magazine that interpreted the average score as the typical score.

Table 3.1: The Average Deviation from the Mean Is 0

	Symmetrical Data		Data Skewed Right		Big Outlier	
	X_i	$X_i - \overline{X}$	X_i	$X_i - \overline{X}$	X_i	$X_i - \overline{X}$
	10	−10	10	−30	10	−90
	20	0	20	−20	20	−80
	30	10	90	50	270	170
Sum	60	0	120	0	300	0
Average	20	0	40	0	100	0

The Joint Economic Committee of Congress once reported that the share of the nation's wealth owned by the richest 0.5 percent of U.S. families had increased from 25 percent in 1963 to 35 percent in 1983 [2]. Politicians made speeches and newspapers reported the story with "Rich Get Richer" headlines. Some skeptics in Washington rechecked the calculations and discovered that the reported increase was due almost entirely to the incorrect recording of one family's wealth as $200,000,000 rather than $2,000,000, an error that raised the average wealth of the rich people who had been surveyed by nearly 50 percent. Someone typed two extra zeros and temporarily misled the nation.

One way to make the mean less sensitive to outliers is to discard the extreme observations. More than a hundred years ago, a renowned statistician, Francis Edgeworth, argued that the mean is improved "when we have thrown overboard a certain portion of our data—a sort of sacrifice which has often to be made by those who sail upon the stormy seas of Probability" [3]. A *trimmed mean* is calculated by discarding a specified percentage of the data at the two extremes and calculating the average of the remaining data. For example, a 10-percent trimmed mean discards the highest 10 percent of the observations and the lowest 10 percent, then calculates the average of the remaining 80 percent. Trimmed means are commonly used in many sports competitions (such as figure skating) to protect the performers from judges who are excessively generous with their favorites and harsh with other competitors.

3.2 Median

The median is the middle value of the data when the data are arranged in numerical order. It does not matter whether we count in from highest to lowest or lowest to highest. If, for example, there are 11 observations, the median is the sixth observation; if there are 10 observations, the median is halfway between the fifth and sixth observations.

In general, at least half of the values are greater than or equal to the median and at least half are smaller than or equal to the median. We have to use the qualifiers *at least* and *or equal to* because more than one observation may be exactly equal to the median, as with these data:

$$1 \quad 2 \quad 2 \quad 2 \quad 7$$

It is not true that half of the observations are greater than 2 and half are less than 2, but it is true that at least half the observations are greater than or equal to 2 and at least half are less than or equal to 2.

The Dow percentage change data in Table 2.1 are arranged in numerical order. Because there are 30 stocks, the median is halfway between the 15th and 16th returns (Citigroup 0.26 and Wal-Mart 0.30):

$$\begin{aligned} \text{median} &= \frac{0.26+0.30}{2} \\ &= 0.28 \end{aligned}$$

Earlier, we calculated the mean of these data to be 0.48. A histogram (Figure 2.10) of the Dow percentage changes shows that the data are roughly symmetrical, but the two large values (Boeing 3.36 and Alcoa 3.78) pull the mean somewhat above the median.

In comparison to the mean, the median is more robust or resistant to outliers. Looking again at the data in Table 3.1, the median stays at 20 while the increase in the value of the third observation from 30 to 90 to 270 pulls the mean upward.

For some questions, the mean is the most appropriate answer. Whether you balance your budget over the course of a year depends on your average monthly income and expenses. Farm production depends on the average yield per acre. The total amount of cereal needed to fill a million boxes depends on the average net weight. For other questions, the median might be more appropriate.

The U.S. Census Bureau reports both the mean and median household income. The mean income tells us how much each person would earn if the total income were equally divided among households. Of course, income is not evenly divided. In 2007, the mean household income in the United States was $69,193 while the median was $50,740. The mean was pulled above the median by a relatively small number of people with relatively high incomes.

3.3 Standard Deviation

An average does not tell us the underlying variation in the data. Sir Francis Galton once commented that "It is difficult to understand why statisticians commonly limit their enquiries to Averages, and do not revel in more comprehensive views. Their souls seem as dull to the charm of variety as that of the native of one of our flat English counties, whose retrospect of Switzerland was that, if its mountains could be thrown into its lakes, two nuisances would be got rid of at once" [4]. Another Englishman with a sense of humor, Lord Justice Matthews, once told a group of lawyers: "When I was a young man practicing at the bar, I lost a great many cases I should have won. As I got along, I won a great many cases I ought to have lost; so on the whole, justice was done" [5].

We might try to measure the variation or spread in the data by calculating the average deviation from the mean. But remember that, as in Table 3.1, the average deviation from the mean is always 0, no matter what the data look like. Because the positive and negative deviations from the mean offset each other, giving an average deviation of 0, we learn nothing at all about the sizes of the deviations. To eliminate this offsetting of positive and negative deviations, we could take the absolute value of each deviation and then calculate the *average absolute deviation*, which is the sum of the absolute values of the deviations from the mean, divided by the number of observations.

The average absolute deviation is easy to calculate and interpret; however, there are two reasons why it is seldom used. First, while it is easily calculated, a theoretical analysis is very difficult. Second, there is an attractive alternative that plays a prominent role in probability and statistics.

Remember that we took absolute values of the deviations to keep the positive and negative deviations from offsetting each other. Another technique that accomplishes this same end is to square each deviation, since the squares of positive and negative deviations are both positive. The average of these squared deviations is the *variance* s^2 and the square root of the variance is the *standard deviation* s:

$$s^2 = \frac{(X_1 - \overline{X})^2 + (X_2 - \overline{X})^2 + \cdots + (X_n - \overline{X})^2}{n-1} \tag{3.2}$$

Notice that the variance of a set of data is calculated by dividing the sum of the squared deviations by $n - 1$, rather than n. In later chapters, we look at data that are randomly selected from a large population. It can be shown mathematically that, if the variance in the randomly selected data is used to estimate the variance of the population from which these data came, this estimate will, on average, be too low if we divide by n, but will, on average, be correct if we divide by $n - 1$. (This is an example of the kinds of general theorems that mathematicians can prove when they work with the standard deviation rather than the average absolute deviation.)

The variance and standard deviation are equivalent measures of the dispersion in a set of data, in that a data set that has a higher variance than another data set also has a higher standard deviation. However, because each deviation is squared, the variance has a scale that is much larger than the underlying data. The standard deviation has the same units and scale as the original data.

3.4 Boxplots

After the data are arranged in numerical order, from smallest to largest, the data can be divided into four groups of equal size, known as quartiles:

First quartile	minimum to 25th percentile
Second quartile	25th percentile to 50th percentile
Third quartile	50th percentile to 75th percentile
Fourth quartile	75th percentile to maximum

These four subsets are called *quartiles* because each encompasses one-fourth of the data. The upper limit of the first quartile is the 25th percentile because this value is larger than 25 percent of the data. Similarly, the upper limit of the second quartile is the 50th percentile, and the upper limit of the third quartile is the 75th percentile.

The 50th percentile is the median, and we have already seen that this is an appealing measure of the center of the data. An appealing measure of the spread of the data is the *interquartile range*, which is equal to the difference between the 25th and 75th percentiles. The interquartile range is (roughly) the range encompassed by the middle half of the data.

A boxplot (also called, more descriptively, a *box-and-whisker diagram*) uses five statistics to summarize the center and spread of the data: the smallest observation, the 25th percentile, the 50th percentile, the 75th percentile, and the largest observation. A boxplot may also identify any observations that are outliers.

Figure 3.1 shows a boxplot using the Dow daily price change data in Table 2.3. A box is used to connect the 25th and 75th percentiles. Because the ends of the box are at the 25th and 75th percentiles, the width of the box is equal to the interquartile range and encompasses the middle half of the data. The median is denoted by the line inside the box. The ends of the two horizontal lines coming out of the box (the "whiskers") show the minimum and maximum values. A boxplot conveys a considerable amount of information, but is less complicated than the histogram in Figure 2.9. It is also relatively robust, in that the box itself (but not the whiskers) is resistant to outliers. A boxplot can be drawn either horizontally or vertically.

In a modified boxplot, outliers that are farther than 1.5 times the interquartile range from the box are shown as separate points, and the whiskers stop at the most extreme points that are not outliers. Figure 3.2 is a modified boxplot for the Dow price changes in Figure 3.1, with the Boeing 3.36 percent and Alcoa 3.78 percent outliers identified.

In practice, the data will usually not divide perfectly into four groups of exactly equal size, and some statisticians (and statistics software) use slightly different rules for calculating the first and third quartiles.

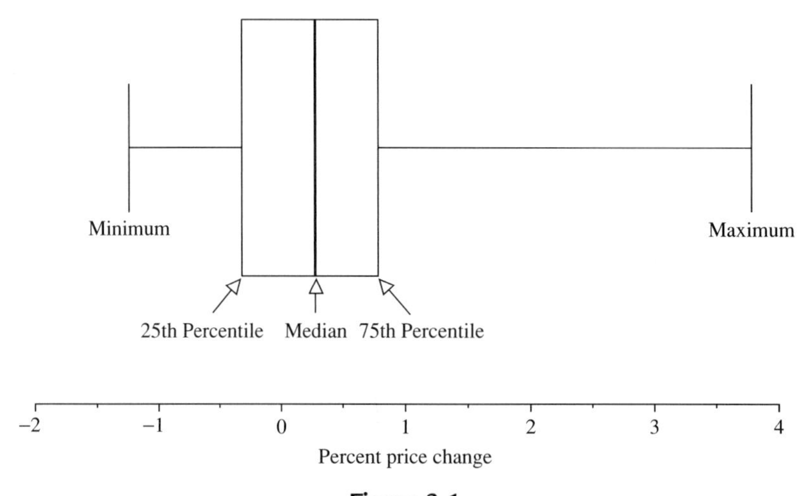

Figure 3.1
A boxplot for the Dow price changes.

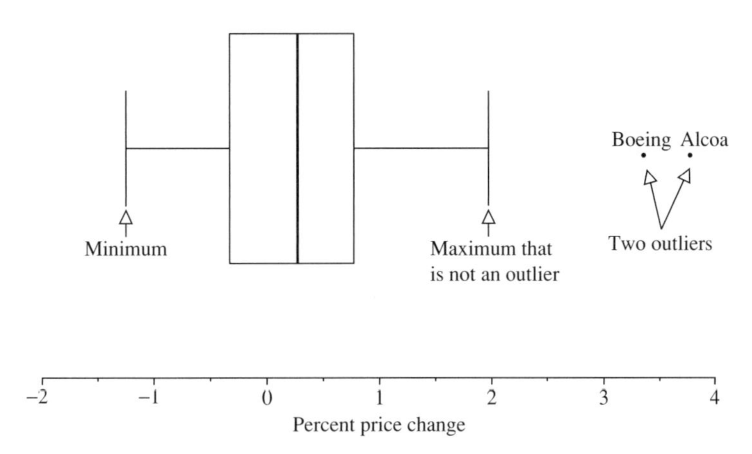

Figure 3.2
A boxplot for the Dow price changes, with two outliers.

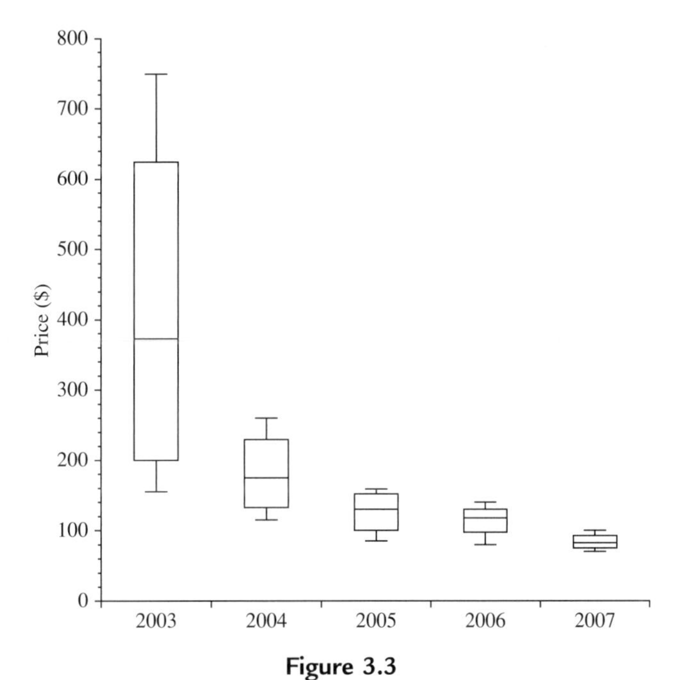

Figure 3.3
Side-by-side boxplots for hard drive prices.

Side-by-side boxplots allow us to compare data sets. Figure 3.3 shows side-by-side boxplots for the hard drive prices in Table 1.6. This figure clearly shows that, not only was there a decline in prices between 2003 and 2007, but the dispersion in prices also declined each year.

In most cases, the standard deviation, the average absolute deviation, and the interquartile range agree in their rankings of the amount of variation in different data sets. Exceptions can occur with outliers because the squaring of deviations can have a large effect on the standard deviation. Just as the median is more resistant to outliers than the mean, so the average absolute deviation and the interquartile range are more resistant to outliers than the standard deviation. Nonetheless, because of its mathematical tractability and importance in probability and statistics, the standard deviation is usually used to gauge the variation in a set of data.

3.5 Growth Rates

Many data are more meaningful when expressed as percentages; for example, we can readily interpret statements such as the CPI increased by 3 percent last year, the Dow went down 2 percent yesterday, or it costs 20 percent more to rent a home than to buy one.

Percentages can also put things in proper perspective. If the Dow falls by 340, is that a lot or a little? When the Dow was at 14,000 in 2007, a 340-point drop was a 2.4 percent decline—sickening but not catastrophic. When the Dow was at 381 in 1929, the 340-point drop that occurred between 1929 and 1932 was a catastrophic 89 percent decline.

In general, the percentage change is computed by dividing the difference between the new and old values by the value *before* the change, then multiplying by 100 to convert a fraction to a percentage:

$$\text{percentage change} = 100\left(\frac{\text{new} - \text{old}}{\text{old}}\right) \tag{3.3}$$

Get the Base Period Right

Percentage changes are sometimes calculated incorrectly by dividing the change by the value *after* the change instead of dividing by the value *before* the change. Thus, when income doubles from \$60,000 to \$120,000, the percentage change might be incorrectly reported as 50 percent,

$$\begin{aligned}\text{incorrect percentage change} &= 100\left(\frac{\text{new} - \text{old}}{\text{new}}\right)\\ &= 100\left(\frac{\$120{,}000 - \$60{,}000}{\$120{,}000}\right)\\ &= 50\end{aligned}$$

instead of 100 percent

$$\begin{aligned}\text{correct percentage change} &= 100\left(\frac{\text{new} - \text{old}}{\text{old}}\right)\\ &= 100\left(\frac{\$120{,}000 - \$60{,}000}{\$60{,}000}\right)\\ &= 100\end{aligned}$$

Newsweek once reported that the salaries of some Chinese government officials had been reduced 300 percent [6]. Suppose that the salary before the reduction were \$60,000. If the salary were eliminated completely, that would be a 100 percent reduction:

$$\begin{aligned} \text{percentage change} &= 100\left(\frac{\text{new} - \text{old}}{\text{old}}\right) \\ &= 100\left(\frac{\$0 - \$60{,}000}{\$60{,}000}\right) \\ &= -100 \end{aligned}$$

To get a 300 percent reduction, the salary would have to be *negative* \$120,000!

$$\begin{aligned} \text{percentage change} &= 100\left(\frac{\text{new} - \text{old}}{\text{old}}\right) \\ &= 100\left(\frac{-\$120{,}000 - \$60{,}000}{\$60{,}000}\right) \\ &= -300 \end{aligned}$$

Since it is unlikely that anyone would pay \$120,000 a year to work for the government, *Newsweek* must have made a mistake when it calculated the percentage change. In fact, it made several mistakes by calculating the percentage change this way:

$$\begin{aligned} \textit{Newsweek} \text{ ``percentage change''} &= -100\left(\frac{\text{old}}{\text{new}}\right) \\ &= -100\left(\frac{\$60{,}000}{\$20{,}000}\right) \\ &= -300 \end{aligned}$$

The negative sign appears because the salary was reduced.

To calculate the percentage change correctly, we divide the change in the salary by the old salary, as in Equation 3.3. If a \$60,000 salary is reduced to \$20,000, this is not a 300 percent decrease, but rather a 66.67 percent decrease:

$$\begin{aligned} \text{percentage change} &= 100\left(\frac{\text{new} - \text{old}}{\text{old}}\right) \\ &= 100\left(\frac{\$20{,}000 - \$60{,}000}{\$60{,}000}\right) \\ &= -66.67 \end{aligned}$$

For a percentage calculation to be informative, it should also use a reasonable base period that is clearly identified. If someone reports that "The price of orange juice increased 50 percent," he or she should tell us the base period. Did the price of orange juice increase 50 percent in a single day or over the past 100 years? If someone does not specify the base period (or gives a peculiar base period), that person may be trying to magnify or minimize the percentage change to support a position, rather than present the facts honestly. To make the price increase seem large, someone may compare it to a period long ago when orange juice was cheap. Someone else, who wants to report a small price increase, may choose a base period when orange juice was expensive, perhaps after bad weather ruined most of the crop. An honest statistician compares the price to a natural base period and identifies that base period: "The price of orange juice has increased by 2 percent during the past year, and by 35 percent over the past ten years." If orange juice prices fluctuate considerably, we can use a time series graph, as explained in Chapter 2, to show these ups and downs.

Because there usually is no room to specify the base period in a brief newspaper headline, we sometimes see accurate, but seemingly contradictory, headlines on the very same day. For example, the *New York Times* once reported that "American Express Net Climbs 16%" [7], while *The Wall Street Journal* reported the same day that "Net at American Express Fell 16% in 4th Quarter" [8]. These were not typographical errors. Both stories were accurate and each explained the base period that they used. The *Times* compared American Express's earnings in the fourth quarter to its earnings in the third quarter; the *Journal* compared American Express's earnings in the fourth quarter to its earnings a year earlier.

Watch out for Small Bases

Because a percentage is calculated relative to a base, a very small base can give a misleadingly large percentage. On your second birthday, your age increased by 100 percent. When you graduate from college, your income may increase by several thousand percent. On July 25, 1946, the average hourly rainfall between 6 A.M. and noon in Palo Alto, California, was 84,816 times the average hourly rainfall in July during the preceding 36 years, dating back to 1910, when the Palo Alto weather station first opened [9]. How much rain fell on this incredibly wet July 25? Only 0.19 inches. During the preceding 36 years, there had been only one July day with measurable precipitation in Palo Alto, and on that day the precipitation was 0.01 inches.

One way to guard against the small-base problem is to give the level as well as the percentage; for example, to note that the rainfall was 0.19 inches on this unusually wet July day in Palo Alto.

The Murder Capital of Massachusetts

Wellfleet is a small town on Cape Cod, renowned for its oysters, artists, and tranquility. There was considerable surprise when the Associated Press reported that Wellfleet had the highest murder rate in Massachusetts in 1993, with 40 murders per 100,000 residents—more than double the murder rate in Boston, which had only 17 murders per 100,000 residents [10]. A puzzled newspaper reporter looked into this statistical murder mystery. She found that there had in fact been no murders in Wellfleet in 1993 and that no Wellfleet police officer, including one who had lived in Wellfleet for 48 years, could remember a murder ever occurring in Wellfleet.

However, a man accused of murdering someone in Barnstable, which is 20 miles from Wellfleet, had turned himself in at the Wellfleet police station in 1993 and the Associated Press had erroneously interpreted this Wellfleet arrest as a Wellfleet murder. Because Wellfleet had only 2,491 permanent residents, this one misrecorded murder arrest translated into 40 murders per 100,000 residents. Boston, in contrast, had 98 murders, which works out to 17 murders per 100,000 residents.

The solution to this murder mystery shows how a statistical fluke can make a big difference if the base is small. A misrecorded murder in Boston would not noticeably affect its murder rate. In Wellfleet, a misrecorded murder changes the reported murder rate from 0 to 40. One way to deal with small bases is to average the data over several years in order to get a bigger base. Wellfleet's average murder rate over the past 50 years is 1 with the misrecorded arrest and 0 without it—either way confirming that it is indeed a peaceful town.

The Geometric Mean (Optional)

The arithmetic mean of n numbers is obtained by adding together the numbers and dividing by n. The geometric mean is calculated by *multiplying* the numbers and taking the nth root:

$$g = (X_1 X_2 \dots X_n)^{1/n}, \; X_i \geq 0 \tag{3.4}$$

For example, the arithmetic mean of 5, 10, and 15 is 10:

$$\begin{aligned} \overline{X} &= \frac{5+10+15}{3} \\ &= 10 \end{aligned}$$

while the geometric mean is 9.09:

$$\begin{aligned} g &= (5 \times 10 \times 15)^{1/3} \\ &= 750^{1/3} \\ &= 9.09 \end{aligned}$$

The geometric mean is not defined for negative values and is simply equal to 0 if one of the observations is equal to 0. One interesting mathematical fact about the geometric mean is that it is equal to the arithmetic mean if all of the numbers are the same; otherwise, as in our numerical example, the geometric mean is less than the arithmetic mean.

Generally speaking, the geometric mean is pretty close to the arithmetic mean if the data are bunched together and far below the arithmetic mean if the data are far apart. For example, for the data 4, 5, 6, the arithmetic mean is 5 and the geometric mean is 4.93; for the data 10, 50, 90, the arithmetic mean is 50 and the geometric mean is 35.57.

The geometric mean is often used for data that change over time because annual rates of change are often more meaningful than the overall percentage change. For example, which of these statements (both of which are true) is more memorable and meaningful?

a. The CPI increased 154 percent between December 1980 and December 2010.
b. The CPI increased 3.2 percent a year between December 1980 and December 2010.

Statement b is based on a geometric mean calculation.

Table 3.2 shows the annual rate of inflation using December U.S. CPI data. The CPI increased by 8.9 percent from December 1980 to December 1981 and by 3.8 percent from December 1981 to December 1982. Since the level of the CPI is arbitrary, we might set the CPI equal to 1 in December 1980. The CPI is consequently 1.089 in December 1981 and (1.089)(1.038) = 1.130 in December 1982. If we continue for all 30 years, the level of the CPI in December 2010 is

$$(1.089)(1.038)(1.038)\ldots(1.015) = 2.54$$

The increase in the CPI from 1 in December 1980 to 2.54 in December 2010 is a 154 percent increase.

The geometric mean tells us that the annual rate of increase over these 30 years was about 3.2 percent:

$$[(1.089)(1.038)(1.038)\ldots(1.015)]^{1/30} = 1.032$$

Table 3.2: Annual Rate of Inflation, December to December

1981	8.9	1996	3.3
1982	3.8	1997	1.7
1983	3.8	1998	1.6
1984	3.9	1999	2.7
1985	3.8	2000	3.4
1986	1.1	2001	1.6
1987	4.4	2002	2.4
1988	4.4	2003	1.9
1989	4.6	2004	3.3
1990	6.1	2005	3.4
1991	3.1	2006	2.5
1992	2.9	2007	4.1
1993	2.7	2008	0.1
1994	2.7	2009	2.7
1995	2.5	2010	1.5

The CPI increased by an average of 3.2 percent a year in that if the CPI increased by 3.2 percent *every* year, it would be 154 percent higher after 30 years:

$$(1.032)^{30} = 2.54$$

In general, if something grows at a rate R_i in year i, then a geometric mean calculation gives us the overall annual rate of growth R:

$$1 + R = [(1 + R_1)(1 + R_2) \ldots (1 + R_n)]^{1/n} \tag{3.5}$$

Equation 3.5 also works for interest rates and other rates of return. If you invest \$100 and earn a 5 percent return the first year, a 20 percent return the second year, and a 10 percent return the third year, your \$100 grows to

$$\$100(1 + 0.05)(1 + 0.20)(1 + 0.10) = \$138.60$$

Using Equation 3.5, the geometric return is 11.5 percent:

$$\begin{aligned} 1 + R &= [(1 + 0.05)(1 + 0.20)(1 + 0.10)]^{1/3} \\ &= 1.386^{1/3} \\ &= 1.115 \end{aligned}$$

in that \$100 invested for three years at 11.5 percent will grow to \$138.60:

$$\$100(1.115)^3 = \$138.60$$

3.6 Correlation

Often, we are more interested in the relationships among variables than in a single variable in isolation. For example, how is household spending related to income? How is GDP related to the unemployment rate? How are the outcomes of presidential elections related to the unemployment rate? How are interest rates related to the rate of inflation?

Chapter 2 discusses the Federal Reserve's stock-valuation model, which hypothesizes that there is a relationship between the earnings/price ratio and the interest rate on 10-year Treasury bonds. A time series graph (Figure 2.23) suggests that these two variables do, indeed, move up and down together. Figure 3.4 shows a scatterplot of the quarterly data used in Figure 2.23. This is another way of seeing that there appears to be a positive relationship between these two variables.

The covariance s_{xy} is a very simple measure of the relationship between two variables:

$$s_{xy} = \frac{(X_1 - \overline{X})(Y_1 - \overline{Y}) + (X_2 - \overline{X})(Y_2 - \overline{Y}) + \cdots + (X_n - \overline{X})(Y_n - \overline{Y})}{n - 1} \tag{3.6}$$

The covariance is analogous to the variance, which is a measure of the dispersion in a single variable. Comparing Equations 3.2 and 3.6, we see that the variance adds up the

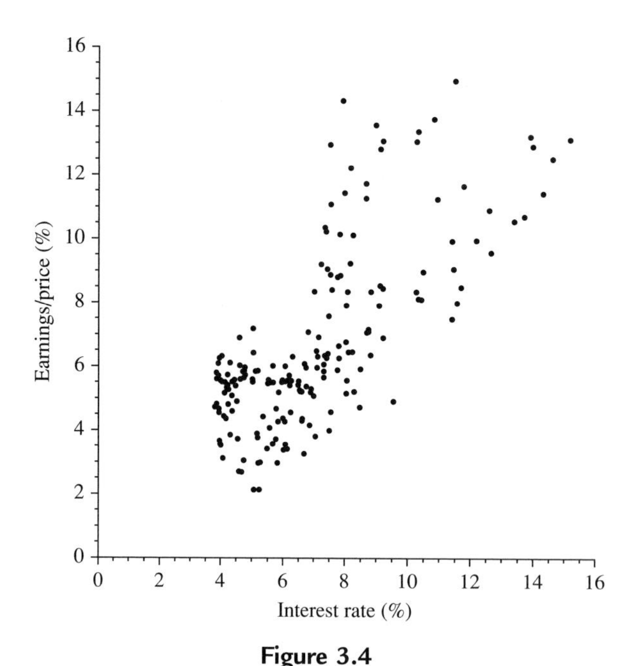

Figure 3.4
S&P 500 earnings/price ratio and 10-year Treasury interest rate, 1960–2009.

squared deviations of a single variable from its mean, while the covariance adds up the products of the deviations of two variables from their means.

Figure 3.5 repeats Figure 3.4, this time showing the mean values of the two variables, so that we can see when each of these variables is above or below its mean. Clearly, when the interest rate is above its mean, the earnings/price ratio tends to be above its mean, and when the interest rate is below its mean, the earnings/price ratio tends to be below its mean.

Now look again at the formula for the covariance. The product of the deviations of X and Y from their respective means is positive when both deviations are positive (the northeast quadrant of Figure 3.5) or when both deviations are negative (the southwest quadrant of Figure 3.5). The product of the deviations of X and Y from their respective means is negative when one deviation is positive and the other deviation is negative (the northwest and southeast quadrants of Figure 3.5). In the case of the 10-year Treasury rate and the S&P 500 earnings/price ratio, the data are overwhelmingly in the northeast and southwest quadrants, and the products of the deviations are overwhelmingly positive. So, the covariance is positive.

This reasoning shows us how to interpret a positive or negative covariance:

Positive covariance: When the value of one variable is above (or below) its mean, the value of the other variable also tends to be above (or below) its mean.
Negative covariance: When the value of one variable is above its mean, the value of the other variable tends to be below its mean, and vice versa.

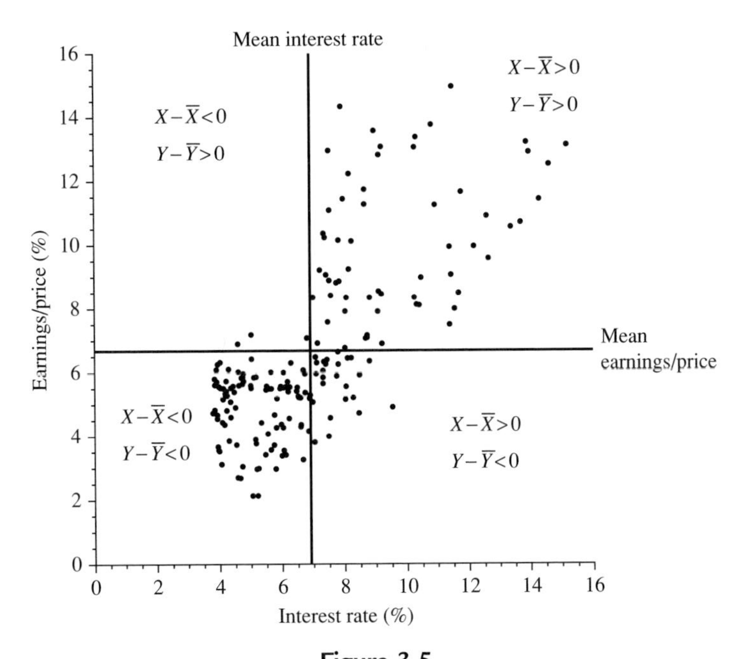

Figure 3.5
S&P 500 earnings/price ratio and 10-year Treasury rate, relative to means.

In our example, the covariance between the 10-year Treasury rate and the S&P 500 earnings/price ratio works out to be 5.14. When the 10-year Treasury rate is above (or below) its average, the S&P 500 earnings/price ratio tends to be above (or below) its average value, too.

Is a 5.14 covariance a big number or a small number? The size of the covariance depends on the scale of the deviations from the means. If the typical deviation is 100, the covariance will clearly be much higher than when the typical deviation is 1. To put the covariance into perspective, we calculate the *correlation* R by dividing the covariance by the standard deviation of X and the standard deviation of Y:

$$R = \frac{(\text{covariance between } X \text{ and } Y)}{(\text{standard deviation of } X)(\text{standard deviation of } Y)} = \frac{s_{XY}}{s_X s_Y} \tag{3.7}$$

Standard deviations are always positive. Therefore, the correlation coefficient R is positive when the covariance is positive and negative when the covariance is negative. A positive correlation means that, when either variable is above its mean value, the other variable tends to be above its mean value. A negative correlation means that, when either variable is above its mean value, the other variable tends to be below its mean value.

It can be shown mathematically that the value of the correlation coefficient cannot be larger than 1 or less than −1. The correlation is equal to 1 if all the data lie on a straight line with a positive slope and is equal to −1 if all the data lie on a straight line with a negative slope. In each of these extreme cases, if you know the value of one of the variables, you can be certain of the value of the other variable. In less extreme cases, we can visualize drawing a positively sloped or negatively sloped straight line through the data, but the data do not lie exactly on the line. The correlation is equal to 0 when there is no linear relationship between the two variables and we cannot visualize drawing a straight line through the data.

The value of the correlation coefficient does not depend on which variable is on the vertical axis and which is on the horizontal axis. For this reason, the correlation coefficient is often used to give a quantitative measure of the direction and degree of association between two variables that are related statistically but not necessarily causally. For instance, the historical correlation between midterm and final examination scores in my introductory statistics classes is about 0.6. I do not believe that either test score has any causal effect on the other test score but rather that both scores are influenced by common factors (such as student aptitude and effort). The correlation coefficient is a numerical measure of the degree to which these test scores are related statistically.

The correlation coefficient does not depend on the units in which the variables are measured—pennies or dollars, inches or centimeters, pounds or kilograms, thousands or millions. For instance, the correlation between height and weight does not depend on whether the heights are measured in inches or centimeters or the weights are measured in pounds or kilograms.

The correlation between the 10-year Treasury rate and the S&P 500 earnings/price ratio turns out to be 0.73, which shows (as suggested by Figures 3.4 and 3.5) that there is a positive, but certainly not perfect, linear relationship between these two variables.

Figure 3.6 shows a scatterplot of annual data on the percent change in real GDP and the change in the unemployment rate. This time, the correlation is −0.85, which shows that there is a strong negative linear relationship between these two variables.

Figure 3.7 shows a scatterplot for two essentially unrelated variables, the rate of return on the S&P 500 and per capita cigarette consumption. The correlation is 0.04, which confirms our visual impression that there is no meaningful linear relation between these two variables. If we try to visualize a straight line fit to these data, we would not know where to draw it.

A correlation of 0 does not mean that there is no relationship between X and Y, only that there is no linear relationship. Figure 3.8 shows how a zero correlation coefficient does not rule out a perfect nonlinear relationship between two variables. In this figure, Y and X are exactly related by the equation $Y = 10X - X^2$. Yet, because there is no linear relationship between Y and X, the correlation coefficient is 0. This is an example of why we should look at a scatterplot of our data.

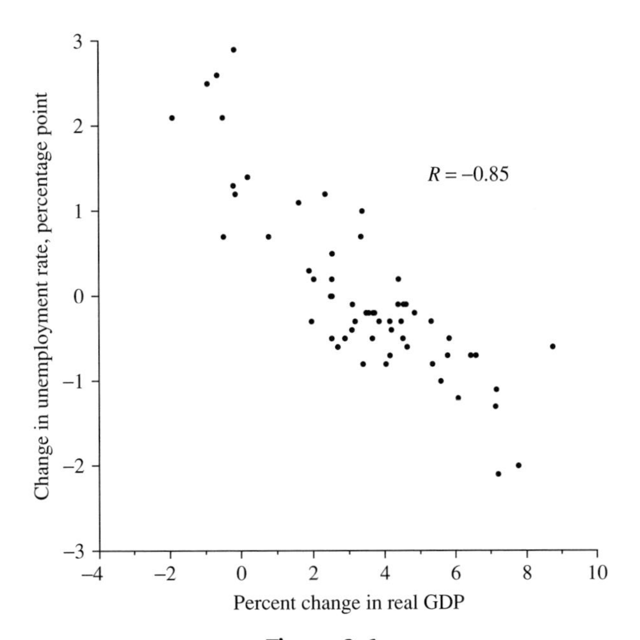

Figure 3.6
Real GDP and unemployment, annual, 1948–2008.

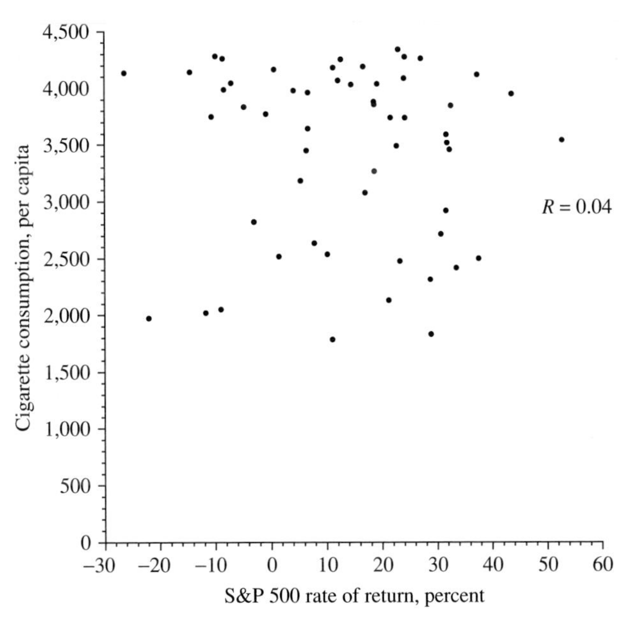

Figure 3.7
S&P 500 annual rate of return and per capita cigarette consumption, 1950–2007.

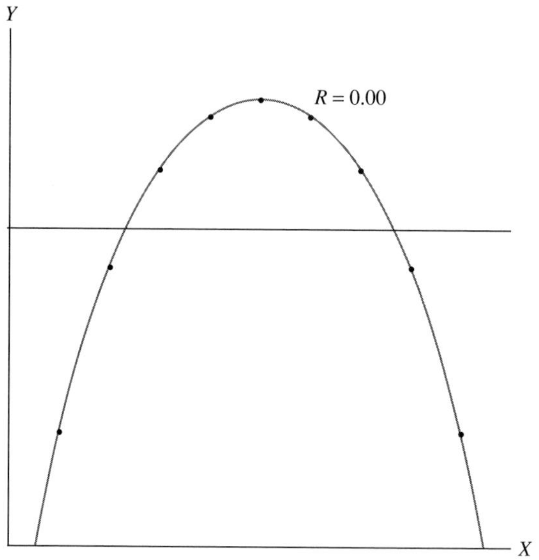

Y

R = 0.00

X

Figure 3.8
A zero correlation does not rule out a nonlinear relationship.

Exercises

3.1 Which of these data sets has a higher mean? Higher median? Higher standard deviation? (Do not do any calculations. Just look at the data.)

X	1	2	3	4	5
Y	5	4	3	2	1

3.2 Which of these data sets has a higher mean? Higher median? Higher standard deviation? (Do not do any calculations. Just look at the data.)

X	1	2	3	4	5
Y	1	2	3	4	6

3.3 Albert Michelson's 1882 measurements of the speed of light in air (in kilometers per second) were as follows [11]:

299,883	299,796	299,611	299,781	299,774	299,696
299,748	299,809	299,816	299,682	299,599	299,578
299,820	299,573	299,797	299,723	299,778	299,711
300,051	299,796	299,772	299,748	299,851	

Calculate the mean, median, and 10-percent trimmed mean. Which is closest to the value 299,710.5 that is now accepted as the speed of light?

3.4 In 1798, Henry Cavendish made 23 measurements of the density of the earth relative to the density of water [12]:

5.10	5.27	5.29	5.29	5.30	5.34	5.34	5.36	5.39	5.42	5.44	5.46
5.47	5.53	5.57	5.58	5.62	5.63	5.65	5.68	5.75	5.79	5.85	

Calculate the mean, median, and 10-percent trimmed mean. Which is closest to the value 5.517 that is now accepted as the density of the earth?

3.5 Twenty-four female college seniors were asked how many biological children they expect to have during their lifetimes. Figure 3.9 is a histogram summarizing their answers: Of these four numbers (0, 1.1, 2, 2.3), one is the mean, one is the median, one is the standard deviation, and one is irrelevant. Identify which number is which.

a. Mean.
b. Median.
c. Standard deviation.
d. Irrelevant.

3.6 Thirty-five male college seniors were asked how many biological children they expect to have during their lifetimes. Figure 3.10 is a histogram summarizing their answers: Of these four numbers (0, 1.4, 2.4, 3.0), one is the mean, one is the median, one is the standard deviation, and one is irrelevant. Identify which number is which.

a. Mean.
b. Median.
c. Standard deviation.
d. Irrelevant.

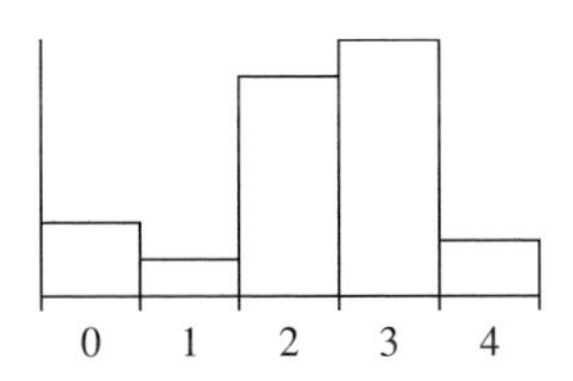

Figure 3.9
Exercise 3.5.

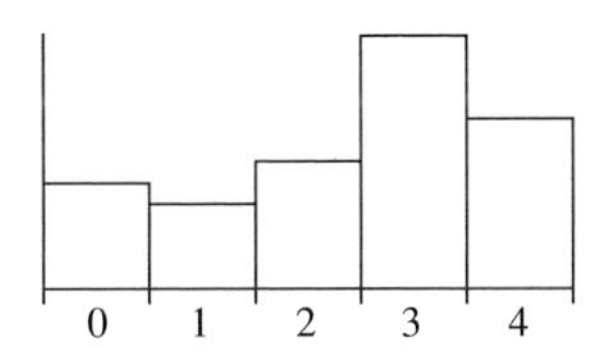

Figure 3.10
Exercise 3.6.

3.7 Suppose that you have ten observations that have a mean of 7, a median of 6, and a standard deviation of 3. If you add 5 to the value of each observation, what are the new values of the
a. Mean?
b. Median?
c. Standard deviation?

3.8 Suppose that you have ten observations that have a mean of 7, a median of 6, and a standard deviation of 3. If you subtract 2 from the value of each observation, what are the new values of the
a. Mean?
b. Median?
c. Standard deviation?

3.9 Suppose that you have ten observations that have a mean of 7, a median of 6, and a standard deviation of 3. If you double the value of each observation, what are the new values of the
a. Mean?
b. Median?
c. Standard deviation?

3.10 Suppose that you have ten observations that have a mean of 7, a median of 6, and a standard deviation of 3. If you halve the value of each observation, what are the new values of the
a. Mean?
b. Median?
c. Standard deviation?

3.11 Identify the apparent statistical mistake in this commentary [13]:

The median cost of a house in [Duarte, California] is a whopping $4,276,462, making it the most expensive housing market in the country. It ranks No. 1 on Forbes' annual ranking of America's Most Expensive ZIP Codes.... [O]nly 12 homes are currently on the market. So a single high-priced listing (like the mammoth nine-bedroom, built this year, that's selling for $19.8 million) is enough to skew the median price skyward.

3.12 Roll four six-sided dice ten times, each time recording the sum of the four numbers rolled. Calculate the mean and median of your ten rolls. Repeat this experiment 20 times. Which of these two measures seems to be the least stable?

3.13 Use the test scores in Exercise 1.8 to calculate the average score for each grade level and then make a time series graph using these average scores.

3.14 An old joke is that a certain economics professor left Yale to go to Harvard and thereby improved the average quality of both departments. Is this possible?

3.15 Ann Landers, an advice columnist, once wrote "Nothing shocks me anymore, especially when I know that 50 percent of the doctors who practice medicine graduated in the bottom half of their class" [14]. Does this observation imply that half of all doctors are incompetent?

3.16 When is the standard deviation negative?

3.17 There was a players' strike in the middle of the 1981 baseball season. Each division determined its season winner by having a playoff between the winner of the first part of the season, before the strike, and the winner of the second part of the season. Is it possible for a team to have the best winning percentage for the season as a whole, yet not win either half of the season and consequently not have a chance of qualifying for the World Series? Use some hypothetical numbers to illustrate your reasoning.

3.18 Use the test scores in Exercise 1.8 two make two boxplots, one using the first-grade scores and the other the eighth-grade scores. Summarize the main differences between these two boxplots.

3.19 Exercise 2.26 shows the percentage returns for the ten Dow stocks with the highest dividend/price (*D*/*P*) ratio and the ten Dow stocks with the lowest *D*/*P* ratio. Display these data in two side-by-side boxplots. Do the data appear to have similar or dissimilar medians and dispersion?

3.20 Table 1.4 shows the U.S. dollar prices of Big Mac hamburgers in 20 countries. Use a boxplot to summarize these data.

3.21 Figure 3.11 shows two boxplots. Which data set has the higher median? The higher standard deviation?

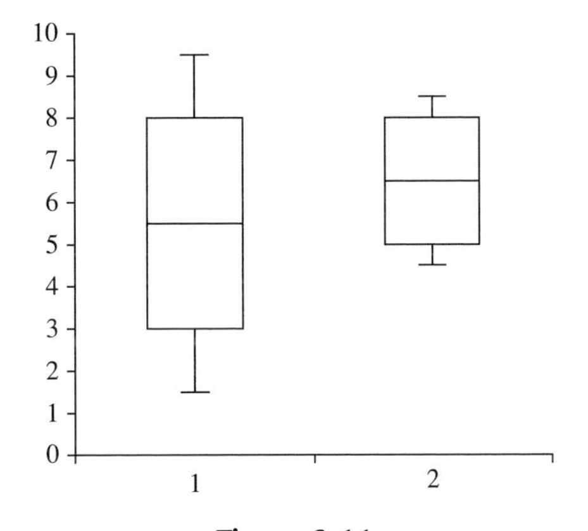

Figure 3.11
Exercise 3.21.

3.22 Figure 3.12 shows two boxplots. Which data set has the higher median? The higher mean? The higher standard deviation?

3.23 Draw a freehand sketch of two side-by-side boxplots, one the boxplot shown in Figure 3.13 and the other the same boxplot if 2 is added to the value of each observation in that boxplot.

3.24 Draw a freehand sketch of two side-by-side boxplots, one the boxplot shown in Figure 3.14 and the other the same boxplot if all the observations in that boxplot are multiplied by 2.

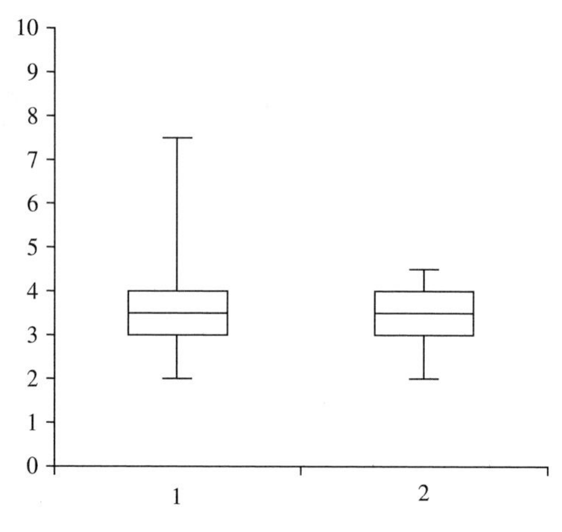

Figure 3.12
Exercise 3.22.

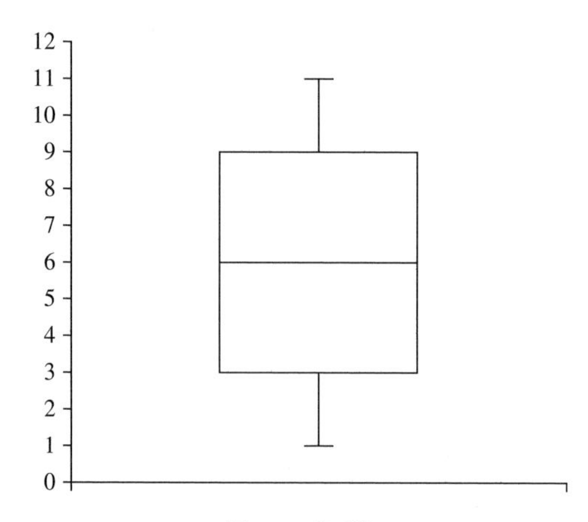

Figure 3.13
Exercise 3.23.

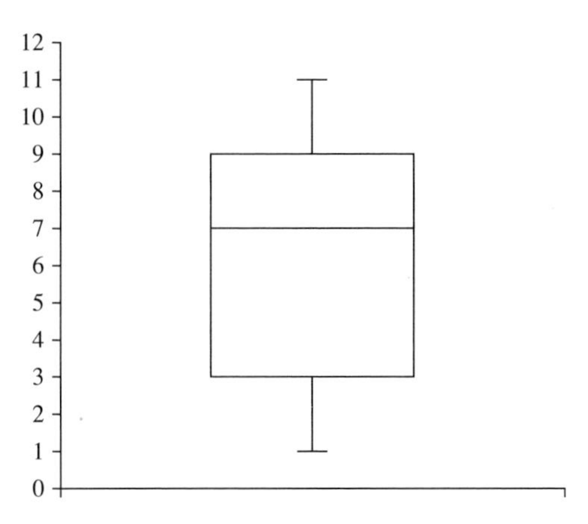

Figure 3.14
Exercise 3.24.

3.25 Here are data on the number of U.S. households (in millions) of different sizes in 2000:

Household size	1	2	3	4	5	6	7 or more
Number of households	31	39	19	16	7	3	1

a. Looking at the 116 million households, ordered by size, what is the size of the median household?
b. Looking at all 291 million individuals, ordered by the size of the household in which they live, what is the size of the household that the median individual lives in?
c. Pomona College reports its average class size as 14 students. However, a survey that asked Pomona College students the size of the classes they were enrolled in found that the average class size was 38 students and that 70 percent of the respondents' classes had more than 14 students. Use the insights gained in the first two parts of this exercise to explain this disparity.

3.26 Joanna is staring work with an annual salary of $40,000. If her salary increases by 5 percent a year, what will her salary be 40 years from now? If prices increase by 3 percent a year, how much higher will prices be 40 years from now than they are today? What will be the real value of the salary Joanna receives 40 years from now in terms of current dollars?

3.27 Treasury zeros pay a fixed amount at maturity. For example, a Treasury zero could be purchased on January 2, 2009, for $445.19 and might be worth $1,000 when it matures 25 years later. What is the annual rate of return on this zero?

3.28 A stock's price/earnings (*P/E*) ratio is the per-share price of its stock divided by the company's annual earnings per share. The *P/E* ratio for the stock market as a whole is used by some analysts as a measure of whether stocks are cheap or expensive, in comparison with other historical periods. Table 3.3 shows some annual *P/E* ratios for the New York Stock Exchange (NYSE). Calculate the mean and standard deviation. The stock market reached a peak in August 1987, when the Dow Jones Industrial Average topped 2,700. The Dow slipped back to 2,500 in October of 1987 and then to 2,250. Then, on a single day, October 19, 1987, the Dow fell by 508 points. At its August 1987 peak, the market's price/earnings ratio was 23. Was this *P/E* value more than two standard deviations above the mean *P/E* for 1970–1986? Was it more than 1.5 times the interquartile range above the median? Draw a box-and-whiskers diagram using these 1970–1986 *P/E* data.

3.29 In the early 1970s, many investors were infatuated by the Nifty 50—a group of 50 growth stocks that were thought to be so appealing that they should be bought and never sold, regardless of price. Table 3.4 shows the price/earnings (*P/E*) ratio in December 1972 and the annualized stock return (*R*) from December 31, 1972, through December 31, 2001, for each of these 50 stocks. Calculate the mean and median values of the returns. Explain why these are not equal.

3.30 Use the data in the preceding exercise to make side-by-side boxplots of the returns for the 25 stocks with the highest *P/E*s and the 25 stocks with the lowest *P/E*s. What differences do you observe in these boxplots?

Table 3.3: Exercise 3.28

Year	*P/E*	Year	*P/E*	Year	*P/E*
1970	15.5	1976	11.2	1982	8.6
1971	18.5	1977	9.3	1983	12.5
1972	18.2	1978	8.3	1984	10.0
1973	14.0	1979	7.4	1985	12.3
1974	8.6	1980	7.9	1986	16.4
1975	10.9	1981	8.4		

Table 3.4: Exercise 3.29

P/E	*R*	*P/E*	*R*	*P/E*	*R*	*P/E*	*R*
90.74	−14.68	50.47	2.45	40.77	9.78	29.34	15.55
85.67	10.50	50.37	13.19	39.00	10.30	28.97	16.99
83.26	−6.84	50.00	12.36	38.85	13.13	27.60	15.35
81.64	8.97	49.45	10.37	38.66	6.68	26.12	15.57
78.52	10.10	48.82	−1.64	38.32	3.19	26.07	12.40
75.75	5.66	48.77	0.89	37.36	9.68	25.93	17.68
65.43	6.04	48.17	1.72	36.89	7.62	25.86	14.12
62.11	−1.37	47.60	13.15	34.10	4.83	25.59	4.91
61.85	13.35	46.28	11.27	33.87	14.21	25.50	10.80
59.97	0.93	46.03	13.14	32.04	11.94	22.37	13.36
54.31	−1.07	45.94	14.27	31.90	13.55	16.28	9.99
53.12	−1.47	41.11	9.95	30.77	6.94		
51.82	11.17	41.00	10.96	30.05	14.66		

3.31 Use the data in Exercise 3.29 to make a scatter diagram with the *P/E* ratio on the horizontal axis and the return on the vertical axis. Does there appear to be a positive relationship, negative relationship, or essentially no relationship?

3.32 Table 3.5 shows ringer percentages for Mary Ann Peninger at the 2000 Women's world horseshoe championships, when she threw first and when she threw second. For example, in the second game, she threw 68.8 percent ringers when she went first and 50.0 percent ringers when she went second. Draw two side-by-side boxplots for her ringer percentages when she threw first and second. What differences do you notice?

3.33 Use the data in Table 3.2 to calculate the annual rate of increase in the CPI over the 10-year period from December 2000 to December 2010.

3.34 Explain the error in the conclusion reached by a security analyst [16]:

The Dow Jones Industrial Average peaked at 381.17 during September 3, 1929. The so-called Great Crash pushed this index down 48% by November 13. But, by April 17, 1930, the index rebounded 48% from the November bottom. In other words, anyone who bought a diversified portfolio of stocks during September 1929 would have experienced no net change in the value of his portfolio by April of 1930.

3.35 Vincent Van Gogh sold only one painting during his lifetime, for about $30. In 1987, a sunflower still life he painted in 1888 sold for $39.85 million, more than three times the highest price paid previously for any work of art. Observers attributed the record price in part to the fact that his other sunflower paintings are all in museums and most likely will never be available for sale. If this painting had been purchased for $30 in 1888 and sold in 1987 for $39.85 million, what would have been the annual rate of return?

Table 3.5: Exercise 3.32 [15]

Game	Threw First	Threw Second
1	86.4	57.1
2	68.8	50.0
3	78.6	68.8
4	61.5	100.0
5	90.0	77.8
6	72.2	79.4
7	75.0	80.8
8	86.7	68.2
9	63.3	80.0
10	83.3	58.9
11	76.7	90.0
12	72.7	66.7
13	80.0	85.0
14	60.7	85.0
15	77.8	77.8

3.36 The world's population was 4.43 billion in 1980 and 6.85 billion in 2010. What was the annual rate of increase during these 30 years? At this rate, what will the world's population be in the year 2050?

3.37 The Dow Jones Industrial Average was 240.01 on October 1, 1928 (when the Dow expanded to 30 stocks), and 14,087.55 on October 1, 2007. The CPI was 51.3 in 1928 and 617.7 in 2007. What was the annual percentage increase in the real value of the Dow over this 79-year period?

3.38 In the 52 years between 1956 and 2008, Warren Buffett's net worth grew from $100,000 to $62 billion. What was his annual rate of return?

3.39 United States real (in 2007 dollars) per capita GDP was $1,504 in 1807 and $45,707 in 2007. What was the annual rate of growth of real per capita GDP over this 200-year period?

3.40 The CPI was 51 in 1800 and 25 in 1900. What was the annual percentage rate of change during this 100-year period?

3.41 Use the data in Table 3.2 to determine the percentage annual rate of increase in the CPI between 1990 and 2010.

3.42 A 1989 radio commercial claimed "If you have a LifeAlert security system, your chances of becoming a victim of a serious crime are 3,000 to 4,000 percent less than your neighbor's." Is this possible?

3.43 Twelve college women and men were asked to read a short article and, when they were finished, estimate how long it took them to read the article. Table 3.6 shows the actual and estimated times (both in seconds) and the percentage difference between the two. Summarize the percentage error data with two side-by-side boxplots, one for women and one for men.

3.44 Explain why this statement is incorrect: "When two assets have a +1.00 correlation, they move up and down at the same time, with the same magnitude."

3.45 Calculate the correlation between the official U.S. poverty thresholds for families of four people and the CPI in Table 3.7.

3.46 After its staff ran more than 5,000 miles in 30 different running shoes, *Consumer Reports* rated the shoes for overall performance on a scale of 0 to 100. Calculate the correlation between price and performance for female shoes and also for male shoes shown in Table 3.8. Briefly summarize your results.

3.47 The data in Table 3.9 were used in the 1964 *Surgeon General's Report* warning of the health risks associated with smoking, where the first variable is annual per capita cigarette consumption and the second variable is deaths from coronary heart disease

Table 3.6: Exercise 3.43

Women			Men		
Actual	Estimated	Error (%)	Actual	Estimated	Error (%)
173	150	−13.3	99	155	56.6
125	165	32.0	143	205	43.4
240	360	50.0	94	120	27.7
115	150	30.4	111	420	278.4
150	180	20.0	119	634	432.8
146	330	126.0	143	96	−32.9
124	190	53.2	77	185	140.3
83	104	25.3	111	240	116.2
76	250	228.9	78	125	60.3
121	240	98.3	115	200	73.9
75	58	−22.7	73	120	64.4
83	120	44.6	133	285	114.3

Table 3.7: Exercise 3.45

	Poverty Threshold	CPI
1960	$3,022	88.7
1970	$3,968	116.3
1980	$8,414	246.8
1990	$13,359	391.4
2000	$17,604	515.8
2010	$22,050	653.2

Table 3.8: Exercise 3.46 [17]

Female Shoes		Male Shoes	
Price	Rating	Price	Rating
124	82	83	85
70	80	70	85
72	78	70	84
70	77	75	82
65	75	70	78
90	73	82	72
60	71	70	72
80	70	125	70
55	70	82	68
70	69	45	68
67	67	70	65
45	64	120	64
55	62	110	64
75	57	133	62
85	46	100	50

Table 3.9: Exercise 3.47 [18]

	Cigarette Use	Heart Disease		Cigarette Use	Heart Disease
Australia	3,220	238.1	Mexico	1,680	31.9
Austria	1,770	182.1	Netherlands	1,810	124.7
Belgium	1,700	118.1	New Zealand	3,220	211.8
Canada	3,350	211.6	Norway	1,090	136.3
Denmark	1,500	144.9	Spain	1,200	43.9
Finland	2,160	233.1	Sweden	1,270	126.9
France	1,410	144.9	Switzerland	2,780	124.5
Greece	1,800	41.2	United Kingdom	2,790	194.1
Iceland	2,290	110.5	United States	3,900	259.9
Ireland	2,770	187.3	West Germany	1,890	150.3
Italy	1,510	114.3			

Table 3.10: Exercise 3.48

Year	Price Index	Change from Previous Year	Change from 2000
2005	112.2	3.4%	112%
2006	115.9	3.2%	116%
2007	119.2	2.8%	119%
2008	123.7	3.8%	134%

per 100,000 persons aged 35 to 64. Calculate the correlation. Does the value of the correlation suggest that there is a positive relationship, negative relationship, or essentially no relationship between cigarette consumption and heart disease?

3.48 Identify the error in Table 3.10, showing consumer prices and the change in prices based on a price index equal to 100 in 2000.

3.49 If $Y = 10 - 3X$, then the correlation between X and Y is

a. 1.
b. 0.
c. −1.
d. Undefined.

3.50 Explain how the following two newspaper headlines that appeared on the same day could both be accurate: "Orders for Machine Tools Increased 45.5% in October" [19]; "October Orders for Machine Tools Decreased 29%" [20].

CHAPTER 4

Probability

Chapter Outline

It is remarkable that a science which began with the consideration of games of chance should become the most important object of human knowledge.... The most important questions of life are, for the most part, really only problems of probability.

—Pierre-Simon, marquis de Laplace

Essential Statistics, Regression, and Econometrics

Historically, the first rigorous application of probability theory involved games of chance. Today, casinos use probabilities when they set payoffs for roulette, craps, and slot machines. Governments use probabilities when they set payoffs for state lotteries. Mathematicians use probabilities to devise optimal strategies for blackjack, backgammon, Monopoly, poker, and many other games. Devoted players can learn these probabilities firsthand from long and sometimes expensive experience.

Another early use of probabilities was in setting insurance rates. This is why life insurance premiums depend on whether a person is 18 or 98, is in good health or using an artificial heart, or is a college professor or a soldier of fortune. Probabilities are used to price medical insurance, car insurance, home insurance, business insurance, and even insurance against a baseball strike or a singer getting laryngitis.

Colleges use probabilities when they decide how many students to admit, how many professors to employ, how many classrooms to build, and how to invest their endowment. Probabilities are involved when an army decides to attack or retreat, when a business decides to expand or contract, and when you decide whether or not to wear a raincoat. Uncertainties are all around us and so are probabilities.

4.1 Describing Uncertainty

The great French mathematician Pierre-Simon Laplace (1749–1827) observed that probabilities are only "common sense reduced to calculation" [1]. Calculations have two very big advantages over common sense. First, common sense is sometimes wrong. Second, common sense can be misunderstood.

Suppose, for example, that during a routine physical examination, a doctor discovers a suspicious lump on a women's breast and orders an X-ray, which turns out positive, indicating that the lump is a malignant tumor. When the patient asks if the lump is malignant, the doctor cannot say yes with certainty because the test is imperfect, but should tell the patient that the results are worrisome.

Words alone are inadequate because the doctor and patient may interpret words differently. When 16 doctors were surveyed and asked to assign a numerical probability corresponding to the diagnosis that a disease is "likely," the probabilities ranged from 20 percent to 95 percent [2]. If, by *likely*, one doctor means 20 percent and another means 95 percent, then it is better for the doctor to state the probability than to risk a disastrous misinterpretation of ambiguous words.

Probabilities can be used not only for life-threatening diseases but for all the daily uncertainties that make life interesting and challenging. We begin with games, because when viewed from a financially safe distance, these are an ideal vehicle for introducing probabilities.

Equally Likely

The *equally likely* approach was devised to handle games of chance—the roll of dice, spin of a roulette wheel, and deal of cards. Each possible result is called an *outcome* and an *event* is a collection of outcomes. If the outcomes are equally likely, then we determine probabilities by counting outcomes. For example, if an ordinary coin is flipped, there are two possible outcomes: heads or tails. If heads and tails are equally likely, each has a 1/2 probability of occurring.

Similarly, when a six-sided die is rolled fairly, there are six possible outcomes. If each is equally likely, each has a 1/6 probability of occurring. What about the probability that an even number will be rolled? Because three of the six equally likely outcomes are even numbers, the probability of an even number is $3/6 = 1/2$.

This principle can be used to determine probabilities whenever the possible outcomes are equally likely: If there are n equally likely outcomes, the probability that any one of m possible outcomes will occur is m/n. In our dice example, there are $n = 6$ possible outcomes and $m = 3$ of these outcomes are even numbers, giving a probability of 3/6.

Some problems require a careful counting of the possible outcomes. When a coin is flipped twice, there are three possible results: two heads, one head and one tail, or two tails. It is tempting to assume that each of these three possibilities has a 1/3 probability of occurring. This was, in fact, the answer given by a prominent eighteenth-century mathematician, Jean d'Alembert. But his answer is wrong!

There is only one way to obtain two heads and only one way to obtain two tails, but there are two different ways to obtain one head and one tail—heads on the first flip and tails on the second or tails on the first flip and heads on the second. To count the possibilities correctly, it is sometimes helpful to draw a *probability tree*, which shows the possible outcomes at each stage. The probability tree in Figure 4.1 (using H for heads and T for tails) shows that there are two equally likely outcomes on the first flip, and that, for each of these outcomes, there are two equally likely outcomes on the second flip.

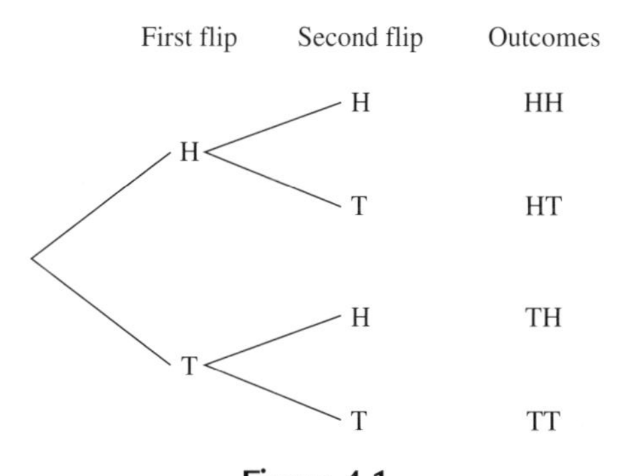

Figure 4.1
Probability tree for two coin flips.

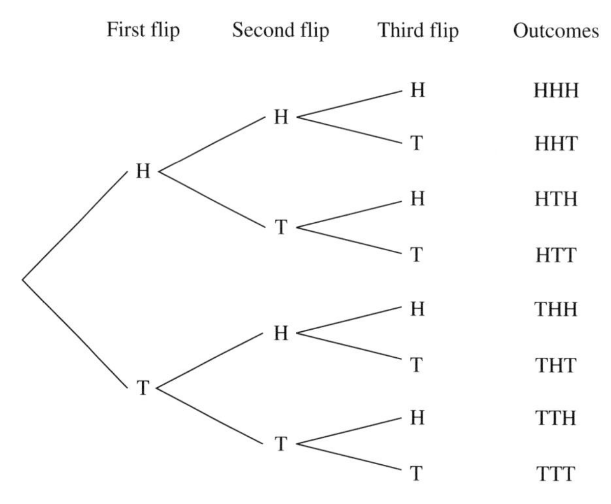

Figure 4.2
Probability tree for three coin flips.

Table 4.1: Probabilities for Three Coin Flips

Events	Number of Ways	Probability
Three heads	1	1/8
Two heads, one tail	3	3/8
One head, two tails	3	3/8
Three tails	1	1/8
Total	8	1

There are four equally likely outcomes, with two involving one head and one tail. Thus, the probability of one head and one tail is 2/4 = 1/2, the probability of two heads is 1/4, and the probability of two tails is 1/4.

What if, instead of flipping one coin twice, we flip two coins simultaneously? It does not matter. These probabilities apply to one coin flipped twice, two coins flipped separately, and two coins flipped simultaneously. A probability tree, in which the coin flips are treated as happening in a sequence, is just a way of helping us count the number of equally likely outcomes.

What about three coin flips? Figure 4.2 shows the probability tree. With three coin flips, there are eight possible outcomes, the probabilities shown in Table 4.1.

A Chimp Named Sarah

Every time a coin is flipped, it has two sides. Every time a die is rolled, it has six sides. Sometimes, we have situations in which the number of possible outcomes changes as we move through a probability tree.

Consider, for example, studies of a chimpanzee named Sarah. In one experiment, Sarah was shown a series of plastic symbols of varying color, size, and shape that formed a question. To answer the question correctly, Sarah had to arrange the appropriate symbols in correct order. In a very simple version, Sarah might be given three symbols—which we label *A*, *B*, and *C*—to arrange in correct order (for example, "bread in bowl"). How many possible ways are there to arrange these three symbols?

As shown in Figure 4.3, the first symbol has three possibilities: *A*, *B*, or *C*. Given the choice of the first symbol, there are only two possibilities for the second symbol. For example, if *A* is selected as the first symbol, the second symbol must be either *B* or *C*. Given the choice of the first two symbols, there is only one remaining possibility for the third symbol. If *A* and *B* are the first two symbols, then *C* must be the third symbol.

Thus the probability tree in Figure 4.3 has three initial branches, followed by two branches and then one final branch. In all, there are 3(2)(1) = 6 possible outcomes. The probability that a random arrangement of three symbols will be in the correct order is 1/6.

In more complicated problems, a probability tree would be very messy but, by remembering how a probability tree is constructed, we can determine the number of possible outcomes without actually drawing a tree. For instance, in one of Sarah's experiments, she had to choose four out of eight symbols and arrange these four symbols in correct order. Visualizing a probability tree, there are eight possible choices for the first symbol and, for each of these choices, there are seven possible choices for the second symbol. For any choice of the first two symbols, there remain six possibilities for the third symbol and then five possibilities for the fourth symbol. Thus, the total number of possible outcomes is 8(7)(6)(5) = 1,680. The probability that four randomly selected and arranged symbols will be correct is 1/1,680 = 0.0006. Sarah's ability to do this provided convincing evidence that she was making informed decisions and not just choosing symbols randomly.

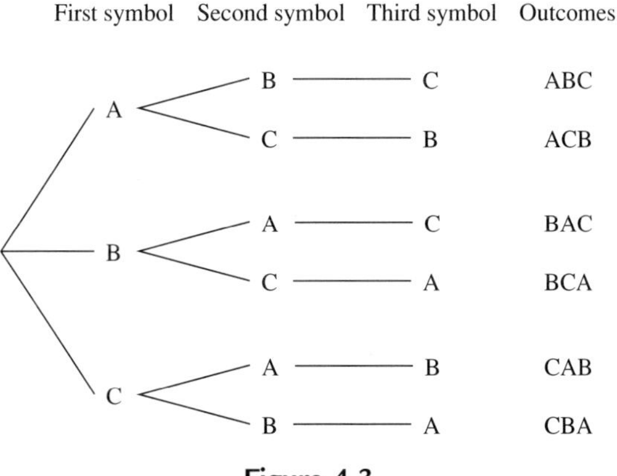

Figure 4.3
Probability tree for Sarah's choices [3].

Long-Run Frequencies

One difficulty with the classical approach to probabilities is that the possible outcomes may not be equally likely. A company selling a 1-year life insurance policy to a 20-year-old woman should not assume that life and death are equally likely outcomes. To handle such cases, we can think about what probabilities imply about long-run frequencies and then use long-run frequencies to estimate probabilities. If a coin has a 1/2 probability of landing heads, we expect that, in a large number of coin flips, heads will come up approximately half the time. Reversing this reasoning, if, in a large number of cases, an outcome occurs half the time, we might estimate its probability to be 1/2. The long-run frequency approach to determining probabilities is to say that if something has occurred m times in n identical situations (where n is a very large number), then its probability is approximately m/n. (More advanced books show how to quantify *approximately*.)

An insurance company can estimate the probability that a healthy 20-year-old woman will die within a year by looking at the recent experiences of millions of other similar women. If they have data on 10 million 20-year-old women, of whom 12,000 died within a year, they can estimate this probability as 12,000/10,000,000 = 0.0012, or slightly larger than one in a thousand.

The equally likely method is most appropriate for games of chance or similar situations in which coins, dice, cards, or other physical apparatus suggest that the possible outcomes are equally likely. The long-run frequency method is most appropriate when we suspect that the outcomes are not equally likely and we have data that can be used to estimate probabilities.

A Scientific Study of Roulette

Outside the United States, roulette wheels usually have 37 slots, numbered 0 to 36. Wheels used in the United States have an additional 00 slot, giving 38 slots in all. If the wheel is perfectly balanced, clean, and fair, a spun ball is equally likely to land in any of the slots. However, imperfections in a wheel can cause some numbers to win more often than other numbers.

In the late 1800s, an English engineer, William Jaggers, took dramatic advantage of these imperfections [4]. He paid six assistants to spend every day for an entire month observing the roulette wheels at Monte Carlo and recording the winning numbers. Jaggers found that certain numbers came up slightly more often than others. He then bet heavily on these numbers and, in a few days, won nearly $325,000—more than $6 million in today's dollars—until the casino caught on and began switching wheels nightly. More recently, in the 1960s, while their fellow students were studying or protesting, a group of Berkeley students were reported to have pulled off a similar feat in Las Vegas. Nowadays, casinos routinely rotate their roulette wheels to frustrate long-run frequency bettors.

Experimental Coin Flips and Dice Rolls

Many people have flipped coins over and over again, to see if heads and tails really do occur equally often. One of the most ambitious was Karl Pearson, a famous statistician, who flipped a coin 24,000 times and recorded 12,012 heads and 11,988 tails, indicating that heads and tails were indeed equally likely. (We do not expect heads and tails to come up exactly half the time, just close to half the time.) Pearson's experiment is dwarfed by the efforts of a Swiss astronomer named Wolf, who rolled dice over a 40-year period, from 1850 to 1893. In one set of experiments, Wolf tossed a pair of dice, one red and one white, 20,000 times. John Maynard Keynes commented on the results [5]:

> *[T]he records of the relative frequency of each face show that the dice must have been very irregular, the six face of the white die, for example, falling 38% more often than the four face of the same die. This, then, is the sole conclusion of these immensely laborious experiments—that Wolf's dice were very ill made.*

Subjective Probabilities

The long-run frequency approach allows us to estimate probabilities when the possible outcomes are not equally likely. However, we need lots of repetitive data to calculate long-run frequencies, and we are often interested in probabilities for virtually unique situations. Consider, for example, a presidential election. Our decisions about choosing a career, buying or selling stocks, expanding or contracting a business, or whether or not to enlist in the military may well be affected by who we think is going to be elected president.

The self-proclaimed Wizard of Odds once calculated presidential probabilities as follows [6]:

> *Without considering the candidates, the odds would be 2 to 1 in favor of a Republican because since 1861 when that party was founded, there have been 12 Republican Presidents and only 7 Democrats.*

The outcome of a presidential election is uncertain and it would be useful to quantify that uncertainty. We would like more than a shrug of the shoulders and a sheepish "Who knows?" However, the equally likely and long-run frequency approaches are not appropriate. The choice of president is not determined by a coin flip every four years. The probabilities change from election to election depending on the candidates and the mood of the electorate.

Bayes's Approach

In the eighteenth century, Reverend Thomas Bayes wrestled with an even more challenging problem—the probability that God exists. The equally likely and long-run frequency approaches are useless, and yet this uncertainty is of great interest to many people, including Reverend

Bayes. Such a probability is necessarily subjective. The best that anyone can do is weigh the available evidence and logical arguments and come up with a personal probability of God's existence. This idea of personal probabilities has been extended and refined by other *Bayesians*, who argue that many uncertain situations can be analyzed only by means of subjective probability assessments. Bayesians are willing to assign probabilities to presidential elections, medical diagnoses, stock prices, legal trials, military strategy, and God's existence.

A subjective probability is based on an intuitive blending of a variety of information and therefore can vary from person to person. Nonetheless, subjective probabilities are useful in that they allow us to communicate our beliefs to others and to make decisions that are consistent with our beliefs.

4.2 Some Helpful Rules

So far, we have focused on probabilities that can be calculated directly by counting equally likely outcomes, calculating long-run frequencies, or engaging in subjective introspection. Some probabilities are better calculated indirectly, by applying standard rules.

We let A identify an event and let $P[A]$ be the probability that A will occur. For example, A might be heads when a coin is flipped and $P[A] = 0.5$. A probability cannot be negative or larger than 1. If A is impossible, $P[A] = 0$; if A is certain to occur, $P[A] = 1$.

The Addition Rule

Our first rule, the addition rule, gives the probability that either of two events (or possibly both) will occur. For example, if we flip two coins, what is the probability that one or the other (or both) coins will land heads? It is tempting to add the 0.5 probability of heads on the first coin to the 0.5 probability of heads on the second coin, but that would give a probability of 0.5 + 0.5 = 1.0, which cannot be correct, since one or more heads is not 100 percent certain.

Simply adding together the probability of heads on the first coin and the probability of heads on the second coin double counts the case of heads on both coins. Our earlier probability tree showed that there are four possible, equally likely, outcomes summarized here in Figure 4.4.

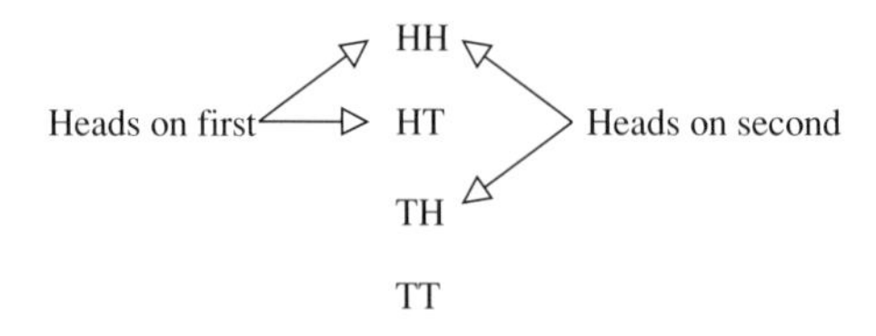

Figure 4.4
Double counting HH.

HH and HT give heads on the first coin. HH and TH give heads on the second coin. If we count all four of these cases, HH is counted twice. The correct probability, counting HH only once, is 3/4.

One way to correct this double counting is to double count and then subtract the outcomes that are double counted. This is the *addition rule*: The probability that either A or B (or possibly both) will occur can be determined by adding the separate probabilities and then correcting for double counting:

$$P[A \text{ or } B] = P[A] + P[B] - P[A \text{ and } B] \tag{4.1}$$

With two coin flips, the probability of heads on the first coin is 1/2, the probability of heads on the second coin is 1/2, and the probability of heads on both coins is 1/4. Therefore, the probability of heads on either the first or second coin is 1/2 + 1/2 – 1/4 = 3/4.

Sometimes $P[A$ and $B]$ is equal to 0, because A and B cannot both occur. If A and B are *mutually exclusive*, so that $P[A$ and $B] = 0$, then the addition rule simplifies to

$$\text{If mutually exclusive, } P[A \text{ or } B] = P[A] + P[B] \tag{4.2}$$

For instance, if a standard six-sided die is rolled and A is the number 1 and B is the number 2, then A and B are mutually exclusive outcomes since the die cannot be both a 1 and a 2. There is no double-counting problem. The probability that a die will be either a 1 or a 2 is 1/6 + 1/6 = 1/3.

Many addition-rule questions are answered more easily by using the subtraction rule, which is discussed later in this chapter. Nonetheless, the addition rule is important because it helps us avoid the common mistake of simply adding together probabilities. For instance, the novelist Len Deighton wrote that a World War II pilot who has a 2 percent chance of being shot down on each mission is "mathematically certain" to be shot down in 50 missions [7]. Deighton obtained this mathematical certainty by adding together 50 2 percent probabilities to get 100 percent. However, the addition rule alerts us to the double-counting problem. (If Deighton's procedure were correct, the probability of being shot down in 51 missions would be a nonsensical 102 percent.) Later in this chapter, when we get to the subtraction rule, we make the correct calculation.

Conditional Probabilities

We began this chapter with the example of worrisome results from a mammogram X-ray. One hundred doctors were asked this hypothetical question [8]:

> *In a routine examination, you find a lump in a female patient's breast. In your experience, only 1 out of 100 such lumps turns out to be malignant, but, to be safe, you order a mammogram X-ray. If the lump is malignant, there is a 0.80 probability that the mammogram will identify it as malignant; if the lump is benign, there is a 0.90 probability*

that the mammogram will identify it as benign. In this particular case, the mammogram identifies the lump as malignant. In light of these mammogram results, what is your estimate of the probability that this lump is malignant?

Of the 100 doctors surveyed, 95 gave probabilities of around 75 percent. However, the correct probability is only 7.5 percent!

To analyze the test results correctly, we can use a *contingency table* in which data are classified in one way by the rows and in another way by the columns. (Because of this two-way classification, a contingency table is sometimes called a *two-way table*.) In our example, the lump has two possible conditions (malignant or benign), which we show as two rows in Table 4.2, and the test has two possible outcomes (positive or negative), which we show as two columns.

The entries in Table 4.2 show the four possible combinations of conditions, for example, having a malignant tumor and testing positive. To determine the entries in the table, we need to assume an arbitrary value for the number of women being tested, say 1,000 women, and assume that the entries correspond to the given probabilities. Because the lumps are malignant in 1 out of 100 cases, we have the entries shown in Table 4.3.

We fill in the rest of the numbers in Table 4.4 by noting that the test gives a positive result in 80 percent of the malignant cases and 90 percent of the benign cases.

Looking at the first numerical row, for those patients with malignant tumors, the test gives a positive result 80 percent of the time, 8/10 = 0.80. Yet, looking at the first numerical

Table 4.2: Mammogram Contingency Table

	Positive	Negative	Total
Malignant			
Benign			
Total			

Table 4.3: Contingency Table for 1,000 Women

	Positive	Negative	Total
Malignant			10
Benign			990
Total			1,000

Table 4.4: Contingency Table for Mammogram Test Results

	Positive	Negative	Total
Malignant	8	2	10
Benign	99	891	990
Total	107	893	1,000

column, of the 107 patients with positive test results, only 7.5 percent actually have malignant tumors: 8/107 = 0.075. Even though 80 percent of the 10 malignant tumors are correctly identified as malignant, these positive test results are far outnumbered by the false positives—the 10 percent of 990 benign lumps that are identified incorrectly. In this example, a small percentage of a large number is bigger than a large percentage of a small number.

It is very easy to misinterpret conditional probabilities, and these doctors evidently misinterpreted them. According to the researcher who conducted this survey [9],

> *The erring physicians usually report that they assumed that the probability of cancer given that the patient has a positive X-ray... was approximately equal to the probability of a positive X-ray in a patient with cancer.... The latter probability is the one measured in clinical research programs and is very familiar, but it is the former probability that is needed for clinical decision making. It seems that many if not most physicians confuse the two.*

The solution is for doctors to become better informed about conditional probabilities.

We calculated the mammogram conditional probabilities by constructing a contingency table and taking the ratio of the appropriate numbers. Conditional probabilities can also be calculated by taking the ratio of two probabilities. The conditional probability that B will occur, given that A has occurred, is

$$P[B \text{ if } A] = \frac{P[A \text{ and } B]}{P[A]} \tag{4.3}$$

In our mammogram example,

$$\begin{aligned} P[\text{malignant if positive}] &= \frac{P[\text{malignant and test positive}]}{P[\text{test positive}]} \\ &= \frac{8/1{,}000}{107/1{,}000} \\ &= 0.075 \end{aligned}$$

Independent Events and Winning Streaks

The probability expressions $P[B]$ and $P[B \text{ if } A]$ answer different questions:

$P[B]$: Considering all possible outcomes, what is the probability that B will occur?
$P[B \text{ if } A]$: Considering only those cases where A occurs, what is the probability that B will also occur?

If the probability that B will occur does not depend on whether or not A occurs, then it is natural to describe A and B as *independent*:

$$A \text{ and } B \text{ are independent if } P[B \text{ if } A] = P[B]$$

In fair games of chance involving coins, dice, roulette wheels, and other physical objects, each outcome is independent of other outcomes, past, present, or future. However, wins are sometimes bunched together, just like heads sometimes comes up three times in a row. Some gamblers mistakenly attach a great deal of significance to coincidences. They apparently believe that luck is like a disease that a player catches then takes a while to get over. For example, Clement McQuaid, who describes himself as a "keen student of gambling games, most of which he has played profitably," offers this advice [10]:

> *There is only one way to show a profit. Bet light on your losses and heavy on your wins. Many good gamblers follow a specific procedure: a.* Bet minimums when you're losing... *b.* Bet heavy when you're winning.

Yes, it is good to win large bets and lose only small bets, But how do we know in advance whether we are going to win or lose our next bet? Suppose you are playing a dice game and have won three times in a row. You remember that you won three times in a row and you are happy about it, but dice have no memories or emotions. Dice do not remember what happened on the last roll and do not care what happens on the next roll. The outcomes are independent and the probabilities are constant, roll after roll. Games of chance are the classic example of independent events.

The Fallacious Law of Averages

If a coin has 0.50 probability of landing heads, the *law of large numbers* says that it will land heads close to 50 percent of the time in the long run. The incorrect *law of averages* (or *gambler's fallacy*) says that, in the long run, the number of heads and tails must be *exactly* equal. For example, some people think that in 1,000 coin flips there must be exactly 500 heads and 500 tails; therefore, if tails come up more often than heads in the first 10, 50, or 100 flips, heads must now come up more often than tails in order to "average" or "balance" things out.

This belief is wrong but widespread. For example, one gambler wrote [11]:

> *Flip a coin 1000 times and it'll come up heads 500 times, or mighty close to it. However, during that 1000 flips of the coin there will be frequent periods when heads will fail to show and other periods when you'll flip nothing but heads. Mathematical probability is going to give you roughly 500 heads in 1000 flips, so that if you get ten tails in a row, there's going to be a heavy preponderance of heads somewhere along the line.*

A coin is an inanimate object that is tossed in the air and examined by curious humans. If the coin is unbent and fairly tossed, then heads and tails are equally likely to appear, no matter what happened on the last flip or the last 999 flips.

The basis for the fallacious law of averages is the mistaken belief that heads and tails must come up equally often; a run of tails must consequently be balanced by a run of heads. The law of large numbers does not say that the *number* of heads must equal the number of tails. The law of large numbers says nothing about the last 10 flips or the last 10 million flips—which have already happened, cannot be changed, and have no effect on future flips. The law of large numbers is concerned with future throws. It says, looking ahead to predict the next 1,000 throws, the probability is high that heads will come up close to half the time.

A *fraction* that is close to 0.5 is not inconsistent with the *number* of heads and tails being unequal. In fact, while we are almost certain that, in the long run, the fraction of the flips that are heads will be very close to 0.5, we are also almost certain that the number of heads will not equal the number of tails!

With two flips, the probability of one head and one tail is 0.5. With four flips, the probability of two heads and two tails drops to 0.375. It can be shown that the probability that the number of heads will exactly equal the number of tails decreases as the number of flips increases: 0.2462 for 10 flips, 0.0796 for 100 flips, 0.0252 for 1,000 flips, and 0.0080 for 10,000 flips.

While the probability is small that there will be exactly 50 percent heads in a large number of flips, the probability is large that there will be close to 50 percent heads. The probability that there will be between 49 percent and 51 percent heads increases as the number of flips increases: 0.2356 for 100 flips, 0.4933 for 1,000 flips, and 0.9556 for 10,000 flips.

This makes sense. With two flips, there is a 50 percent chance of one head and one tail, and the only other possible outcomes are no heads or all heads. With 1,000 flips, there are lots of possible outcomes that are close to 50 percent heads: 499, 501, 498, 502, 497, 503, and so on. Each of these possibilities has approximately the same probability, but these probabilities must be small—otherwise the sum of the probabilities would be more than 1. Hence, there is a small probability of exactly 500 heads but a large probability that the number of heads will be close to 500.

Many gamblers believe in the fallacious law of averages, hoping to find a profitable pattern in the chaos created by random chance. When a roulette wheel turns up a black number several times in a row, there are always people eager to bet on red, counting on the law of averages. Others will rush to bet on black, trying to catch a "hot streak." The casino cheerfully accepts wagers from both, confident that future spins do not depend on the past.

One of the most dramatic roulette runs occurred at Monte Carlo on August 18, 1913. At one table, black came up again and again. After about ten blacks in a row, the table was surrounded by excited people betting on red and counting on the law of averages to reward them. But black kept coming. After 15 straight blacks, there was near panic as people tried to reach the table so that they could place even heavier bets on red. Still black came up. By the 20th black, desperate bettors were wagering every chip they had left on red, hoping to recover a fraction of their losses. When this memorable run ended, black had come up 26 times in a row and the casino had won millions of francs.

In honest games, at Monte Carlo and elsewhere, betting strategies based on the law of averages do not work. In fact, the few gambling systems that have paid off have been based on the opposite principle—that physical defects in a roulette wheel or other apparatus cause some events to occur more often than others. If the number 6 comes up an unusually large number of times (not twice in a row, but maybe 30 times in 1,000 spins—which is 11 percent more often than expected), then a mechanical irregularity may be responsible. Instead of betting against 6, expecting its success to be balanced out, we might bet on 6, hoping that a physical imperfection will cause 6 to continue to come up more often than it would if the wheel were perfect.

Sports are a fertile source of law-of-averages fallacies. When a baseball player goes hitless in 12 times at bat, a commentator may announce that the player "has the law of averages on his side" or that he "is due for a hit." The probability of a base hit does not increase just because a player has not had one lately. The 12 outs in a row may have been bad luck, 12 line drives hit right at fielders. If so, this bad luck does not make the player any more likely to have good luck his next time at bat. If it is not bad luck, then maybe a physical problem is causing the player to do poorly. Psychology may also play a role because, unlike coins, players have memories and emotions and care about wins and losses.

The manager of a baseball player who is 0 for 12 should be concerned—not confident that the player is due for a hit. Similarly, a player who had four hits in his last four at bats should not be benched because he is "due for an out." Yet that is exactly what happened when a manager once had a pinch hitter bat for a player who already had four hits that day, explaining that players seldom get five hits in a row. (They never will if they never get the opportunity!)

I once watched a Penn State kicker miss three field goals and an extra point in an early-season football game. The television commentator said that the Penn State coach should be happy about those misses as he looked forward to some tough games in coming weeks. The commentator explained that every kicker will miss some over the course of the season and it is good to get these misses "out of the way" early in the year. This unjustified optimism is the law-of-averages fallacy again. Those misses need not be balanced by successes. If anything, the coach should be worried because this poor performance suggests that his kicker is not very good.

The Multiplication Rule

Equation 4.3 tells us how to calculate a conditional probability. Often, we are interested in the reverse calculation: We know a conditional probability and want to calculate $P[A \text{ and } B]$, the probability that both A and B will occur. We can make this calculation by rearranging Equation 4.3 to obtain the *multiplication rule*: The probability that both A and B will occur is

$$P[A \text{ and } B] = P[A]\, P[B \text{ if } A] \tag{4.4}$$

The probability of both A and B occurring is equal to the probability that A will occur multiplied by the probability that B will occur given that A has occurred. The multiplication rule can be extended indefinitely to handle more than two events. The probability that A and B and C will occur is equal to the probability that A will occur, multiplied by the probability that B will occur given that A has occurred, multiplied by the probability that C will occur given that A and B have occurred: $P[A \text{ and } B \text{ and } C] = P[A]\ P[B \text{ if } A]\ P[C \text{ if } (A \text{ and } B)]$.

For example, what is the probability that four cards drawn from an ordinary deck of playing cards will all be aces? Because there are four aces among the 52 cards, the probability that the first card is an ace is 4/52. Given that the first card is an ace, there are three aces among the 51 remaining cards and the probability that the second card will also be an ace is 3/51. If the first two cards are aces, two aces are left among 50 cards and the probability of a third ace is 2/50. The probability of a fourth ace, given that the first three cards are aces, is 1/49. Multiplying these probabilities,

$$P[\text{four aces}] = \frac{4}{52}\frac{3}{51}\frac{2}{50}\frac{1}{49} = 0.0000037$$

We noted previously that, if two events are independent, then $P[B \text{ if } A] = P[B]$. If A and B are independent, the multiplication rule simplifies to this:

$$\text{If } A \text{ and } B \text{ are independent, } P[A \text{ and } B] = P[A]\,P[B] \tag{4.5}$$

If we roll two dice, for example, the results are independent and the probability of two 1s is simply the product of each die's probability of rolling a 1:

$$P[\text{double ones}] = \frac{1}{6}\frac{1}{6} = \frac{1}{36} = 0.0278$$

Legal Misinterpretations of the Multiplication Rule

In the 1960s, a white woman with blond hair tied in a ponytail was seen fleeing a Los Angeles robbery in a yellow car driven by a black man with a beard and a moustache. Four days later, the police arrested Malcolm Collins, a black man with a beard, moustache, and yellow Lincoln, and his common-law wife, a white woman with a blond ponytail. A mathematics professor calculated the probability that two people picked at random would have this combination of characteristics by estimating the probability of each characteristic: $P[\text{black man with beard}] = 1/10$, $P[\text{man with moustache}] = 1/4$, $P[\text{owning a yellow car}] = 1/10$, $P[\text{interracial couple}] = 1/1{,}000$, $P[\text{blond woman}] = 1/3$, $P[\text{wearing a ponytail}] = 1/10$. Using the multiplication rule,

$$P[\text{all six characteristics}] = \frac{1}{10}\frac{1}{4}\frac{1}{10}\frac{1}{1{,}000}\frac{1}{3}\frac{1}{10} = \frac{1}{12{,}000{,}000}$$

The small value of this probability helped convict Collins and his wife, the jurors apparently believing, in the words of the California Supreme Court, that "there could be but one chance in 12 million that the defendants were innocent and that another equally distinctive couple committed the robbery" [12].

The professor's calculation implicitly assumes independence, and the California Supreme Court questioned the appropriateness of this assumption. The probability of having a moustache is not independent of having a beard: While only 25 percent of all men have moustaches, perhaps 75 percent of men with beards do. Similarly, being a black man, a blond woman, and an interracial couple are not independent, nor perhaps are ponytails and blond hair. The court also raised the possibility that the assumed characteristics might have been incorrect, for example, the man may have worn a false beard as a disguise. The court decided that this was not proof beyond a reasonable doubt and reversed the Collins' conviction.

The Subtraction Rule

An event will either occur or not occur; therefore, $P[A] + P[\text{not } A] = 1$. From this, we derive the *subtraction rule*: The probability that A will not occur is

$$P[\text{not } A] = 1 - P[A] \tag{4.6}$$

If the probability of rolling a 1 with a die is 1/6, then the probability of not rolling a 1 is $1 - 1/6 = 5/6$.

The subtraction rule is so obvious that it seems hardly worth mentioning. However, it is a very useful rule because sometimes the easiest way to determine the probability that something will happen is to calculate the probability that it will not happen. For instance, what is the probability that, among four fairly flipped coins, there will be at least one heads? The easiest approach is to calculate the probability of no heads (four tails) and then use the subtraction rule to determine the probability of at least one heads. The multiplication rule tells us that the probability of four tails is $(1/2)^4$; therefore, the probability of at least one heads is $1 - (1/2)^4 = 0.9375$.

We can use the subtraction rule whenever we want to calculate the probability that something will happen "at least once." For example, earlier in this chapter, we looked at Len Deighton's incorrect assertion that a pilot with a 2 percent chance of being shot down on each mission is "mathematically certain" to be shot down in 50 missions. To determine the correct probability, we use the subtraction rule to turn the question around: What is the probability that a pilot will complete 50 missions without being shot down? We can then subtract this value from 1 to determine the probability that a pilot will not complete 50 missions successfully.

To calculate the probability that a pilot with a 2 percent chance of being shot down on each mission will complete 50 missions successfully, we need to assume that the mission outcomes are independent. This assumption is implicit in Deighton's constant 2 percent probability, though it is not completely realistic, as pilots no doubt improve with experience. (Probabilities also vary with the difficulty of individual missions; the 2 percent figure must be a simplifying average.) Assuming independence, the probability of not being shot down in 50 missions is equal to 98 percent multiplied by itself 50 times:

$$P[\text{not shot down in 50 missions}] = 0.98^{50}$$

The subtraction rule then tells us that the probability of being shot down is equal to 1 minus the probability of not being shot down:

$$\begin{aligned} P[\text{shot down in 50 missions}] &= 1 - P[\text{not shot down in 50 missions}] \\ &= 1 - 0.98^{50} \\ &= 0.6358 \end{aligned}$$

Instead of Deighton's erroneous 100 percent, the correct probability is about 64 percent.

Bayes's Rule

Reverend Thomas Bayes wanted to use information about the probability that the world would be the way it is if God exists (P[world if God exists]) to make inferences about the probability that God exists, given the way the world is (P[God exists if world]). Bayes was unable to prove the existence of God, but his analysis of how to go from one conditional probability to its reverse has turned out to be extremely useful and is the foundation for the modern Bayesian approach to probability and statistics.

Although Bayes worked out some calculations for several games of chance, his few published writings do not contain a general formula for reversing conditional probabilities. It was Laplace who wrote down the general expression and called it *Bayes's theorem*:

$$P[A \text{ if } B] = \frac{P[A]P[B \text{ if } A]}{P[A]P[B \text{ if } A] + P[\text{not } A]P[B \text{ if not } A]} \tag{4.7}$$

You do not need to memorize Bayes's theorem. In most cases, the simplest and most intuitive procedure is to construct a contingency table for a hypothetical population.

For instance, we began this chapter with the example of worrisome results from a mammogram test. If the lump is malignant, this test has a 0.8 probability of a positive reading: P[test positive if malignant] = 0.8. The doctor and patient are interested in the reverse conditional probability—the probability that a patient who tests positive has a malignant tumor. Table 4.4 shows that P[malignant if test positive] = 8/107 = 0.075.

Bayes's theorem is commonly used in two ways, both of which are illustrated in Table 4.4. The first use is to go from one conditional probability, $P[A \text{ if } B]$, to the reverse, $P[B \text{ if } A]$. Here, we went from P[test positive if malignant] = 0.80 to P[malignant if test positive] = 0.075.

The second use of Bayes's theorem is to revise a probability $P[A]$ in light of additional information B. Here, before the mammogram, the probability that a randomly selected lump is malignant is $P[A] = 0.01$. After the positive mammogram reading, the revised probability that the lump is malignant is $P[A \text{ if } B] = 0.075$. The probability that the lump is malignant increases from 1 percent to 7.5 percent, but is still far from certain.

A Bayesian Analysis of Drug Testing

An editorial in the *Journal of the American Medical Association* on mandatory urine drug tests argued that "An era of chemical McCarthyism is at hand, and guilty until proven innocent is the new slogan" [13]. The editorial noted that it would cost $8 billion to $10 billion annually to test every employee in the United States once a year and that the accuracy of these tests (measured in the fraction of the people who are diagnosed correctly) ranged from 75 percent to 95 percent for some drugs and from 30 percent to 60 percent for others.

If an employee is given a drug test, two kinds or errors can happen. A false-positive result occurs when the test incorrectly indicates the presence of a drug; a false-negative result occurs when the test fails to detect the presence of drugs. These mistakes can occur for a variety of reasons, including the mislabeling of samples, the use of contaminated laboratory equipment, and the technician's misreading of subjective criteria regarding chemical color, size, and location.

To illustrate the potential seriousness of the false-positive problem, consider a test administered to 10,000 persons, of whom 500 (5 percent) use the drug that the test is designed to detect and 9,500 (95 percent) do not. Suppose further that the test is 95 percent accurate: 95 percent of the drug users will be identified as drug users and 95 percent of those who are drug free will be identified as drug free.

Table 4.5 shows that, of the 500 people who use this drug, 475 (95 percent) will have a positive test result and 25 (5 percent) will not. Of the 9,500 who do not use this drug, 475 (5 percent) will have a positive result and 9,025 (95 percent) will not. Using these numbers, we can calculate the fractions of the diagnoses that are incorrect:

$$P[\text{drug user if negative reading}] = \frac{25}{9{,}050} = 0.0028$$

$$P[\text{drug free if positive reading}] = \frac{475}{950} = 0.50$$

Table 4.5: A Contingency Table for Random Drug Testing

	Test Positive	Test Negative	Total
Drug User	475	25	500
Drug Free	475	9,025	9,500
Total	950	9,050	10,000

Less than 1 percent of those who test negative are in fact drug users. However, an astounding 50 percent of those who test positive do not use the drug. This is another example of how important it is to interpret conditional probabilities correctly. In this example, 95 percent of all drug users test positive, but only 50 percent of those who test positive are drug users.

4.3 Probability Distributions

A *random variable* X is a variable whose numerical value is determined by chance, the outcome of a random phenomenon.[1] For example, X might be the number of heads when three coins are flipped or the number that comes up when a six-sided die is rolled. A *discrete random variable* has a countable number of possible outcomes: The die can be 1, 2, 3, 4, 5, or 6. A *continuous random variable*, in contrast, has a continuum of possible values: In theory, distance or time can be measured with infinite precision and the number of possible values cannot be counted.[2]

Table 4.1 shows the probabilities for the four possible values of X, the number of heads when three coins are flipped: zero, one, two, or three heads. A graph of the probabilities for all possible values is called a *probability distribution*. Figure 4.5 shows the probability distribution for the number of heads when three coins are flipped.

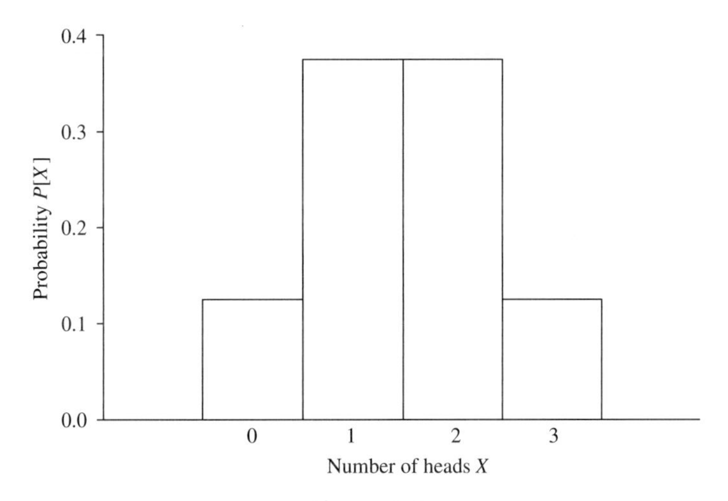

Figure 4.5
Probability distribution for three coin flips.

[1] Statisticians often use upper case and lower case letters to distinguish between a random variable, which can take on different values, and the actual values that happen to occur. We use upper case notation for simplicity.

[2] A discrete random variable can have an infinite number of possible values, for example, the number of times a coin is flipped before obtaining a heads. What matters is that the possible values can be counted.

Table 4.6: Probabilities and Frequencies for Three Coin Flips

Number of Heads	Probability	Observed Frequency
0	0.125	0.120
1	0.375	0.350
2	0.375	0.400
3	0.125	0.130
Total	1.000	1.000

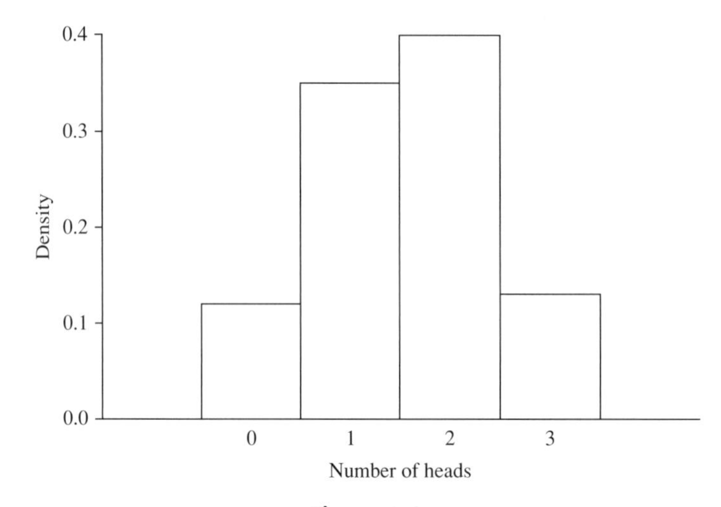

Figure 4.6
Histogram for three coin flips.

A probability distribution is similar to the histograms described in Chapter 2, in that the total area is equal to 1, but these graphs are conceptually different. A probability distribution shows the theoretical probabilities while a data histogram shows the actual frequencies with which the outcomes happen to occur in a particular data set.

For example, I flipped three coins 100 times and recorded how often I got no heads (12 times), one head and two tails (35 times), two heads and one tail (40 times), and three heads (13 times). Table 4.6 shows that the observed frequencies are close to, but not exactly equal to, the theoretical probabilities.

Figure 4.6 shows a histogram for these 100 coin flips. Remember, a probability distribution is similar to a histogram, but a probability distribution shows theoretical probabilities while a histogram shows actual frequencies.

Expected Value and Standard Deviation

Sometimes, a few simple numbers can summarize effectively the important characteristics of a probability distribution. The *expected value* (or *mean*) of a discrete random variable X

is a weighted average of all possible values of X, using the probability of each X value as weights:

$$\mu = E[X] = \sum X_i P[X_i] \tag{4.8}$$

The Greek symbol μ is pronounced "mew," and the notation $E[X]$ denotes the expected value of the random variable X. The summation sign Σ means that we add up the products $X_i P[X_i]$.

Suppose, for example, that X is equal to the number of heads that occur when three coins are flipped. The probabilities are given in Table 4.6. Equation 4.8 shows that the expected value of X is an average of the possible outcomes weighted by their respective probabilities:

$$\begin{aligned}\mu &= 0(0.125) + 1(0.375) + 2(0.375) + 3(0.125) \\ &= 1.5\end{aligned}$$

The expected value is not necessarily the most likely value of X. You will never flip 1.5 heads! The expected value is the anticipated long-run average value of X if this experiment is repeated a large number of times. If, as anticipated, no heads occurs 12.5 percent of the time, one head occurs 37.5 percent of the time, two heads occurs 37.5 percent of the time, and three heads occurs 12.5 percent of the time, the average value of X will be 1.5.

Suppose that your instructor offered to let you play a game where you flip three coins and are paid $1 for every head that appears. The coins are fair and honest and so is your professor. On any single play, you might receive nothing, $1, $2, or $3. If you play a large number of times, you can expect to receive, on average, $1.50 per play. If you have to pay $1 each time you play this game, you can expect to make an average profit of $0.50 per play in the long run. If it costs you $2 to play this game, you can expect to lose $0.50 per play in the long run.

The mean μ of a probability distribution is analogous to the mean $\overline{X}$ of observed data. The mean of a probability distribution weights the possible outcomes by the probability that they will occur. The mean of observed data uses the frequency with which outcomes happen to occur.

In the same way, the variance σ^2 of a probability distribution is analogous to the variance s^2 of observed data, again using probabilities in place of observed frequencies. The *variance* of a discrete random variable X is a weighted average of the squared deviations of the possible values of X from the mean, using the probability of each X value as weights:

$$\sigma^2 = E[(X - \mu)^2] = \sum (X_i - \mu)^2 P[X_i] \tag{4.9}$$

The Greek symbol σ is pronounced "sigma." The square root of the variance is the standard deviation σ.

Table 4.7 shows the calculation of the variance for our example of three coin flips. The variance works out to be $\sigma^2 = 0.75$, so that the standard deviation is $\sigma = \sqrt{0.75} = 0.866$. The variance of a probability distribution is analogous to the variance of observed data, in that both use the squared deviations from the mean to gauge dispersion. The difference is

Table 4.7: Variance for the Probability Distribution for Three Coin Flips

Number of Heads X_i	$X_i - \mu$	$(X_i - \mu)^2$	Probability $P[X_i]$	$(X_i - \mu)^2 P[X_i]$
0	−1.5	2.25	0.125	0.28125
1	−0.5	0.25	0.375	0.09375
2	0.5	0.25	0.375	0.09375
3	1.5	2.25	0.375	0.28125
Total			1.000	0.75000

that the variance of a probability distribution uses probabilities to weight the possible outcomes while the variance of a data set uses the observed frequencies.

The Normal Distribution

In Chapter 2, we saw that histograms are often shaped like a bell: a single peak with the bars declining symmetrically on either side of the peak—gradually at first, then more steeply, then gradually again. W. J. Youdon, a distinguished statistician, described this familiar shape this way [14]:

The
normal
law of error
stands out in the
experience of mankind
as one of the broadest
generalizations of natural
philosophy. It serves as the
guiding instrument in researches
in the physical and social sciences and
in medicine, agriculture, and engineering.
It is an indispensable tool for the analysis and the
interpretation of the basic data obtained by observation and experiment.

This bell-shaped curve describes data on such disparate phenomena as the heights of humans, the weights of tomatoes, scores on the SAT test, baseball batting averages, and the location of molecules.

Probability Density Curves

We noted earlier that a continuous random variable can have a continuum of possible values. For example, Figure 4.7 shows a spinner for randomly selecting a point on a circle. We can imagine that this is a clean, well-balanced device in which each point on the circle is equally likely to be picked. But, how many possible outcomes are there? How many points are there on the circle? In theory, there are an uncountable infinity of points because between any two points on the circle, there are still more points.

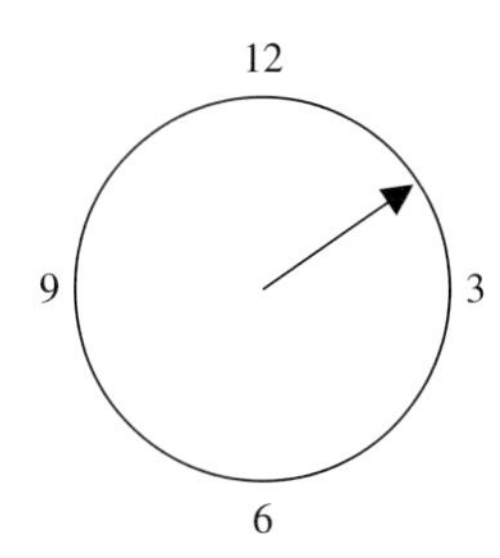

Figure 4.7
A continuous random number.

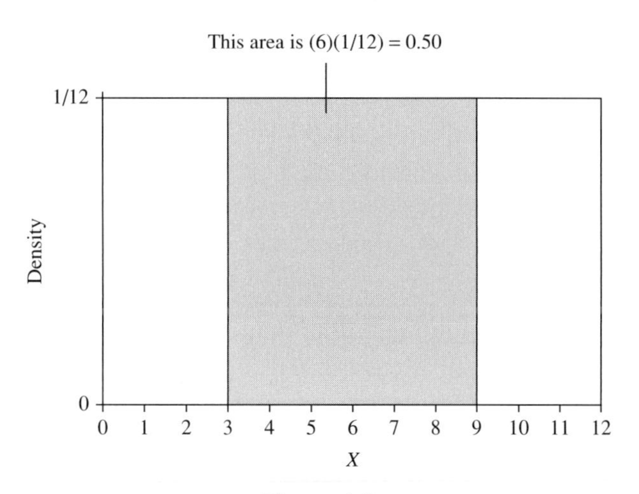

Figure 4.8
A continuous probability distribution.

Weight, height, and time are other examples of continuous variables. Even though we might say that Rachel Smith is 19 years old, a person's age can, in theory, be specified with infinite precision. Instead of saying that she is 19 or 20, we could say that she is 19 years and 220 days, or 19 years, 220 days, and 10 hours. With continuous variables, we can specify finer and finer gradations within any interval.

How can we specify probabilities when there are an uncountable number of possible outcomes? If we give each point a positive probability, the sum of this uncountable number of probabilities will be infinity, not 1. Mathematicians handle this vexing situation by assigning probabilities to intervals of outcomes, rather than to individual outcomes.

We display interval probabilities by using a probability density curve, as in Figure 4.8, in which the probability that the outcome will be in a specified interval is given by the area under the curve. The shaded area shows the probability that the random number will be between 3 and 9. The area of this rectangle is (base)(height) = (6)(1/12) = 1/2. What is the probability that the random number will be between 0 and 12? This probability is the entire area under the curve: (base)(height) = (12)(1/12) = 1. In fact, the height of the probability density curve, 1/12, was derived from the requirement that the total area must be 1.

A rectangular distribution, like Figure 4.8, is called the *uniform distribution*. Continuous probability distributions can have all sorts of shapes, as long as the densities are never negative and the total area under the curve is equal to 1.

The smooth density curve for a continuous random variable is analogous to the jagged probability distribution for a discrete random variable. The population mean and the standard deviation consequently have the same interpretation. The population mean is the anticipated long-run average value of the outcomes; the standard deviation measures the extent to which the outcomes are likely to differ from the mean. The population mean is at the center of a symmetrical density function; in Figure 4.8, for example, the mean is 6. More generally, however, the mean and standard deviation of a continuous random variable cannot be calculated without using advanced mathematics.

Standardized Variables

The mean and standard deviation are two important tools for describing probability distributions. One appealing way to standardize variables is to transform them so that they have the same mean and the same standard deviation. This reshaping is easily done in the statistical beauty parlor. The standardized value Z of a random variable X is determined by subtracting the mean of the probability distribution and dividing by the standard deviation:

$$Z = \frac{X - \mu}{\sigma} \tag{4.10}$$

Observed data can be similarly transformed by subtracting the mean of the data and then dividing by the standard deviation:

$$Z = \frac{X - \overline{X}}{s} \tag{4.11}$$

A standardized variable Z measures how many standard deviations X is above or below its mean. If X is equal to its mean, Z is equal to 0. If X is one standard deviation above its mean, Z is equal to 1. If X is two standard deviations below its mean, Z is equal to –2.

Suppose that the distribution of the heights of U.S. women between the ages of 25 and 34 has a mean of 66 inches and a standard deviation of 2.5 inches. If so, we can translate the height of an individual woman from inches to standardized Z values, as illustrated in Table 4.8.

Instead of saying that a woman is 71 inches tall (which is useful for some purposes, such as clothing sizes), we can say that her height is two standard deviations above the mean (which is useful for other purposes, such as comparing her height with the heights of other women).

Table 4.8: Standardized Heights

X	$Z = \frac{X - 66}{2.5}$
61.0	−2
63.5	−1
66.0	0
68.5	+1
71.0	+2

The Central Limit Theorem

Karl Gauss (1777–1855) collected a great deal of data measuring the shape of the earth and the movements of planets and typically found that histograms of his data were roughly bell shaped. Other researchers found that histograms of many other physical and social data are often bell shaped. You can imagine the excitement they must have felt when they first discovered this remarkable regularity. These researchers were analyzing very different situations governed by unpredictable chance, and yet a regular pattern emerged. No wonder Sir Francis Galton called this phenomenon a "wonderful form of cosmic order" [15].

Even more remarkably, mathematicians were eventually able to prove a mathematical theorem that helps explain this recurring pattern in the data: The *central limit theorem* states that if Z is the sum of n independent, identically distributed random variables with a finite, nonzero standard deviation, then the probability distribution of Z approaches the normal distribution as n increases.

For example, if we flip a coin n times, the number of heads is the sum of the results of these n flips. Figure 4.3 shows the probability distribution for the number of heads when $n = 3$ coins are flipped. Figure 4.9 shows the probability distribution for $n = 10$ and

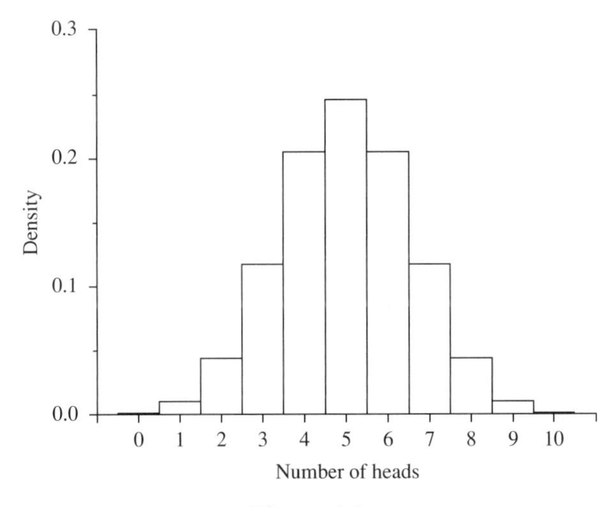

Figure 4.9
Probability distribution for ten coin flips.

Figure 4.10 shows the probability distribution for $n = 50$. As n increases, the probability distribution does indeed become more bell shaped.

Figure 4.11 shows the perfect normal distribution that emerges when n becomes infinitely large.

Many random variables are the cumulative result of a sequence of random events. For instance, a random variable giving the sum of the numbers when eight dice are rolled can be viewed as the cumulative result of eight separate random events. The percentage change

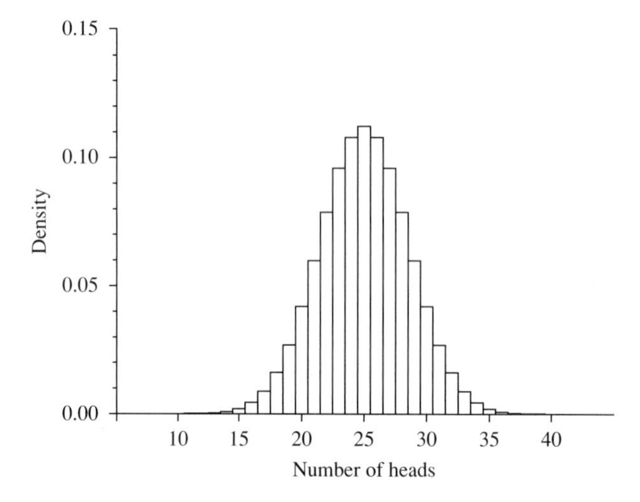

Figure 4.10
Probability distribution for 50 coin flips.

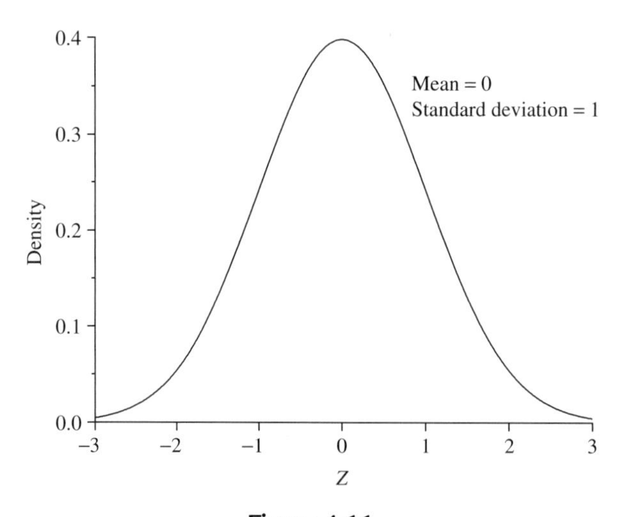

Figure 4.11
The standardized normal distribution.

in a stock's price over a 6-month period is the cumulative result of a large number of random events during that interval.

As remarkable as it is, the central limit theorem would be of little practical value if the normal curve emerged only when n is extremely large. The normal distribution is important because it so often appears even when n is quite small. Look again at the case of $n = 10$ coin flips in Figure 4.9; for most purposes, a normal curve would be a completely satisfactory approximation of that probability distribution. If the underlying distribution is reasonably smooth and symmetrical (as with dice rolls and coin flips), the approach to a normal curve is very rapid and values of n larger than 20 or 30 are sufficient for the normal distribution to provide an acceptable approximation.

Furthermore, the central limit theorem would be just a mathematical curiosity if the assumption that the cumulated variables are independent and identically distributed had to be satisfied strictly. This assumption is appropriate for dice rolls and other repetitive games of chance but in practical affairs is seldom exactly true. Probabilities vary from one trial to the next as conditions change and, in many cases, because of the outcomes of earlier trials. For example, height and weight depend on heredity and a lifetime of diet and exercise—factors that are not independent and do not have identical probability distributions. Yet, histograms of height and weight are bell shaped. A baseball player's chances of getting a base hit depend on the player's health, the opposing pitcher, the ballpark, and other factors. Yet, histograms for the total number of base hits over a season are bell shaped. A person's ability to answer a particular test question depends on the question's difficulty and the student's ability and alertness, but histograms of test scores are bell shaped.

This is why the normal distribution is so popular and the central limit theorem so celebrated. However, do not be lulled into thinking that all probability distributions and histograms are bell shaped. Many outcomes are approximately, but not perfectly, normal; and many probability distributions are not normal at all. Figure 4.8 shows a uniform distribution; the distribution of household income is skewed right. My purpose is not to persuade you that there is only one probability distribution but to explain why so many phenomena are well described by the normal distribution.

Finding Normal Probabilities

The density curve for the standardized normal distribution is graphed in Figure 4.11. The probability that the value of Z will be in a specified interval is given by the corresponding area under this curve. These areas can be determined by complex numerical procedures, with the results collected in tables, such as Table A.1 in the Appendix of this book. Such calculations are now standard in statistical software, too, and are especially useful for Z values that are not shown in Table A.1.

The Normal Probability Table

Table A.1 shows the area in the right-hand tail of a standardized normal distribution. The left-hand column of Table A.1 shows values of Z in 0.1 intervals, such as 1.1, 1.2, and 1.3. For finer gradations, the top row shows 0.01 intervals for Z. By matching the row and column, we can, for instance, determine that the probability that Z is larger than 1.25 is 0.1056.

Table A.1 gives all the information we need to determine the probability for any specified interval. For example, to determine the probability that Z is less than –1.25, we note that, because of symmetry, the area in a left-hand tail is equal to the corresponding area in the right-hand tail:

$$P[Z < -1.25] = P[Z > 1.25] = 0.1056$$

For areas more complicated than the right-hand or left-hand tail, it is usually safest to make a quick sketch, so that the probabilities that must be added or subtracted are readily apparent. For example, Figure 4.12 shows that, to determine the probability that Z is between 0 and 1.25, we use the fact that the total area in the right-hand tail is 0.5.

Nonstandardized Variables

So far, we calculated probabilities for a normally distributed variable Z that is standardized to have a mean of 0 and a standard deviation of 1. Nonstandardized variables can have positive or negative means and standard deviations that are larger or smaller than 1 (but not negative). The mean tells us where the center of the distribution is, and the standard deviation gauges the spread of the distribution about its mean. Together, the mean and standard deviation describe a normal distribution.

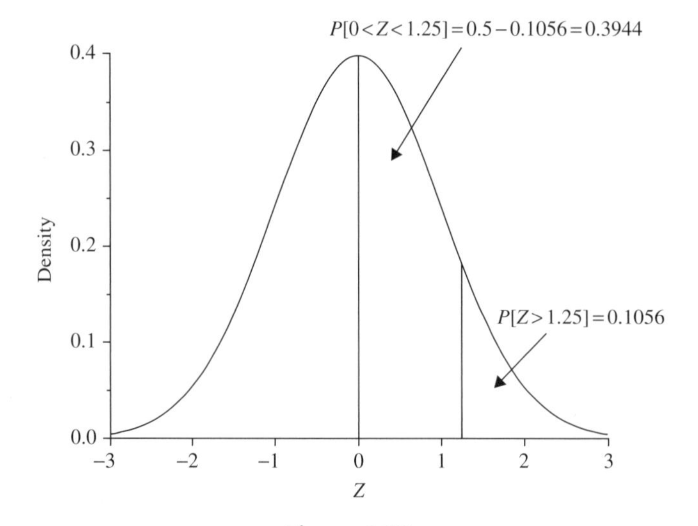

Figure 4.12
The probability that Z is between 0 and 1.25.

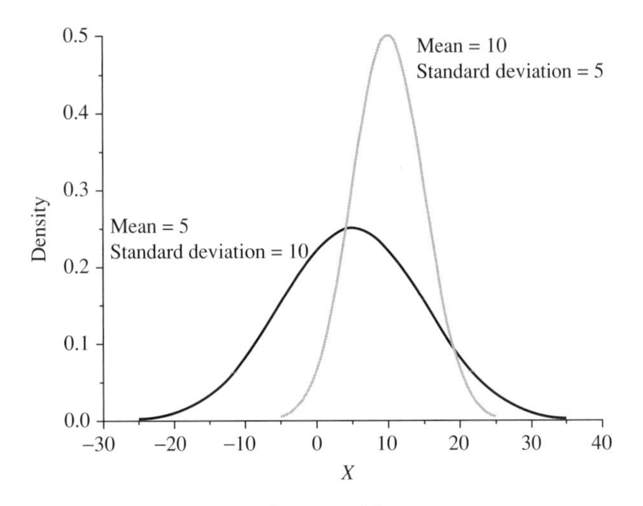

Figure 4.13
Two normal distributions.

Figure 4.13 shows two distributions with different means and standard deviations. Perhaps, these random variables are the annual percentage returns on two investments, one with a 5 percent mean and 10 percent standard deviation and the other with a 10 percent mean and 5 percent standard deviation. The distribution with the higher mean is centered to the right of the one with the lower mean. The distribution with the higher standard deviation is more spread out than the one with the lower standard deviation.

It is natural to describe these two investments by saying that the one with the higher expected return also has the more certain return. If offered a choice, most investors would choose the second investment.

How do we determine probabilities for a normally distributed variable X if it does not have a mean of 0 and standard deviation of 1? We determine the corresponding value of the standardized Z value and find the probability in Table A.1 (or with statistical software). Suppose, for example, X is normally distributed with a mean of 5 and a standard deviation of 10, for which we use the shorthand notation $X \sim N[5, 10]$. If we want to determine the probability that the value of X is larger than 15, the corresponding Z value is

$$\begin{aligned} Z &= \frac{X - \mu}{\sigma} \\ &= \frac{15 - 5}{10} \\ &= 1.0 \end{aligned}$$

The standardized variable Z is equal to 1 because, for a normally distributed variable with a mean of 5 and a standard deviation of 10, the value $X = 15$ is one standard deviation above the mean. Table A.1 shows the probability to be 0.1587.

One, Two, Three Standard Deviations

The three rules of thumb in Table 4.9 can help us estimate probabilities for normally distributed random variables without consulting Table A.1.

A normally distributed random variable has about a 68 percent (roughly two-thirds) chance of being within one standard deviation of its mean, a 95 percent chance of being within two standard deviations of its mean, and better than a 99.7 percent chance of being within three standard deviations. Turning these around, a normally distributed random variable has a 32 percent chance of being more than one standard deviation from its mean, roughly a 5 percent chance of being more than two standard deviations from its mean, and less than a 0.3 percent chance of being more than three standard deviations from its mean.

For example, a number of tests are designed to measure a person's IQ (intelligence quotient). The first IQ test consisted of 54 mental "stunts" that two psychologists, Alfred Binet and Theodore Simon, devised in 1904 to identify students in the Paris school system who needed special assistance. IQ tests today are designed to measure general intelligence, including an accurate memory and the ability to reason logically and clearly.

Because an individual's score on an IQ test depends on a very large number of hereditary and environmental factors, the central limit theorem explains why the distribution of IQ scores is approximately normally distributed. One of the most widely used tests today is the Wechsler Adult Intelligence Scale, which has a mean IQ of 100 and a standard deviation of 15. Therefore, a score of 100 indicates that a person's intelligence is average. About half the people tested score above 100, while half score below 100.

1. The one-standard-deviation rule implies that about 32 percent of the population have IQ scores more than 15 points away from 100: 16 percent above 115 and 16 percent below 85.
2. The two-standard-deviations rule implies that about 5 percent of the population have IQ scores more than 30 points away from 100: 2.5 percent above 130 and 2.5 percent below 70.
3. The three-standard-deviations rule implies that about 0.27 percent of the population have IQ scores more than 45 points away from 100: 0.135 percent above 145 and 0.135 percent below 55.

Table 4.9: One, Two, Three Standard Deviations

The probability that a normally distributed variable will be within:	
One standard deviation of its mean is	0.6826
Two standard deviations of its mean is	0.9544
Three standard deviations of its mean is	0.9973

Exercises

4.1 A prominent French mathematician, Jean d'Alembert, argued that because there are four possible outcomes when coins are flipped three times (no heads, a head on the first flip, a head on the second flip, or a head on the third flip), the probability of no heads is 1/4. Explain why you either agree or disagree with his reasoning.

4.2 A very successful football coach once explained why he preferred running the ball to passing it: "When you pass, three things can happen [completion, incompletion, or interception], and two of them are bad [16]. Can we infer that there is a 2/3 probability that something bad will happen when a football team passes the ball?

4.3 In the game odd-or-even, two players simultaneously reveal a hand showing either one or two fingers. If each player is equally likely to show one or two fingers and their choices are made independently, what is the probability that the sum will be even?

4.4 The traditional Indian game Tong is similar to the odd-or-even game described in the preceding exercise, except that each player can show one, two, or three fingers. If each possibility is equally likely and the choices are made independently, what is the probability that the sum will be an even number?

4.5 A television weather forecaster once said that there was a 50 percent chance of rain on Saturday and a 50 percent chance of rain on Sunday, and therefore a 100 percent chance of rain that weekend. Explain the error in this reasoning. If there is a 40 percent chance that it will rain on both Saturday and Sunday, what is the probability that it will rain on at least one of these two days?

4.6 Jurors are not supposed to discuss a case with anyone until after all of the evidence has been presented. They then retire to the jury room for deliberations, which sometimes begin with a secret-ballot vote of the jurors. Suppose that we have a case where each selection of a person for the jury has a 0.9 probability of picking someone who, after hearing all of the evidence, would initially vote guilty and a 0.1 probability of picking someone who would initially vote not guilty. What is the probability that the initial vote will be unanimously guilty if there is a

a. 12-person jury?

b. 6-person jury?

4.7 Suppose that a computer user chooses six random characters (either letters or digits) for a website password. (For example, I just chose ZDQMF2.) Characters can be used more than once (as with ZDZMF2) and no distinction is made between upper case and lower case letters. Suppose that a password attacker tries to enter your account by trying passwords with six randomly chosen characters but gets only three tries before the website locks the account for 24 hours. Because the attack program uses randomly selected characters, it may try the same six characters more than once. What is the probability that the three tries will be enough to access your account?

4.8 What are the chances of winning the following game? Put two red marbles and five blue marbles in a bag. The player puts his hand in the bag and pulls out three marbles. He wins if none of these marbles is red.

4.9 The ancient Enneagram (Figure 4.14) has nine personality types represented by the numbers 1 through 9 drawn around a circle. Three equilateral triangles can be drawn by connecting the points 9-3-6, 4-7-1, and 5-8-2. These three groupings correspond to three emotional states identified by modern psychological theory: attachment (9-3-6), frustration (4-7-1), and rejection (5-8-2).

If the numbers 1–9 were randomly separated into three groups of three numbers, what is the probability that one group would contain the numbers 9, 3, and 6 (not necessarily in that order), another group would contain the numbers 4, 7, and 1, and the third group would contain the numbers 5, 8, and 2?

4.10 Answer this letter to newspaper columnist Marilyn vos Savant [17]: "I have a really confusing one for you. Let's say my friend puts six playing cards face-down on a table. He tells me that exactly two of them are aces. Then I get to pick up two of the cards. Which of the following choices is more likely? (a) That I'll get one or both of the aces, or (b) That I'll get no aces?"

4.11 You are playing draw poker and have been dealt a pair of fours, a jack, a seven, and a three. You keep the pair of fours and discard the other three cards. What is the probability that at least one of the three new cards you receive will be a four?

4.12 A police department recorded the clothing colors worn by pedestrians who died after being struck by cars at night [18]. They found that four-fifths were wearing dark clothes and concluded that it is safer to wear light clothes. Use some hypothetical numbers to explain why these data do not necessarily justify this conclusion.

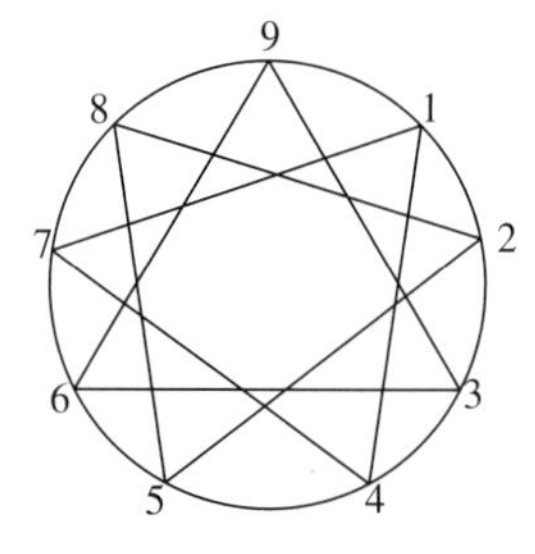

Figure 4.14
Exercise 4.9.

4.13 A book [19] calculated the probabilities of "being injured by various items around your house" and concluded "As the figures show, our homes are veritable booby traps. Even your cocktail table is out to get you. These accounted for almost 64,000 injuries, more than all ladders [62,000 injuries]." With 74,050,000 U.S. households, they calculated the probability of being injured by a cocktail table as 64,000/74,050,000 = 0.00086 and the probability of being injured by a ladder as 62,000/74,050,000 = 0.00084. Explain why these data do not really show that it is more dangerous to use a cocktail table than a ladder.

4.14 A Temple University mathematics professor used these data to show that most Americans have an exaggerated fear of terrorists [20]:

Without some feel for probability, car accidents appear to be a relatively minor problem of local travel while being killed by terrorists looms as a major risk of international travel. While 28 million Americans traveled abroad in 1985, 39 Americans were killed by terrorists that year, a bad year—1 chance in 700,000. Compare that with the annual rates for other modes of travel within the United States—1 chance in 96,000 of dying in a bicycle crash, 1 chance in 37,000 of drowning, and 1 chance in only 5,300 of dying in an automobile accident.

How do you suppose the author calculated the probability of dying in a car accident? Do these calculations prove that it is more dangerous to drive to school than to fly to Paris?

4.15 In an interview for a banking job, a student was asked to calculate the following: A can contains 20 coins, 19 of which are ordinary coins and 1 of which has heads on both sides. A coin is randomly selected from the can and flipped five times. It lands heads all five times. What is the probability that it is the two-headed coin?

4.16 In the United States, criminal defendants are presumed innocent until they are proven guilty beyond a reasonable doubt, because it is thought better to let nine guilty people go free than to send one innocent person to prison. Assume that 90 percent of all defendants are guilty, 90 percent of the guilty defendants are convicted, and 90 percent of the innocent defendants are set free.
a. Of those people convicted, what percent are innocent?
b. Of those people set free, what percent are guilty?

4.17 Suppose that there are two kinds of households, the Careless and the Careful; 99 percent of households are Careful and 1 percent are Careless. In any given year, a home inhabited by a Careless household has a 0.010 probability of being destroyed by fire, and a home occupied by a Careful household has a 0.001 probability of being destroyed by fire. If a home is destroyed by fire, what is the probability that it was occupied by a Careless household?

4.18 Use the information in Exercise 4.17 to answer this question. Suppose that every home is worth \$100,000 and that an insurance company sells 1-year policies to every household for \$500 that will pay either \$100,000 or nothing, depending on whether the home is destroyed by fire.

a. Is the expected value of the amount the insurance company has to pay larger or smaller than \$500?
b. For a Careless household, is the expected value of the payoff larger or smaller than \$500?
c. For a Careful household, is the expected value of the payoff larger or smaller than \$500?
d. What potential problem do you see for this insurance company?

4.19 In three careful studies [21], lie detector experts examined several persons, some known to be truthful and others known to be lying, to see if the experts could tell which were which. Overall, 83 percent of the liars were pronounced "deceptive" and 57 percent of the truthful people were judged "honest." Using these data and assuming that 80 percent of the people tested are truthful and 20 percent are lying, what is the probability that a person pronounced "deceptive" is in fact truthful? What is the probability that a person judged "honest" is in fact lying? How would these two probabilities be altered if half of the people tested are truthful and half are lying?

4.20 For a woman who gives birth at age 35, the probability of having a baby suffering from Down syndrome is 1/270 [22]. A test of the amniotic fluid in the mother's uterus is virtually 100 percent accurate in predicting Down syndrome but is expensive and can cause a miscarriage. A study of the effectiveness of an inexpensive blood test that does not risk miscarriage found that in 89 percent of the Down syndrome cases, the test gave a positive reading, while in 75 percent of the cases without Down syndrome, the test gave a negative reading. Of those cases where there is a positive reading, what fraction are false positives?

4.21 The random walk model of stock prices states that stock market returns are independent of the returns in other periods; for example, whether the stock market does well or poorly in the coming month does not depend on whether it has done well or poorly during the past month, the past 12 months, or the past 120 months. On average, the monthly return on U.S. stocks has been positive about 60 percent of the time and negative about 40 percent of the time. If monthly stock returns are independent with a 0.6 probability of a positive return and a 0.4 probability of a negative return, what is the probability of

a. 12 consecutive positive returns?
b. 12 consecutive negative returns?
c. A positive return, if the return the preceding month was negative?

4.22 The Romeros are planning their family and want an equal number of boys and girls. Mrs. Romero says that their chances are best if they plan to have two children. Mr. Romero says that they have a better chance of having an equal number of boys and girls if they plan to have ten children. Assuming that boy and girl babies are equally likely and independent of previous births, which probability do you think is higher? Explain your reasoning.

4.23 Answer this question that a reader asked Marilyn vos Savant, who is listed in the Guinness Book of World Records Hall of Fame for "Highest IQ" [23]: "During the last year, I've gone on a lot of job interviews, but I haven't been able to get a decent job.... Doesn't the law of averages work here?"

4.24 Answer this question that a reader asked Marilyn vos Savant [24]:

> *At a lecture on fire safety that I attended, the speaker said: "One in 10 Americans will experience a destructive fire this year. Now, I know that some of you can say you have lived in your homes for 25 years and never had any type of fire. To that I would respond that you have been lucky. But it only means that you are moving not farther away from a fire, but closer to one."*

Is this last statement correct? Why?

4.25 An investment advisor claims that 60 percent of the stocks she recommends beat the market. To back up her claim, she recommends n stocks for the coming year. In fact, each of her recommendations has a 50 percent probability of beating the market and a 50 percent probability of doing worse than the market. Assuming independence, is she more likely to have at least a 60 percent success rate if she recommends 10 stocks or 100 stocks?

4.26 A woman wrote to Dear Abby, saying that she had been pregnant for 310 days before giving birth [25]. Completed pregnancies are normally distributed with a mean of 266 days and a standard deviation of 16 days. What is the probability that a completed pregnancy lasts at least 310 days?

4.27 A nationwide test has a mean of 75 and a standard deviation of 10. Convert the following raw scores to standardized Z values: $X = 94$, 83, 75, and 67. What raw score corresponds to a Z value of 1.5?

4.28 After Hurricane Katrina, Congress mandated that hurricane protection for New Orleans be improved to a once-in-a-100-years standard, meaning that there is only a 1/100 probability of failing in any given year. Suppose you take out a 30-year mortgage to buy a home in this area. If the probability of failing is 1/100, and assuming independence, what is the probability of at least one failure in the next 30 years?

4.29 Explaining why he was driving to a judicial conference in South Dakota, the chief justice of the West Virginia State Supreme Court said "I've flown a lot in my life. I've used my statistical miles. I don't fly except when there is no viable alternative" [26]. What do you suppose the phrase *used my statistical miles* means? Explain why you either agree or disagree with the judge's reasoning.

4.30 Yahtzee is a family game played with five normal six-sided dice. In each turn, a player rolls the five dice and can keep some of the numbers and roll the remaining dice. After the second roll, the player can choose to keep more numbers and roll the remaining dice a final time. At the end of a Yahtzee game recently, Smith needed to roll at least one 5 to get a 35-point bonus and win the game. He rolled all five dice once, twice, and then a third time and did not get any 5s. What are the chances of this happening?

4.31 At the midpoint of the 1991 Cape Cod League baseball season, Chatham was in first place with a record of 18 wins, ten losses, and one tie [27]. (Ties are possible because games are sometimes stopped due to darkness or fog.) The Brewster coach, whose team had a record of 14 wins and 14 losses, said that his team was in a better position than Chatham: "If you're winning right now, you should be worried. Every team goes through slumps and streaks. It's good that we're getting [our slump] out of the way right now." Explain why you either agree or disagree with his reasoning.

4.32 Answer this letter to Ask Marilyn [28]:

> *You're at a party with 199 other guests when robbers break in and announce that they are going to rob one of you. They put 199 blank pieces of paper in a hat, plus one marked "you lose." Each guest must draw, and the person who draws "you lose" will get robbed. The robbers offer you the option of drawing first, last, or at any time in between. When would you take your turn?*

4.33 Limited Liability is considering insuring Howard Hardsell's voice for $1,000,000. They figure that there is only a 0.001 probability that they will have to pay off. If they charge $2,000 for this policy, will it have a positive expected value for them? What is the expected value of the policy to Howard? If Howard's expected value is negative, why would he even consider buying such a policy?

4.34 Each student who is admitted to a certain college has a 0.6 probability of attending that college and a 0.4 probability of going somewhere else. Each student's decision is independent of the decisions of other students. Compare a college that admits 1,000 students with a larger college that admits 2,500 students. Which college has the higher probability that the percentage of students admitted who decide to attend the college will be

a. Exactly equal to 60%?
b. Between 50% and 70%?
c. More than 80%?

Table 4.10: Exercise 4.35

		Profit (millions of dollars)	
Rental Market	Probability	1-Story	2-Story
Strong	0.2	4	10
Medium	0.5	2	3
Weak	0.3	−1	−2

4.35 A business is considering building either a one-story or two-story apartment complex. The value of the completed building will depend on the strength of the rental market when construction is finished. The subjective probability distributions for the net profit with each option are shown in Table 4.10. Use the mean and standard deviation to compare these two alternatives.

4.36 The annual returns on U.S. corporate stock and U.S. Treasury bonds over the next 12 months are uncertain. Suppose that these returns can be described by normal distributions with U.S. corporate stock having a mean of 15% and standard deviation of 20%, and U.S. Treasury bonds having a mean of 6% and standard deviation of 9%. Which asset is more likely to have a negative return? Explain your reasoning.

4.37 In his 1868 work, Carl Wunderlich concluded that temperatures above 100.4° Fahrenheit should be considered feverish. In a 1992 study, Maryland researchers suggested that 99.9° Fahrenheit was a more appropriate cutoff [29]. If the oral temperatures of healthy humans are normally distributed with a mean of 98.23 and a standard deviation of 0.67 (values estimated by the Maryland researchers), what fraction of these readings are above 100.4? Above 99.9?

4.38 The Australian Bureau of Meteorology uses the monthly air-pressure difference between Tahiti and Darwin, Australia, to calculate the Southern Oscillation Index: SOI = $10(X - \mu)/\sigma$, where X is the air-pressure difference in the current month, μ is the average value of X for this month, and σ is the standard deviation of X for this month. Negative values of the SOI indicate an El Niño episode, which is usually accompanied by less than usual rainfall over eastern and northern Australia; positive values of the SOI indicate a La Niña episode, which is usually accompanied by more than usual rainfall over eastern and northern Australia. Suppose that X is normally distributed with a mean of μ and a standard deviation of σ. Explain why you believe that the probability of an SOI reading as low as −22.8, which occurred in 1994, is closer to 1.1×10^{-15}, 0.011, or 0.110.

4.39 If the age at death of a 20-year-old U.S. woman is normally distributed with a mean of 76.4 years and a standard deviation of 8.1 years, what is the probability that she will live past 80?

4.40 Galileo wrote a short note on the probability of obtaining a sum of 9, 10, 11, or 12 when three dice are rolled [30]. Someone else had concluded that these numbers are equally likely, because there are six ways to roll a 9 (1-4-4, 1-3-5, 1-2-6, 2-3-4, 2-2-5, or 3-3-3), six ways to roll a 10 (1-4-5, 1-3-6, 2-4-4, 2-3-5, 2-2-6, or 3-3-4), six ways to roll an 11 (1-5-5, 1-4-6, 2-4-5, 2-3-6, 3-4-4, or 3-3-5), and six ways to roll a 12 (1-5-6, 2-4-6, 2-5-5, 3-4-5, 3-3-6, or 4-4-4). Yet Galileo observed "from long observation, gamblers consider 10 and 11 to be more likely than 9 or 12." How do you think Galileo resolved this conflict between theory and observation?

4.41 The noon temperature on July 4 in two cities, *A* and *B*, is normally distributed with respective means μ_A and μ_B and standard deviations σ_A and σ_B. For each of these three cases, identify the city that has the higher probability of a noon July 4 temperature above 100°.

a. $\mu_A = 80$, $\sigma_A = 20$, $\mu_B = 80$, $\sigma_B = 30$.
b. $\mu_A = 80$, $\sigma_A = 20$, $\mu_B = 70$, $\sigma_B = 20$.
c. $\mu_A = 80$, $\sigma_A = 20$, $\mu_B = 70$, $\sigma_B = 30$.

4.42 Mark believes that there is a 4/5 probability that it will snow in Boston on New Year's Eve 2015; Mindy believes that the probability is only 2/3. What bet could they make that would give each a positive expected value? For example, Mark pays Mindy $2 if it snows and Mindy pays Mark $5 if it does not.

4.43 Consider a multiple-choice question that has *n* possible answers. A person who does not answer the question gets a score of 0. A person who answers the question gets +1 if the answer is correct and –*X* if the answer is incorrect. What value of *X* would make the expected value of the score equal to 0 for someone who randomly selects an answer?

4.44 On long automobile trips, Mrs. Jones drives and Mr. Jones gives directions. When there is a fork in the road, his directions are right 30 percent of the time and wrong 70 percent of the time. Having been misled many times, Mrs. Jones follows Mr. Jones's directions 30 percent of the time and does the opposite 70 percent of the time. Assuming independence, how often do they drive down the correct fork in the road? If Mrs. Jones wants to maximize the probability of choosing the correct road, how often should she follow Mr. Jones's directions?

4.45 A carnival game has four boxes (Figure 4.15), into which the contestant tosses four balls. Each box is deep enough to hold all four balls and the contestant is allowed to toss each ball until it lands in a box. The contestant wins the prize if each box has one ball. Assuming that balls are equally likely to land in any box (this is a game of chance, not skill), what is the probability of winning the game?

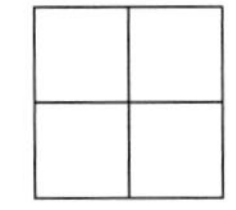

Figure 4.15
Exercise 4.45.

4.46 Answer this question to Ask Marilyn [31]:

I recently returned from a trip to China, where the government is so concerned about population growth that it has instituted strict laws about family size. In the cities, a couple is permitted to have only one child. In the countryside, where sons traditionally have been valued, if the first child is a son, the couple may have no more children. But if the first child is a daughter, the couple may have another child. Regardless of the sex of the second child, no more are permitted. How will this policy affect the mix of males and females?

4.47 A standard die is painted green on four sides and red on two sides and will be rolled six times. You can choose one of these three sequences and will win $25 if the sequence you choose occurs. Which do you choose and why?

a.	red	green	red	red	red	either
b.	red	green	red	red	red	green
c.	green	red	red	red	red	red

4.48 A car was ticketed in Sweden for parking too long in a limited time zone after a policeman recorded the positions of the two tire air valves on one side of the car (in the one o'clock and eight o'clock positions) and returned hours later to find the car in the same spot with the tire valves in the same positions [32]. The driver claimed that he had driven away and returned later to park in the same spot, and it was a coincidence that both tire valves stopped in the same positions as before. The court accepted the driver's argument, calculating the probability that both valves would stop at their earlier positions as (1/12)(1/12) = 1/144 and feeling that this was not a small enough probability to preclude reasonable doubt. The court advised, however, that had the policeman noted the position of all four tire valves and found these to be unchanged, the very slight $(1/12)^4 = 0.0005$ probability of such a coincidence would be accepted as proof that the car had not moved. As defense attorney for a four-valve client, how might you challenge this calculation?

4.49 Smith will win a prize if he can win two consecutive squash games in a three-game match against Andrabi and Ernst alternately, either Andrabi-Ernst-Andrabi or Ernst-Andrabi-Ernst. Assume that Andrabi is a better player than Ernst and that Smith's chances of winning a game against either player are independent of the order in which the games are played and the outcomes of other games. Which sequence should Smith choose and why?

4.50 Here is a probability variant of a three-card Monte game used by street hustlers. You are shown three cards: one black on both sides, one white on both sides, and one white on one side and black on the other. The three cards are dropped into an empty bag and you slide one out; it happens to be black on the side that is showing. The operator of the game says, "We know that this is not the double-white card. We also know that it could be either black or white on the other side. I will bet $5 against your $4 that it is, in fact, black on the other side." Can the operator make money from such bets without cheating?

CHAPTER 5

Sampling

Chapter Outline

"Data! data! data!" he cried impatiently, "I can't make bricks without clay."
—Sherlock Holmes

We use probabilities to describe how likely it is that something will happen. Data, in contrast, are a record of things that actually have happened. Here are three examples:

A fairly flipped coin
Probability: There is a 50 percent chance of landing tails.
Data: A coin is flipped ten times and lands heads three times.
A fairly rolled six-sided die
Probability: There is a 1/6 probability of rolling a 3.
Data: A die is rolled 20 times and 3 comes up five times.
A stock's price
Probability: There is a 0.51 probability that the price will be higher tomorrow than today.
Data: The stock's price has gone up 18 of the past 30 days.

Essential Statistics, Regression, and Econometrics

Probabilities and data are closely related, because we use probabilities to anticipate what the data might look like and we use data to estimate probabilities.

For example, life insurance companies use mortality data to estimate the probability that a healthy 21-year-old woman will die within a year and use this estimated probability to price a 1-year life insurance policy for a healthy 21-year-old woman.

In this chapter, we look at some general principles for deciding whether empirical data are likely to yield reliable information.

5.1 Populations and Samples

The starting point is to distinguish between a population and a sample. To illustrate, we look at a study of the adult daughters of women who had immigrated to California from another country [1]. The *population* consists of 1,111,533 immigrant women who gave birth to daughters in California between 1982 and 2007. A *sample* might be 30 women selected from this population. We often use samples because it is impractical (or impossible) to look at the entire population. If we burn every lightbulb to see how long bulbs last, we have a large electricity bill and a lot of burnt-out lightbulbs. Many studies are not destructive but simply too expensive to apply to the entire population. Instead, we sample. A lightbulb manufacturer tests a sample of its bulbs. The government tests water samples. Medical researchers give drugs to a sample of patients. Here, a researcher with limited time and resources might look at a sample of 30 women instead of trying to analyze all 1,111,533.

Figure 5.1 illustrates how sample data come from a population. The key insight is that, when we analyze data, we need to think about the population from which the data came. Once we have that insight, we can go back another step in our reasoning. *Before* we gather

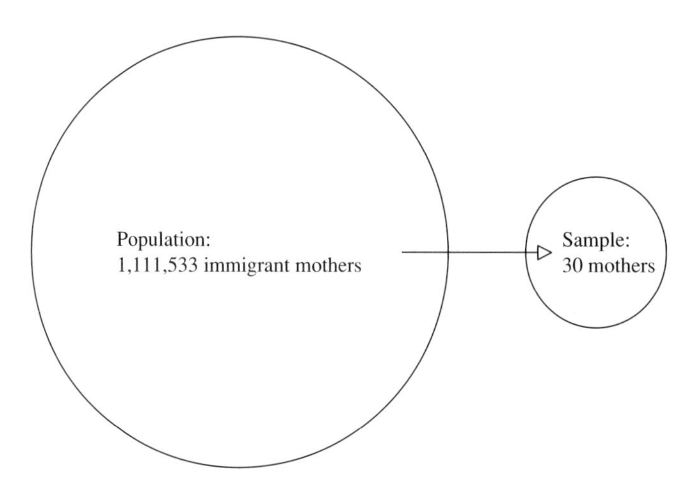

Figure 5.1
Sample data come from a population.

data, we should think about the population we are interested in studying, *then* think about how we can obtain a sample that is representative of this target population. Well-designed research begins by defining a population that is appropriate for our research question, then figuring out how to get useful data from this population.

In a lightbulb study, the population is the bulbs made by a certain manufacturer, and the sample is the bulbs tested. In a medical study, the population is the people suffering from a certain disease, and the sample is those who try the medication. In the study of immigrant women, the population is 1,111,533 mothers; the sample is the 30 mothers we choose to analyze.

5.2 The Power of Random Sampling

A *convenience sample* is a sample that we select because it is quick and easy, for example, immigrant mothers we know personally. There are obviously problems that outweigh convenience. Our friends may be atypical (maybe they are unusually well educated) or so similar that they do not reflect the diversity of women in the population (maybe they are all Cuban).

Alternatively, we could try to select "typical" immigrant mothers, for example, by standing in a mall and interviewing shoppers who seem likely to be immigrant mothers. Handpicking strangers is probably better than choosing our friends but still has problems. When we handpick people who seem typical, we exclude people who are unusual; and we consequently end up with a sample that is less varied than the population. Even worse, our sample reflects our beliefs. If we think that immigrant women are mostly from Mexico, our sample will confirm this preconceived idea. We have no way of knowing if our beliefs are wrong.

This is like the story of the Chinese researcher who wanted to estimate the height of the emperor but, by custom, was not allowed to look at the emperor. Instead, he got guesses from many other people, none of whom had ever seen the emperor. His research did not tell us the height of the emperor but rather people's beliefs about the emperor's height. It is similar here. If we select women we think are typical immigrant women, we will get a sample that does not tell us what immigrant women are like but rather what we think immigrant women are like.

Random Samples

If we do not want our personal beliefs to influence our results, we need a sampling procedure that does not depend on our personal beliefs. A random sample does that. Think about how cards are dealt in a poker game. The dealer does not turn the cards face up and pick out typical cards. Instead, the cards are turned face down, shuffled thoroughly, and dealt without trickery. The dealer's beliefs have no effect on the cards that are dealt.

A random sample is just like a fair deal in a card game. In a fairly dealt 5-card hand, each of the 52 cards is equally likely to be included in the hand and every possible 5-card hand is equally likely to be dealt. In a *simple random sample*, each member of the population is equally likely to be included in the sample and every possible sample is equally likely to be selected.[1]

Because a random sample is chosen by chance, the sample could turn out to be unrepresentative of the population. A random sample of college students could turn out to all be science majors. A random sample of voters could turn out to all be Republicans. A random sample of immigrant women could turn out to all be from Canada. This slight risk is needed to prevent the sample from being tainted by the researcher's subjective beliefs. Of course, the larger the sample, the less chance there is of obtaining a lopsided sample.

Another advantage of random sampling is that the mathematical properties of random samples are known. We will be able to make statements like this: "We are 95 percent confident that the sample mean will be within 3 percentage points of the population mean." If the researcher handpicks the sample, there is no basis for gauging the reliability of the results.

Sampling Evidence in the Courtroom

The legal system used to be deeply skeptical of probability and statistics. Something is either true or not; a 99 percent probability that something is true is not *proof* that it is true. Therefore, it was argued that sample data can never prove anything about a population.

A 1955 case involved a sales tax levied by the city of Inglewood, California, on purchases made by Inglewood residents in Inglewood stores [2]. Sears Roebuck discovered that, over an 850-day period, its Inglewood store had been paying the city a tax on all its sales, including sales to people who did not live in Inglewood. Sears sued the city, asking for a refund of the tax that it had mistakenly paid on sales to nonresidents. To estimate the size of the refund, Sears took a sample of 33 days and calculated that 36.69 percent of the sales had been to nonresidents. Applying this sample estimate to the entire 850-day period, Sears estimated that it had overpaid Inglewood $28,250.

At the trial, the Inglewood attorney asked only one question: Did Sears know for a fact that *exactly* 36.69 percent of its sales over the full 850-day period were to nonresidents? Because the answer was no, the judge ruled that Sears had to look at the receipts for all 850 days. It took Sears employees more than 3,000 hours to look over nearly 1 million sales receipts and obtain a final tally of $27,527. The sample estimate based on 33 days was within $750 (2.6 percent) of the result based on 850 days. In addition, surely mistakes were made during the mind-numbing scrutiny of 1 million sales receipts. Even

[1] A *systematic random sample* chooses one of the first *k* items in a population and then every *k*th item thereafter. If the population is not randomly arranged, but is in some sort of systematic order (for example, in alphabetical or chronological order), then each member of the population is equally likely to be included in the sample, but every possible sample is not equally likely to be selected.

after a complete tabulation, no one knew for a fact exactly how many sales had been to nonresidents.

Over time, the legal system has come to accept sampling as reliable evidence. In 1975, the Illinois Department of Public Services used a sample of 353 Medicaid claims that a doctor had filed to estimate that the doctor had been overpaid an average of $14.215 per claim. As in the Sears case, the doctor's lawyer argued that all the claims should have been examined. Unlike the Sears case, the court rejected this argument, noting that "the use of statistical samples has been recognized as a valid basis for findings of fact" [3]. The court also noted that it would be impractical and inefficient to examine all the claims.

Besides the additional expense, complete tabulations are seldom perfect. Think about contested election results where the votes are counted and recounted and give a different answer every time. Surely, you would make mistakes if you had to work hour after hour grinding through thousands of virtually identical documents. In fact, studies have shown that estimates based on sample data can be more accurate than tabulations of population data.

Choosing a Random Sample

How, in practice, do we select a simple random sample? Think again of a fair deal from a deck of cards. Every card has a unique identifier: 2 of spades, 4 of clubs, jack of diamonds, and so on. The cards are turned face down and well shuffled, so that each card has an equal chance of being the top card, or the second card, or the card on the bottom of the deck. The cards are then dealt and revealed.

In the same way, consider the random selection of ten students in your statistics class. Each student in the class can be given a unique identifier, for example, a number. The numbers can be written on cards which are shuffled, and ten cards dealt. Each person in the class has an equal chance of being included in the sample and all possible ten-person samples have an equal chance of being selected. It is a simple random sample.

In practice, cards are seldom used to select a random sample. The point is that the procedure should be equivalent to using cards. In our study of immigrant women, the population consisted of 1,111,533 immigrant women who gave birth to daughters in California between 1982 and 2007. These data came from records compiled from California birth certificates. A computer program was used to go through these records and identify women who were born outside the United States and gave birth to a daughter in California. A total of 1,111,533 women were found who met these requirements.

Information about these women was stored in a computer file and each mother was assigned a number from 1 to 1,111,533. A random number generator was used to select 30 numbers between 1 and 1,111,533. These numbers are shown in Table 5.1. Thus, the random sample consisted of mother number 389,912, mother number 681,931, and so on.

Table 5.1: Thirty Random Numbers

389,912	681,931	599,576	217,409	630,699	457,855
518,033	682,942	489,708	1,093,656	204,834	379,121
26,406	1,024	95,013	647,767	165,253	753,295
791,140	653,993	348,917	82,528	593,648	356,526
841,279	768,383	688,497	317,824	680,126	904,444

Random Number Generators

Random numbers can be found in tables of random numbers constructed from a physical process in which each digit from 0 to 9 has a one-tenth chance of being selected. The most famous of these tables contains 1 million random digits obtained by the RAND Corporation from an electronic roulette wheel using random frequency pulses.

With modern computers, random numbers can be obtained from a computer algorithm for generating a sequence of digits. One popular algorithm takes a starting number (called the *seed*), multiplies the seed by a constant b, and adds another constant a. The result is divided by a third constant c and the remainder is the random number. For example, suppose $a = 5$, $b = 7$, $c = 32$, and the seed is 12. We have $5 + 7 \times 12 = 89$, and $89/32 = 2$ with a remainder of 25. So, 25 is the first random number. The second random number is obtained by repeating the process, using 25 in place of 12. This process generates all 32 numbers between 1 and 32, and then repeats with the 33rd number being 25 again. A very long cycle can be obtained by choosing very large values for b and c. Other algorithms use different procedures but are similar, in that they use mathematical formulas to generate long strings of numbers.

Modern computers can make such calculations extremely quickly. Karl Pearson once spent weeks flipping a coin 24,000 times. Now, with a computer program for generating random numbers, 24,000 coin flips can be simulated in a fraction of a second.

Computerized random number generators produce what are called *pseudo-random numbers*, in that the numbers are produced by a deterministic process that gives exactly the same sequence of numbers every time. To randomize the process, we need to select an unpredictable starting position (the seed); one popular procedure is to use the computer's internal clock to determine the seed, since there is virtually no chance that two users taking samples from the same population will use the random number generator at exactly the same time.

5.3 A Study of the Break-Even Effect

Our study of immigrant daughters has a well-defined population: 1,111,533 immigrant women who gave birth to daughters between 1982 and 2007. In many studies, the population is not so well defined, but we nonetheless use the concept of a population in order to assess the implications of our sample data. The next study illustrates this.

We use an example from behavioral economics, which is concerned with how economic decisions are influenced by various psychological and emotional factors, in contrast to classical economic theory, which assumes dispassionate, rational decision making. One example of the difference between behavioral and classical economic theory relates to how people react to losing money in the stock market. Classical economic theory says that we should ignore our losses, which are in the past and cannot be undone. Behavioral economic theory says that people who have lost money may try to "break even" by winning back what they have lost. Thus, Daniel Kahneman and Amos Tversky, two behavioral economics pioneers, argued that a "person who has not made peace with his losses is likely to accept gambles that would be unacceptable to him otherwise." As evidence, they observe that at racetracks, betting on long shots tends to increase at the end of the day because people are looking for a cheap way to win back what they lost earlier in the day.

A formal study [4] of this break-even theory looked at how experienced poker players reacted to losing $1,000. High-stakes poker games are a good source of data because they avoid some problems with artificial experiments that are sometimes conducted in college classrooms. For example, students may not understand the instructions or may not think about their answers carefully because there is little or no reward for being careful.

A *convenience sample* might be people with whom we play poker or friends we know who play poker. However, the people we play with may not be experienced high-stakes players. Also, our friends may use similar strategies that do not reflect the variety of styles used by the target population.

Alternatively, we could select "typical" players who we have seen on television or at a local casino. As with the study of immigrant daughters, when we handpick people who seem typical, we exclude people who are unusual and we consequently end up with a sample that is less varied than the population. Even worse, our sample reflects our beliefs. If we think typical poker players use a certain style of play, our sample confirms this preconceived idea. We have no way of knowing if our beliefs are wrong.

In the poker study, the target population was experienced high-stakes Texas hold 'em poker players (Figure 5.2). It is not possible to monitor private poker games played all over the world, so the population was narrowed to people who play high-stakes Texas hold 'em games on the Internet.

It is still not feasible to look at every player at every Internet site for eternity. So, a sample is used. First, the researchers had to choose a time period; they selected January 2008 through May 2008, simply because that was when they did the study. The researchers assumed that there is no reason why this particular time period should be systematically different from other time periods. If there is reason to doubt this assumption, then the study's credibility is in doubt.

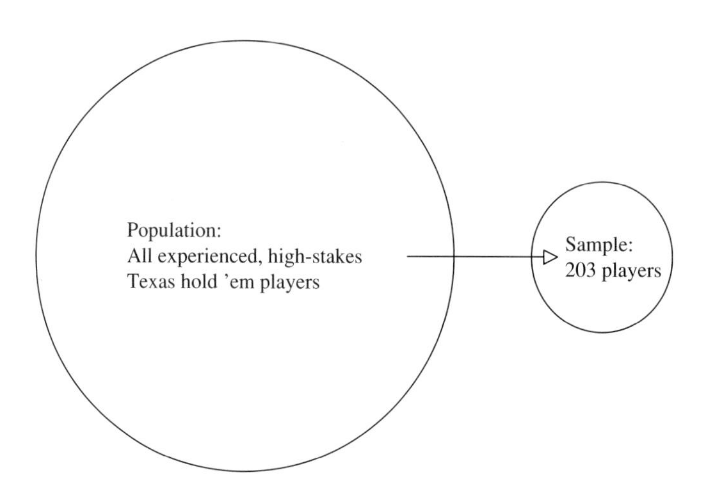

Figure 5.2
Sample data come from a population.

Next, they had to choose an Internet site. The researchers selected a well-known online poker room that is reported to be the favorite site of professional poker players and attracts many experienced high-stakes players.

Finally, the players were selected. The researchers looked at people who played at least 100 high-stakes hands because these were evidently serious, experienced players.

Imprecise Populations

The data for this poker study were not obtained by taking a random sample from an explicit population. Instead, as in many empirical studies, a target population was identified and data were collected from that population. A formal random sample could not be selected because the members of the population could not be identified and numbered in preparation for a random sample. When we analyze data as if it were a random sample, we should think about whether the data might be biased in some way. Is there anything special about the time period, January 2008 through May 2008? Does this Internet site attract atypical players? Are there any potential problems with choosing 100 hands as a cutoff?

The study would certainly be suspect if it only used data for May 2, 2009 (Why this particular day?), from an obscure Internet site where high school kids use play money (They are neither experienced nor serious!), and used 83 hands as a cutoff (Why 83 hands?). There is a plausible explanation for the time period, Internet site, and number of hands used by this study. If a different study using different data comes to different conclusions, then we have to rethink the question of whether our data are biased.

Texas Hold 'Em

Texas hold 'em poker uses a standard 52-card deck of playing cards. Before the cards are dealt, the player on the dealer's left puts a "small blind" into the pot, and the player two seats to the left of the dealer puts in a "big blind." In this study, the small blind was $25 and the big blind was $50. Each player is dealt two cards (which only the player sees) and must "call" by putting enough money into the pot to make their bet equal to the highest bet (which is initially the big blind), raise the bet, or fold. The bets go clockwise around the table until the highest bet has been called by all remaining players or only one player remains.

If more than one player is still in, three cards ("the flop") are dealt face up in the middle of the table and are considered part of each player's hand. Another round of betting occurs, starting with the player to the left of the dealer. Then a fourth community card is dealt face up and there is another round of betting; finally, a fifth community card is dealt and there is a final round of betting. The player with the best five-card hand, using the two private cards and five community cards, wins the pot. The dealer position shifts clockwise around the table after each hand.

Texas hold 'em is a strategic game because there are several rounds of betting and the five community cards are visible to all players. However, the outcome of each individual hand depends on the luck of the draw; for example, a player who is dealt two aces may lose to a player with two 3s if another 3 shows up in the community cards. A player's long-run success depends on making good betting decisions—"knowing when to hold 'em and when to fold 'em." Experts generally play tight-aggressive in that they play relatively few hands but make large bets when they do play a hand. The opposite strategy is loose-passive: playing many hands but betting the smallest amount needed to stay in. In Chapter 6, we use our sample data to characterize the strategies used by experienced, high-stakes players.

5.4 Biased Samples

Ideally, we would like our data to be a random sample from the target population. In practice, samples can be tainted by a variety of biases. If we are aware of these potential biases, we may be able to obtain more reliable data and be alert to pitfalls in other studies.

Selection Bias

Selection bias occurs when the results are distorted because the sample systematically excludes or underrepresents some elements of the population. For example, an airline once advertised that 84 percent of the frequent business travelers from New York to Chicago preferred their airline to another airline [5]. The target population was frequent business travelers from New York to Chicago. The puzzling thing about this ad was that, in practice, only 8 percent of the people flying from New York to Chicago flew on this airline. If 84 percent prefer this airline, why were only 8 percent flying on it?

The answer to this puzzle is that the sample had considerable selection bias in that it was not randomly drawn from the target population but, instead, was a survey of passengers on one of this airline's flights from New York to Chicago. It is not surprising that frequent business travelers who choose to fly this airline prefer it. It is surprising that 16 percent of the passengers prefer another airline. But, we can hardly expect this airline to advertise "Sixteen percent of the people flying on our planes wish they weren't."

This particular kind of selection bias is known as *self-selection bias* because people choose to be in the sample. There is also self-selection bias when people choose to go to college, marry, have children, or immigrate to another country. In each case, we should be careful making comparisons to people who made different choices. For example, college graduates may have higher incomes, in part, because they are more ambitious. If so, it would be a mistake to attribute their higher income solely to college attendance. For a fair comparison, we would have to do something like the following: among all those who want to go to college, a random sample is allowed to go.

This was actually done in the early 1960s in Ypsilanti, Michigan, where black children from low socioeconomic households were selected or rejected for an experimental preschool program based on the outcome of a coin flip. Those who attended the preschool program were subsequently more likely to complete high school, less likely to be arrested, and more likely to have jobs [6]. It may seem heartless to those who lost the coin flip, but this experiment demonstrated the value of the preschool program.

Survivor Bias

In a *prospective* study, the researcher selects a sample and then monitors the sample as time passes. For example, a researcher might select 1,000 people and give them medical checkups every year to identify the causes of health problems. Or, 1,000 children might be selected and interviewed each year to see whether playing competitive sports affects their chances of going to college. Or, 1,000 companies might be selected from companies listed on the New York Stock Exchange (NYSE) in 1990 and followed over the next 20 years to see whether there is a relationship between a company's executive compensation program and the performance of its stock.

In a *retrospective* study, in contrast, a sample is selected and the researcher looks back at the history of the members of this sample. A researcher might examine the medical histories of 1,000 elderly women to identify the causes of health problems. Or, a researcher might select 1,000 college students and see how many played competitive sports. Or, a researcher might select 1,000 companies from companies listed on the NYSE in 2011 and look back over the preceding 20 years to see whether there is a relationship between the company's executive compensation program and its stock performance.

Retrospective studies often have *survivor bias* in that, when we choose a sample from a current population to draw inferences about a past population, we leave out members of the past population who are not in the current population: We look at only the survivors. If we examine the medical records of the elderly, we overlook those who did not live long enough to become elderly. If we look at college students, we exclude people who did not go to college. If we look at companies on the NYSE in 2011, we exclude companies that went bankrupt before 2011.

Does Anger Trigger Heart Attacks?

The *New York Times* once reported that researchers had found that, "People with heart disease more than double their risk of heart attacks when they become angry, and the danger lasts for two hours" [7]. This conclusion was based on interviews with 1,623 people who survived heart attacks. In these interviews, conducted a few days after the attacks, 36 people said they had been angry during the 2 hours before the attack and 9 people said they were angry the day before the attack.

It is certainly a common perception that anger can trigger a heart attack; however, it is also said to be healthier to express anger than to "keep it bottled up inside." It would be interesting to know which of these folk myths is closer to the truth. Unfortunately, the data in this study tell us nothing at all about the relationship between anger and heart attacks. More people were angry 2 hours before the attack than were angry the day before; however, 97 percent of these heart attack victims did not report being angry at all, either on the day of the attack or the day before.

In addition, this is a retrospective study that involves two kinds of survivor bias. Those who died of heart attacks and those who had no heart attacks are both excluded from the study. To assess the effect of anger on heart attacks, we need two groups (people who were angry and people who were not) and we need data on both possible outcomes (having a heart attack and not having a heart attack). To exaggerate the point, suppose that a complete picture of the data looks like Table 5.2.

Table 5.2: Anger and Heart Attacks

	Heart Attack	No Heart Attack	Total
Angry	45	1,955	2,000
Not angry	1,578	0	1,578
Total	1,623	1,955	3,578

In these hypothetical data, 2,000 people were angry and 1,578 were not angry. As in the study, 45 of the 1,623 people who suffered heart attacks reported being angry. These hypothetical data show that 97 percent of the people who were angry did not have heart attacks and 100 percent of those who were not angry had heart attacks. People who were not angry had a much greater risk of heart attack—exactly the reverse of the conclusion drawn by the researchers who examined only people who suffered heart attacks. A scientific study would surely not yield data as one-sided as this caricature. It might even show that angry people are more likely to experience heart attacks. The point is that the survivor bias in the study does not allow us to determine whether or not anger increases the risk of a heart attack.

5.5 Observational Data versus Experimental Data

In many scientific studies, researchers can collect *experimental data* in laboratories by varying some factors (for example, temperature, velocity, or salinity) while holding other factors constant and seeing what happens. However, in economics, sociology, and other behavioral sciences, researchers typically cannot do experiments involving human subjects (for example, making some people lose their jobs, divorce their spouses, or have four children) and see how they react. In these cases, researchers have to make do with *observational data*, in which they observe people in different circumstances over which the researcher has no control: people who have lost their jobs, divorced, or had four children.

Both studies cited in this chapter (immigrant mothers and high-stakes poker players) use observational data. Researchers who use observational data can learn from laboratory scientists about the kinds of data that yield useful, reliable conclusions and recognize the limitations of data that are gathered in more natural settings.

Conquering Cholera

When it is not practical or ethical to do controlled laboratory experiments, researchers must use observational data. For example, in the 1800s, London was periodically hit by terrible cholera epidemics. John Snow, a distinguished doctor, suspected that cholera might be caused by drinking water contaminated with sewage, but he could not ethically do a controlled experiment in which some persons were forced to drink contaminated water. Instead, he used observational data by comparing the cholera mortality rates in an area of London where the water was supplied by two different companies: Southwark & Vauxhall (which pumped water from a nearby section of the Thames river that was contaminated with sewage) and Lambeth (which obtained its water far upstream, from a section of the Thames that had not been contaminated by London sewage). During the first seven weeks of the 1854 cholera epidemic, there were 1,263 cholera deaths in 40,046 houses supplied by

Southwark and 98 deaths in 26,107 houses supplied by Lambeth [8]. These data provided convincing evidence of a relationship between drinking contaminated water and the incidence of cholera.

Confounding Factors

A well-known physics experiment involves attaching an object to a string, letting it swing back and forth as a pendulum, and recording the time it takes to complete one cycle. This experiment becomes more meaningful when there is some sort of comparison, for example, seeing how the cycle time is related to the weight of the object. However, cycle times are also affected by other factors (such as the length of the string and the angle from which the object is dropped) that need to be taken into account. When *confounding* factors are present, the effects we are interested in are mixed up with the effects of extraneous factors.

To isolate the effects of the factors of interest, experiments can be done *under controlled conditions*, holding constant the confounding factors. If all other relevant factors are held constant, then the one factor that is allowed to vary is evidently causing the observed differences in outcomes.

In a properly done pendulum experiment, the objects might have different weights but be attached to strings of the same length and dropped from the same angle. When all confounding factors are held constant, physicists can demonstrate the remarkable fact that the weight of the object has *no* effect on the cycle time.

When working with observational data, where we cannot control confounding factors, we can look for data where the confounding factors happen to be constant. If we are doing a study of the effect of smoking on health and we think gender might be a confounding factor, we can do one study of males and a separate study of females. If we are doing a study of the effect of socioeconomic status on child bearing and we think age and religious beliefs might be confounding factors, we can look at people of the same age with similar religious beliefs. In his cholera study, Snow chose an area of London in which adjacent houses were served by different water companies to reduce the confounding effects of socioeconomic factors.

Alternatively, we can use *multiple regression* (Chapter 10), a remarkable statistical procedure that estimates what the effects of each factor would be if the other factors were held constant. In our physics example, the researcher could record the cycle times for a large number of experiments where weight, string length, and angle all vary, then use multiple regression to estimate how cycle time would be affected by weight if string length and angle had been held constant.

The important lesson is that researchers should always think about whether there are possibly confounding factors, then either hold these factors constant or use a statistical procedure that takes them into account. Of course, confounding factors are sometimes overlooked initially and conclusions are sometimes modified when researchers eventually take these overlooked factors into account. For example, a 1971 study found that people who drink lots of coffee get bladder cancer more often than do people who do not drink coffee [9]. However, a confounding factor was that people who drink lots of coffee were also more likely to smoke cigarettes. In 1993, a rigorous analysis of 35 studies concluded that once this confounding factor was taken into account, there was no evidence that drinking coffee increased the risk of bladder cancer [10].

Exercises

5.1 What data might be used to make a reasonable estimate of the probability that a 65-year-old man dies within the next 5 years?

5.2 What data might be used to make a reasonable estimate of the probability that a newborn baby will be female?

5.3 Explain how you would obtain a simple random sample from this population: varsity athletes at your school.

5.4 Explain how you would obtain a simple random sample from this population: stocks traded on the New York Stock Exchange (NYSE).

5.5 Identify an appropriate population for each of these samples:
   a. The price of Cheerios in ten stores is recorded.
   b. Ten Toyota Camrys are driven 1,000 miles and the miles per gallon are recorded.
   c. One thousand people are asked if they agree with a presidential decision.

5.6 Identify an appropriate population for each of these samples:
   a. A coin is flipped 100 times and lands heads 54 times.
   b. A baseball player gets 86 base hits in 300 times at bat.
   c. Two thousand people are asked who they intend to vote for in the next presidential election.

5.7 In the 1920s, a vaccine called BCG was tested to see if it could reduce deaths from tuberculosis [11]. A group of doctors gave the vaccine to children from some tubercular families and not to others, apparently depending on whether parental consent could be obtained easily. Over the subsequent 6-year period, 3.3% of the unvaccinated children and 0.67% of the vaccinated children died from tuberculosis. A second study was then conducted in which doctors vaccinated only half of the children who had parental consent, using the other half as a control group. This time 1.42% of the unvaccinated children and 1.52% of the vaccinated children died from tuberculosis. Provide a logical explanation for these contradictory findings.

5.8 Shere Hite sent detailed questionnaires (which, according to Hite, took an average of 4.4 hours to fill out) to 100,000 women and received 4,500 replies, 98 percent saying that they were unhappy in their relationships with men [12]. A *Washington Post-ABC News* poll telephoned 1,505 men and women and found that 93 percent of the women considered their relationships with men to be good or excellent [13]. How would you explain this difference?

5.9 A study of the effects of youth soccer programs on self-esteem found that children who played competitive soccer had more self-esteem than other children. Identify the most important statistical bias in this sample and describe a (hypothetical) controlled experiment that would get rid of this problem.

5.10 A study found that, among youths between the ages of 8 and 18 who were playing competitive soccer, those who had been playing for several years liked soccer more than did those who had been playing for only a few years, suggesting that "the more you play soccer, the more you like it." Identify the most important statistical bias in this sample and describe a (hypothetical) controlled experiment that would get rid of this problem.

5.11 A study looked at a random sample of 100 companies whose stocks were traded on the NYSE. The companies were divided into quintiles (the 20 largest companies, the 20 next largest, and so on) and the rates of return over the preceding 50 years were calculated for each quintile. This study found that the quintile with the smallest companies ("small cap" stocks) had done better than the quintile with the largest companies ("large cap" stocks). Identify the most important statistical bias in this sample and explain how the study could be redone to get rid of this problem.

5.12 A study of the 30 largest U.S. companies found that their average growth rate over the preceding 20 years had been well above the average growth rate for all companies, suggesting that big companies grow faster than the average company. Identify the most important statistical bias in this sample and explain how the study could be redone to get rid of this problem.

5.13 A study of the geographic mobility looked at how often adult daughters moved outside the ZIP code in which they were born [14]. The study found that the daughters of immigrants to the United States were more likely to move than the daughters of women born in the United States, suggesting that U.S. women have less geographic mobility than women in other countries. Identify the most important statistical bias in this sample.

5.14 The daughters of women who immigrate to the United States have more upward economic mobility than the daughters of women born in the United States, suggesting that U.S. women have less economic mobility than women in other countries. Identify the most important statistical bias in this sample.

5.15 Average grades are higher in senior-level college chemistry courses than in introductory chemistry courses, suggesting that students who get bad grades in an introductory chemistry course should not give up taking chemistry courses. Identify the most important statistical bias in this sample and describe a (hypothetical) controlled experiment that would get rid of this problem.

5.16 In 1993, a member of a presidential task force on food assistance stated "If you think that blacks as a group are undernourished, look around at the black athletes on television—they're a pretty hefty bunch" [15]. Identify the most important statistical problem with this conclusion.

5.17 The data in Table 5.3 show the fraction of all divorces based on the month in which the marriage ceremony was performed [16]. For example, 6.4 percent of all divorces involve couples who were married in January. Do these data show that persons who marry in June are more likely to get divorced than people who marry in January?

5.18 A doctor notices that more than half of her teenage patients are male, while more than half of her elderly patients are female. Does this mean that males are less healthy than females when young, but healthier when elderly?

5.19 Explain any possible flaws in this conclusion [17]:

A drinker consumes more than twice as much beer if it comes in a pitcher than in a glass or bottle, and banning pitchers in bars could make a dent in the drunken driving problem, a researcher said yesterday. Scott Geller, a psychology professor at Virginia Polytechnic Institute and State University in Blacksburg, Va., studied drinking in three bars near campus.... [O]n average, bar patrons drank 35 ounces of beer per person when it came from a pitcher, but only 15 ounces from a bottle and 12 ounces from a glass.

5.20 Red Lion Hotels ran full-page advertisements claiming "For every 50 business travelers who try Red Lion, a certain number don't come back. But 49 of 50 do." The basis for this claim was a survey of people staying at Red Lion, 98 percent of whom said "they would usually stay in a Red Lion Hotel when they travel." Use a numerical example to help explain why the survey results do not prove the advertising claim.

Table 5.3: Exercise 5.17

Month Married	Fraction of Divorces	Month Married	Fraction of Divorces
January	0.064	July	0.087
February	0.068	August	0.103
March	0.067	September	0.090
April	0.073	October	0.078
May	0.080	November	0.079
June	0.117	December	0.087

5.21 A medical insurance survey found that more than 90 percent of the plan's members are satisfied. Identify two kinds of survivor bias that may affect these results. Is the reported satisfaction rate biased upward or downward?

5.22 A survey found that the average employee at the WD Company had worked for the company for 20 years, suggesting that the average person who takes a job with WD works for the company for 20 years. Explain why the survey data might be consistent with a situation is which the average person who takes a job with WD works for the company for
  a. Fewer than 20 years.
  b. More than 20 years.

5.23 A study of students at a very selective liberal arts college found that varsity athletes had, on average, higher grades than nonathletes. The researcher concluded that the college could increase the grades of nonathletes by requiring them to participate in sports. What problems do you see with this study?

5.24 A *New York Time*s article argued that one reason many students do not graduate from college within 6 years of enrolling is that they choose to "not attend the best college they could have" [18]. For example, many students with a 3.5 high school grade point average could have gone to the University of Michigan Ann Arbor, which has an 88% graduation rate, but chose instead to go to Eastern Michigan, which has only a 39% graduation rate.
  a. What flaws do you see in the implication that these students would have a better chance of graduating if they went to the University of Michigan instead of Eastern Michigan?
  b. What kind of data would we need to draw a valid conclusion?

5.25 A major California newspaper [19] asked its readers the following question: "Do you think the English-only law should be enforced in California?" Of the 2,674 persons who mailed in responses, 94 percent said yes. Why might this newspaper's readers be a biased sample of the opinions of California residents?

5.26 A doctor told a meeting of the Pediatric Academic Societies that a U.S. government survey of 4,600 high school students found that both boys and girls were much more likely to have smoked cigarettes, used alcohol and marijuana, had sex, and skipped school if they had body piercings [20]. Does his study demonstrate that we could reduce these activities by making body piercings illegal?

5.27 Explain why these data are not convincing evidence that the way to seize more illegal drugs is to conduct fewer searches: "In 2000, [the U.S. Customs service] conducted 61 percent fewer searches than in 1999, but seizures of cocaine, heroin and Ecstasy all increased" [21].

5.28 Reuters Health [22] reported that

In a study of more than 2,100 secondary school students, researchers found that boys who used computers to do homework, surf the Internet and communicate with others were more socially and physically active than boys who did not use computers at all.

On the other hand, boys who used computers to play games tended to exercise less, engage in fewer recreational activities, and have less social support than their peers....

The findings, published in the October issue of the Journal of Adolescent Health, *suggest that parents should monitor how their sons use their computers—and not just how much time they spend in front of the screen.*

What statistical problem do you see with this conclusion?

5.29 A study of college seniors found that economics majors had above-average scores on a test of mathematical ability, suggesting that the study of economics improves mathematical ability. Identify the most important statistical bias in this sample and describe a (hypothetical) controlled experiment that would get rid of this problem.

5.30 A study of college graduates found that economics majors had higher average starting salaries than art history majors, suggesting that the study of economics increases your chances of getting a good job. Identify the most important statistical bias in this sample and describe a (hypothetical) controlled experiment that would get rid of this problem.

5.31 A study found that college students who take more than one course from a professor give the professor higher course evaluations than do students who take only one course from the professor, suggesting that taking multiple courses from a professor causes students to appreciate the professor more. Identify the most important statistical bias in this sample and describe a (hypothetical) controlled experiment that would get rid of this problem.

5.32 A study found that professors tend to give higher grades to students who take more than one course from the professor than to students who take only one course from the professor, suggesting that professors reward students who come back for more. Identify the most important statistical bias in this sample and describe a (hypothetical) controlled experiment that would get rid of this problem.

5.33 A Volvo advertisement said that the average American drives for 50 years and buys a new car, on average, every 3¼ years—a total of 15.4 cars in 50 years. Since Volvos last an average of 11 years in Sweden, Volvo owners need to buy only 4.5 cars in 50 years. Carefully explain why these data are very misleading.

5.34 A study looked at the 50 states in the United States and compared educational spending per student in each state with the average score on college entrance examination tests (the SAT and ACT). There is a positive correlation across states in spending per student and in the percentage of students taking college entrance examination tests.

a. Explain why high school seniors who take the SAT (or ACT) are not a random sample of all high school seniors.

b. Do you think that the average test score in a state would tend to increase or decrease if a larger percentage of the state's students took college entrance examination tests?

c. Suppose that your answer to (b) is correct and, also, that educational spending has no effect at all on test scores. Explain why we would observe either a positive or negative statistical relationship between spending and average test scores.

5.35 Studies have shown that people who marry and divorce have, on average, fewer children than do people who stay married, suggesting that having more children makes divorce less likely.

a. Explain why there might be sampling bias in this study.

b. Suppose that the chances of getting divorced are not affected by the number of children you have. Explain how it could still be the case that people who marry and divorce have, on average, fewer children than do people who stay married.

c. What would have to be done to have a controlled experiment?

5.36 Studies have shown that students who go to highly selective colleges have, on average, higher lifetime incomes than do students who go to less selective colleges.

a. Explain why there might be sampling bias in this study.

b. Suppose that the college you go to has no effect on your lifetime income, but that highly selective colleges are more attractive because they have much better dormitories, food, and extracurricular activities. Explain how it could still be the case that students who go to highly selective colleges have, on average, higher lifetime incomes than do students who go to less selective colleges.

c. What would have to be done for a controlled experiment?

5.37 A 1950s study found that married men were in better health than men of the same age who never married or were divorced, suggesting that the healthiest path is for a man to marry and never divorce [23].

a. Explain why there might be sampling bias in this study.

b. Suppose that marriage is generally bad for a man's health. Explain how it could still be the case that men who marry and stay married are in better health than (i) men who never marry and (ii) men who marry and then divorce.

c. What would have to be done for a controlled experiment?

5.38 A discrimination case compared the 1976 salaries of Georgia Power Company black and white employees who had been hired by the company in 1970 to similar entry-level positions [24].

a. Explain why there is survivor bias in this comparison.

b. Suppose that there are two types of employees, productive and unproductive, and that Georgia Power hired 100 black employees and 100 white employees in 1970 at the exact same salary, with half the new hires productive and half unproductive, regardless of race. Each year, every unproductive employee got a 2 percent annual raise and every productive employee got a 7 percent annual raise. Explain how it could still turn out that the average salary of white employees in 1976 was higher than the average salary of black employees.

5.39 "The most dangerous place to be is in bed, since more people die in bed than anywhere else. From now on, I'm sleeping on the floor." What problems do you see with this conclusion?

5.40 A study found that the grade point averages (GPAs) of women at all-women colleges tend to be higher than the GPAs of women at coeducational colleges, and concluded that women do better when they attend all-women colleges.

a. Explain why there might be sampling bias in this study.

b. Suppose that every woman's GPA does not depend on the college she attends. Explain how it still might be true that women at all-women colleges tend to have higher GPAs than do women at coeducational colleges.

c. What would have to be done to have a controlled experiment?

5.41 During World War II, 408,000 U.S. American military personnel died while on duty and 375,000 civilian Americans died because of accidents in the United States. Should we conclude that fighting a war is only slightly more dangerous than living in the United States?

5.42 To obtain a sample of college students, the Educational Testing Service divided all schools into groups, including large public universities, small private colleges, and so on. Then they asked the dean at a representative school in each group to recommend some students to participate in a study [25]. Identify three distinct reasons why this sample might not be representative of the college student population.

5.43 Newspaper columnist Ann Landers once asked her readers, "If you had it to do over again, would you have children?" [26]. Ten thousand people responded, of which 70 percent said *no*.

a. Explain why there might be sampling bias in this study.

b. Suppose that 90 percent of all parents would answer *yes* to this question. Explain how the Ann Landers survey might get 70 percent *no* answers.

5.44 A study looked at the annual number of days of restricted activity (for example, missing work or staying in bed for more than half a day because of injury or illness) for men with different cigarette-smoking histories [27]:

Never smoked	14.8 days
Currently smoking	22.5 days
Quit smoking	23.5 days

These data suggest that it is best never to smoke, but once you start smoking, it's better not to stop.

a. Explain why there might be sampling bias in this study.
b. Suppose that smoking is bad for your health. Explain how it could still be the case that people who quit smoking are in worst health than people who keep smoking.
c. What would have to be done to have a controlled experiment?

5.45 A "study" asked people to identify their favorite game and then calculated the mortality rate (deaths per 1,000 people that year) for each game category [28]. For example, of those people whose favorite game is bingo, 41.3 of 1,000 people died that year. The mortality rates were 27.2 for bridge, 0.05 for poker, and 0.04 for the video game Mortal Kombat. The study concluded that

The numbers reveal an alarming truth: bingo may not be as safe as some people have assumed. "This is the first evidence we have seen that violent video games actually reduce the death rate," says PlayGear CEO Pete Elor, "It comes as a blow to the head for people who advocate less violent forms of entertainment." Lawyer Gerald Hill thinks the church and community need to take action: "When you look at the numbers, there's just no way you can get around it. Bingo is claiming lives right and left...." We can only hope that this study will cause people to think twice before engaging in such risky behavior.

a. Explain why there might be sampling bias in this study.
b. Suppose that all four of these games are completely safe in that playing the game has no effect at all on one's health. Explain how it could still be the case that the mortality rate is much higher for bingo players than for Mortal Kombat players.
c. What would have to be done to have a controlled experiment?

5.46 A television commercial said that 70 percent of the cars made by one automaker during the past 12 years were still in use. Does this mean that 70 percent of this company's cars last 12 years?

5.47 A study found that Harvard freshmen who had not taken SAT preparation courses scored an average of 63 points higher on the SAT than did Harvard freshmen who had taken such courses. Harvard's admissions director said that this study suggested that SAT preparation courses are ineffective [29].

a. Explain why there might be sampling bias in this study.

b. Suppose that SAT preparation courses increase SAT scores by 25 points. Explain how it could still be the case that students who do not take such courses score higher, on average, than do students who do take these courses.
c. What would have to be done to have a controlled experiment?

5.48 A survey conducted by American Express and the French Tourist Office found that most visitors to France do not think that the French are unfriendly [30]. The sample consisted of "1,000 Americans who have visited France more than once for pleasure over the past two years."
a. Explain why there might be sampling bias in this study.
b. Suppose that 90 percent of all people who visit France think that the French are very unfriendly. Explain how it could still be the case that most of the people in this survey do not think the French are unfriendly.
c. Who should have been surveyed in order to obtain more useful results?

5.49 Some universities with budget problems have offered generous payments to any tenured professor who agrees to leave the university.
a. How does this plan help the university's budget?
b. Explain why the professors who accept this offer are not a random sample of the university's professors.
c. Explain why this plan could lower the average quality of the university's professors.

5.50 Congressmen Gerry Studds (a Massachusetts Democrat) and Daniel Crane (an Illinois Republican) were both censured by the House of Representatives for having had sex with teenage pages. A questionnaire in the *Cape Cod Times* asked readers whether Studds should "resign immediately," "serve out his present term, and not run for reelection," or "serve out his present term, run for reelection" [31]. Out of a Sunday circulation of 51,000 people, 2,770 returned this questionnaire:

Resign immediately	1,259 (45.5%)
Serve out term, and not run for reelection	211 (7.5%)
Serve out term, run for reelection	1,273 (46.0%)
Undecided	27 (1.0%)

a. Explain why there might be sampling bias in this study.
b. If a true random sample had been conducted, which of these four percentages would you expect to be higher than in this newspaper survey?

CHAPTER 6

Estimation

Chapter Outline

While the individual man is an insolvable puzzle, in the aggregate he becomes a mathematical certainty. You can, for example, never foretell what any one man will do, but you can say with precision what an average number will be up to.

—Sherlock Holmes

Businesses use data to estimate the reliability of their products and predict demand. Doctors use data to evaluate a patient's condition and predict whether a treatment will be effective. Governments use data to estimate the unemployment rate and predict the effects of government policies on the economy.

Chapter 5 explains why random samples are often the best way to collect data. But how can we gauge the reliability of a random sample? If the data are randomly selected, then pretty much anything can happen, right? Surprisingly, the fact that the data come from a random sample is what allows us to gauge reliability!

If a fair die is rolled once, we can get any number from 1 to 6. But if a fair die is rolled 1,000 times, we can be pretty certain that the *average* number will be close to 3.5. In fact, it can be shown that there is a 95 percent chance that the average will turn out to be between 3.3 and 3.7. We can make this probability calculation because we know the probabilities when dice are rolled fairly (a random sample). We would not know the probabilities if we were personally

Essential Statistics, Regression, and Econometrics

choosing numbers between 1 and 6. It is the randomness in a random sample that allows us to make probability statements about the outcomes.

6.1 Estimating the Population Mean

The starting point for any estimation procedure is to specify what we are trying to estimate. Returning to the poker example in Chapter 5, the researchers wanted a numerical measure of each person's style of play so that they could see if people change their style of play after a big win or loss. One generally accepted measure of a player's style is *looseness*: the percentage of hands in which a player voluntarily puts money into the pot. At the six-player tables used in this study, people are typically considered to be very tight players if their looseness is below 20 percent and to be extremely loose players if their looseness is above 50 percent.

Thus, each player's style of play is characterized by his or her looseness, and one summary measure of the playing style of the population of experienced high-stakes poker players is their average looseness. We use the Greek letter μ, pronounced "mew," to represent the population mean:

$$\mu = \text{population mean}$$

The population mean μ is a *parameter* because it is a given characteristic of the population. The players in the population are what they are, no matter which players are selected for the sample. In our poker example, μ is the average looseness of the population of all experienced high-stakes poker players.

The sample used in this study consists of 203 players. They played an average of 1,972 hands during a 4-month period. Half of the players won or lost more than \$36,000; 10 percent won or lost more than \$355,000. We assume that these players are effectively a random sample from the population of all the people who play lots of high-stakes Texas hold 'em games on the Internet.

Looseness statistics X were calculated for each of these 203 players. The sample mean $\overline{X}$ is the average value:

$$\overline{X} = \frac{X_1 + X_2 + \cdots + X_n}{n} \tag{6.1}$$

In our poker example, $n = 203$ and the sample mean works out to be $\overline{X} = 25.53$. This is a reasonable number since professional poker players have a reputation for relatively tight play.

But, remember, a sample mean is not necessarily equal to the population mean. This sample mean came from one sample of 203 players. If we were to take another sample, we would almost certainly get a somewhat different mean—maybe a little higher, maybe a little lower, but almost certainly not the same. So, how seriously can we take the results of one particular sample? It turns out that we can actually use the sample itself to estimate how much confidence we have in our sample mean.

6.2 Sampling Error

The difference between the sample mean and the population mean is called *sampling error.* Sampling error is not due to mistakes made by the researcher, such as recording data incorrectly. Sampling error is the inevitable variation from sample to sample caused by the luck of the draw in choosing a random sample. Sampling error is positive in some samples and negative in other samples. If many samples are taken, the average sampling error will be close to 0 and the average value of the sample mean will be close to the population mean.

In contrast, if there is a systematic error, the average value of the sample mean will not equal the population mean. In our poker example, a software program was written to calculate the looseness coefficient (the percentage of hands in which the player makes a voluntary bet). Suppose that there is a programming error and the blind bets (which are involuntary) are counted as voluntary bets. Each player's calculated looseness coefficient will be too high and so is the average value of the sample mean. This is a systematic error that can be detected and corrected. We can fix the programming error. Sampling error, in contrast, cannot be corrected, because no one made a mistake.

The important point about sampling error is that we are aware of it. When we take a random sample and calculate the sample mean, we recognize that this sample is one of many samples that might have been selected, and this sample mean is one of many sample means that might have been obtained.

We can never know whether a sample mean is above or below the population mean, because we do not know the population mean. However, we can estimate the probability that a sample mean will be close to the population mean. For example, we might be able to say that there is 0.95 probability that a sample mean will be within 7 percentage points of the population mean.

Here is how we do it. Suppose that the probability distribution of the looseness coefficient X is as depicted in Figure 6.1. We do not know the probability distribution, but we pretend that we do so that we can see why our procedure makes sense.

The probability distribution in Figure 6.1 shows the probability that a randomly selected player's looseness coefficient will be in a specified interval. For this particular distribution, there is a 30 percent chance that we will select a person whose looseness coefficient X is less than 20, a 65 percent chance that X will be between 20 and 40, and a 5 percent chance that X will be larger than 40. This probability distribution is skewed right. The average value of the looseness coefficient is $\mu = 25$, but the peak of the density function is at 20 and the median is 23.4. The standard deviation of X is $\sigma = 10$.

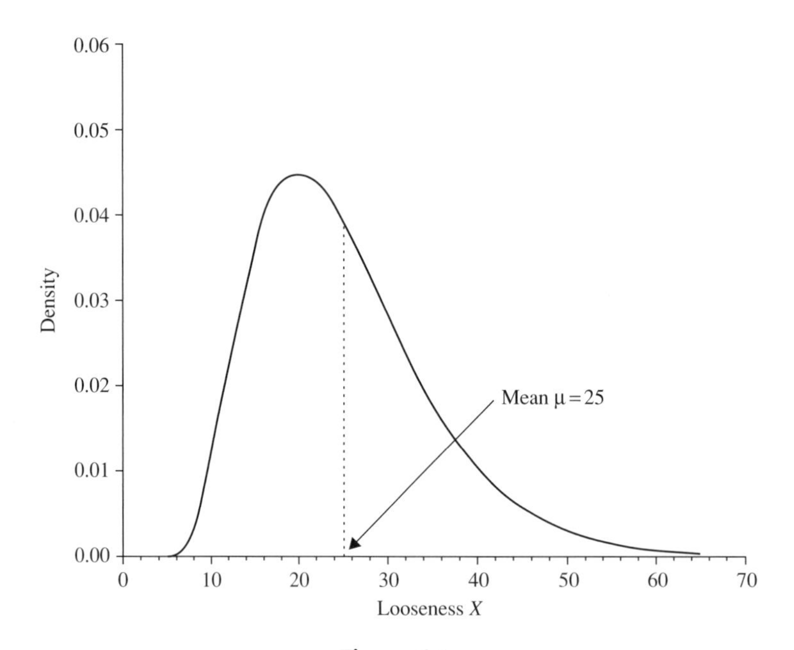

Figure 6.1
The population distribution of the looseness coefficient X.

6.3 The Sampling Distribution of the Sample Mean

Figure 6.1 applies when just one person is selected. What can we say about the sample mean when, say, ten people are selected? We do not know what the sample mean will turn out to be, but we do know that some values of the sample mean are more likely than others.

Look again at Figure 6.1. If we choose ten people randomly from this distribution, it is very unlikely that the sample mean will be below 10 or above 50. It could happen, but it would take an extremely improbable run of luck to choose ten people so far above or below the population mean. It is much more likely that some people will be above the population mean and some below and that these observations offset each other, giving us a sample mean that is not far from the population mean.

There is, in fact, a probability distribution for the sample mean, just as there is a probability distribution for individual players. The probability distribution for the sample mean is called the *sampling distribution* for the sample mean. It tells us the probability that the sample mean will turn out to be in a specified interval, for example, the probability that the sample mean will be between 25 and 30.

The Shape of the Distribution

What can we say about the shape and location of the sampling distribution? First, for reasonably sized samples, the shape of the sampling distribution is approximately normal.

Here is why. Each observation in a random sample is an independent draw from the same population. The sample mean is the sum of these n outcomes, divided by n. Except for the unimportant division by n, these are the same assumptions used in the central limit theorem. Therefore, the sampling distribution for the mean of a random sample from *any* population approaches a normal distribution as n increases.

The central limit theorem explains why histograms for many different measurements, such as human heights, are often bell shaped. Even more importantly, the central limit theorem tells us that the sampling distribution for the mean of a reasonably sized random sample is bell shaped. This is very important because it implies that we do not need to make any assumptions about the probability distribution for the population from which the random sample is taken. Here, the distribution of individual looseness coefficients could be normal, skewed, rectangular, or U shaped. It does not matter. The sampling distribution for the sample mean is still normal.

The only caution is that the sample should be large enough for the central limit theorem to work its magic. If the underlying population is itself approximately normal, a sample of ten observations is large enough. If the underlying distribution is not normal but roughly symmetrical, a sample of size 20 or 30 is generally sufficient. If the underlying distribution is very asymmetrical, then 50, 100, or more observations may be needed.

Figure 6.2 shows a population distribution for X that is very asymmetrical, with a large chance of values close to 0, and smaller and smaller probabilities for large values of X. This figure also shows the sampling distribution for the sample mean for samples of size 2 and 30. Even though the probability of a single value close to 0 is very large, the average of two values is unlikely to be close to 0. As the sample size increases to 30, there is virtually no chance that the average value will be close to 0. Not only that, for a sample of size 30, the sampling distribution is, for all practical purposes, a normal distribution. The damping effects of averaging many draws gives the sampling distribution a symmetrical bell shape.

This example shows the important distinction between the probability distribution for X and the sampling distribution for the sample mean $\overline{X}$. The probability distribution for a single value of X shows which *individual* values of X are likely and unlikely. This distribution can have a wide variety of shapes. The sampling distribution for the sample mean shows which sample *means* are likely and unlikely. For reasonably sized samples, it has only one possible shape—the bell-shaped normal distribution.

Unbiased Estimators

What do you suppose will be the average value of a very large number of sample means? The average of, say, a 1 million means of samples of size 25 is equal to the average of all 25 million observations. The average of 25 million observations will almost certainly be very, very close to the population mean μ. If we took 1 billion samples, we could be even

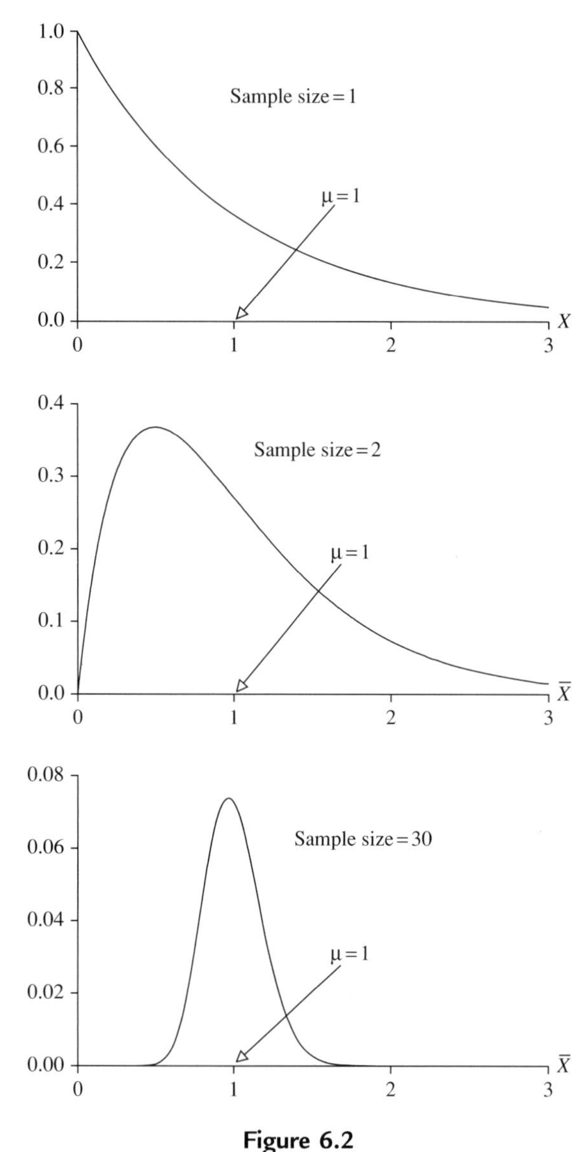

Figure 6.2
The central limit theorem in action.

more certain. The probability distribution for the sample mean assumes that we take an infinite number of samples, and, no surprise, the mean of this sampling distribution is equal to the population mean μ:

$$\text{mean of } \overline{X} = \mu \tag{6.2}$$

In Figure 6.2, the mean of the sampling distribution of $\overline{X}$ is equal to 1, which is the mean of the probability distribution of X.

A sample statistic is an *unbiased* estimator of a population parameter if the mean of the statistic's sampling distribution is equal to the value of the population parameter. In our example, the sample mean is an unbiased estimator of μ, since the mean of the sampling distribution is μ.

Other things being equal, it is better to use an unbiased estimator than one that gives estimates that are systematically too high or too low. A statistician who uses unbiased estimators can anticipate sampling errors that, over a lifetime, average close to 0. Of course, average performance is not the only thing that matters. Remember the lawyer who summarized his career by saying "When I was a young man practicing at the bar, I lost a great many cases I should have won. As I got along, I won a great many cases I ought to have lost; so on the whole justice was done" [1]. A humorous statistician might say, "Half of my estimates were much too high and half were far too low; on average, my estimates were great." A serious statistician thinks about how good the estimates are on average and also thinks about how accurate they are in individual cases.

Sampling Variance

The standard deviation of the sampling distribution gauges whether the value of the estimator is likely to vary greatly from sample to sample. An estimator that is very likely to be either much too high or far too low has a large standard deviation. In contrast, an estimator that is unbiased and has a small standard deviation is very likely to yield estimates that are consistently close to the population mean.[1]

It can be shown mathematically that the standard deviation of the sampling distribution for the sample mean is equal to σ (the standard deviation of X) divided by the square root of the sample size, n:

$$\text{standard deviation of } \overline{X} = \frac{\sigma}{\sqrt{n}} \tag{6.3}$$

Notice that the standard deviation of the sample mean declines as the sample size n increases. This makes sense. There is much less uncertainty about what the sample mean will turn out to be if the sample size is increased from 10 to 100. At the limit, if the sample size were infinitely large, there would be no uncertainty about the value of the sample mean. It would equal μ, the population mean.

Putting It All Together

We can summarize the sampling distribution of the sample mean as follows:

$$\overline{X} \sim N\left[\mu, \frac{\sigma}{\sqrt{n}}\right] \tag{6.4}$$

[1] A small standard deviation may not be desirable if the estimator is quite biased. In this case, you can be very confident that the estimate will be far from the population mean.

The sample mean has a sampling distribution that is (approximately) normal with a mean equal to the population mean for X and a standard deviation equal to standard deviation of X divided by the square root of the sample size.

We can apply Equation 6.4 to our poker example. The probability distribution for X in Figure 6.1 has a mean of 25 and a standard deviation of 10:

$$\text{mean of } X\text{: } \mu = 25$$
$$\text{standard deviation of } X\text{: } \sigma = 10$$

If we select a random sample of ten players, the probability distribution for the sample mean $\overline{X}$ is (approximately) normal with a mean of 25 and a standard deviation of 3.16:

$$\text{mean of } \overline{X}\text{: } \mu = 25$$
$$\text{standard deviation of } \overline{X}\text{: } \frac{\sigma}{\sqrt{n}} = \frac{10}{\sqrt{10}} = 3.16$$

Figure 6.3 shows this probability distribution. It is an approximately bell-shaped normal curve, centered at the population mean $\mu = 25$.

In accord with the two standard deviations rule of thumb for normal distributions, there is a 0.95 probability that the value of the sample mean will be within (approximately) two standard deviations of its mean. The exact number is 1.96 standard deviations; so, there is a 0.95 probability that the value of the sample mean will be in this interval:

$$\mu - 1.96\frac{\sigma}{\sqrt{n}} < \overline{X} < \mu + 1.96\frac{\sigma}{\sqrt{n}}$$

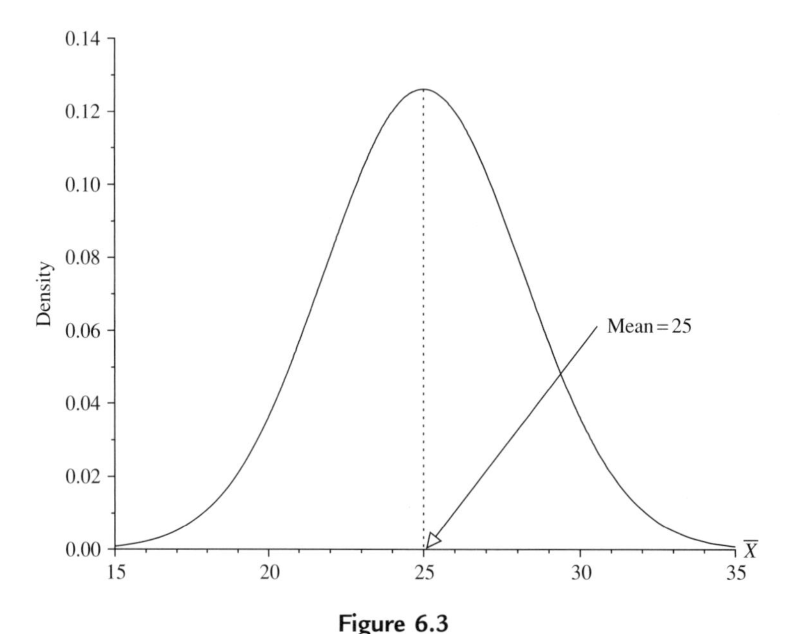

Figure 6.3
The population distribution of $\overline{X}$, for samples of size 10.

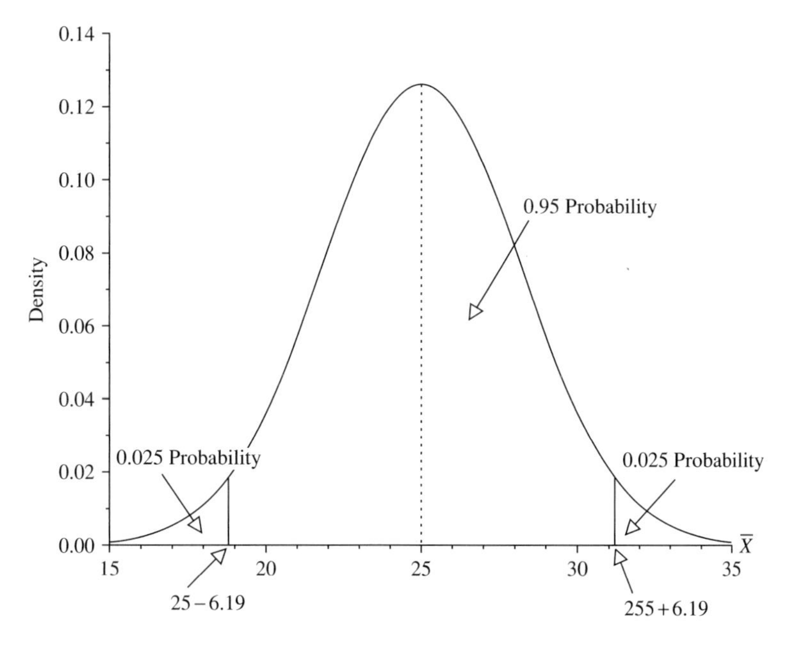

Figure 6.4
There is a 0.95 probability that the sample mean $\overline{X}$ will be within 6.19 of the population mean $\mu = 25$.

In our example, $\frac{\sigma}{\sqrt{n}} = 3.16$ and

$$25 - 1.96(3.16) < \overline{X} < 25 + 1.96(3.16) = 25 - 6.19 < \overline{X} < 25 + 6.19$$

There is a 0.95 probability that the sample mean will be within 6.19 of the population mean. Figure 6.4 shows this probability.

If we use a larger sample, we can be even more confident that our estimate will be close to μ. Figure 6.5 shows how the population distribution for the sample mean narrows as the size of the sample increases. If we base our estimate of the population mean on a sample of ten poker players, there is a 0.95 probability that our estimate will miss by less than 6.19. With a random sample of 100 players, the standard deviation is 1:

$$\text{standard deviation of } \overline{X}\text{: } \frac{\sigma}{\sqrt{n}} = \frac{10}{\sqrt{100}} = 1$$

and there is a 0.95 probability that the sample mean will be within 1.96(1) = 1.96 of μ.

Confidence Intervals

There is a 0.95 probability that the sample mean will turn out to be within 1.96 standard deviations of the population mean. This implies a 0.95 probability that the sample mean plus or minus 1.96 standard deviations will include the value of μ:

$$\overline{X} - 1.96\frac{\sigma}{\sqrt{n}} \text{ to } \overline{X} + 1.96\frac{\sigma}{\sqrt{n}}$$

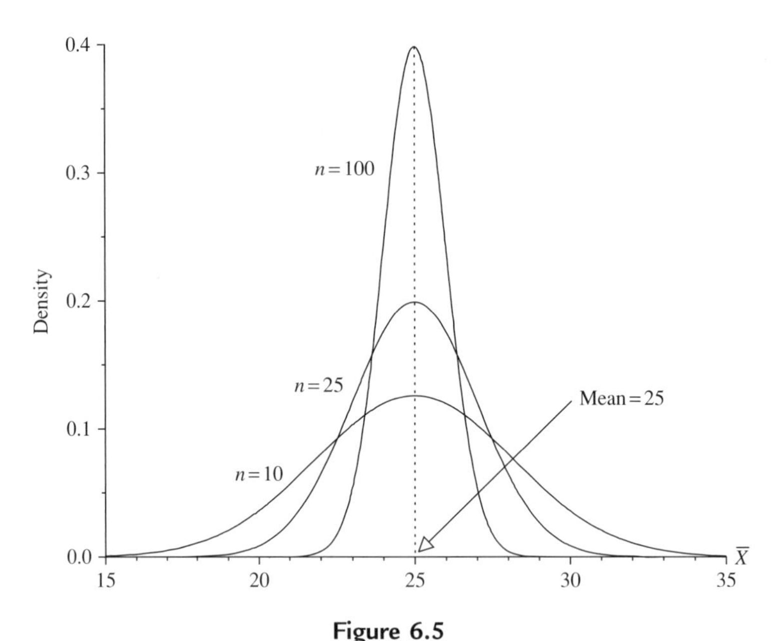

Figure 6.5
The distribution of $\overline{X}$ for different sample sizes n.

Such an interval is called a *confidence interval* and the 0.95 probability is the interval's *confidence level*. The shorthand formula for a 95 percent confidence interval for the population mean is

$$95\% \text{ confidence interval for } \mu\text{: } \overline{X} \pm 1.96\frac{\sigma}{\sqrt{n}} \tag{6.5}$$

It is important to recognize that the sample mean varies from sample to sample, but the population mean is fixed. A 95 percent confidence interval is interpreted as follows: "There is a 0.95 probability that the sample mean will be sufficiently close to μ that my confidence interval includes μ." A clever analogy is to a game of horseshoes. The population mean is the target stake and the confidence interval is a horseshoe thrown at the stake. There is a 0.95 probability that we will throw a ringer, not a 0.95 probability that the stake will jump in front of our horseshoe.

6.4 The *t* Distribution

Equation 6.3 shows that the standard deviation of our estimator $\overline{X}$ depends on σ, the standard deviation of X. So far, we have been assuming that we know the value of σ. In practice, it is unlikely that we know the value of σ, but we can estimate it from our random sample, since the standard deviation of the observed values of X is a reasonable estimate of the standard deviation of the probability distribution for X.

The formula for the sample variance s^2 is given in Equation 3.2 in Chapter 3 and repeated here:

$$s^2 = \frac{(X_1 - \overline{X})^2 + (X_2 - \overline{X})^2 + \cdots + (X_n - \overline{X})^2}{n-1} \tag{6.6}$$

The standard deviation s is the square root of the variance.

When we calculate a confidence interval using Equation 6.5, we can replace the population standard deviation σ (which is unknown) with the estimated standard deviation s. However, we need to adjust our confidence interval. Sometimes, the confidence interval will miss μ, not because the sample mean is off target but because our estimate of the standard deviation is too low and our confidence interval is too narrow. To compensate for the fact that our estimate of the standard deviation sometimes might be too low, we need to widen our confidence interval.

In 1908, W. S. Gosset figured out the correct width for confidence intervals when the population distribution of X is normal [2]. Gosset was a statistician who worked for the Irish brewery Guinness to improve its barley cultivation and brewing. Because of an earlier incident in which confidential trade secrets had been revealed, Guinness did not allow its employees to publish papers. Gosset got around this restriction by publishing his work using the pseudonym *Student* and the probability distribution he derived became known as *Student's t distribution*.

When a random sample is taken from a normal distribution and the sample mean is standardized by subtracting the mean μ and dividing by the standard deviation of the sample mean, the resulting Z statistic,

$$Z = \frac{\overline{X} - \mu}{\sigma/\sqrt{n}}$$

has a normal distribution with a mean of 0 and a standard deviation of 1.

Gosset figured out the sampling distribution of the variable that is created when the unknown standard deviation σ is replaced by sample standard deviation s:

$$t = \frac{\overline{X} - \mu}{s/\sqrt{n}} \tag{6.7}$$

The exact distribution of t depends on the sample size, because the larger the sample, the more confident we are that the sample standard deviation s is a good estimate of the population standard deviation σ. Figure 6.6 compares the t distribution for a sample of size 10 with the normal distribution. The t distribution has a somewhat smaller probability of being close to 0 and a somewhat larger probability of being in the tails.

With smaller samples, the t distribution is more dispersed. With very large samples, the t distribution is indistinguishable from the normal distribution. The probabilities for various t distributions are shown in Table A.2 in the Appendix. Instead of identifying

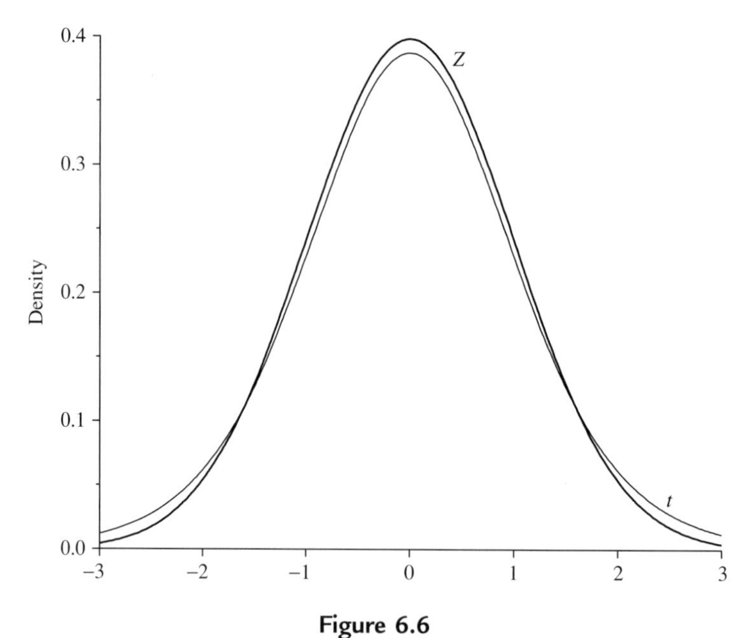

Figure 6.6
The normal distribution and t distribution with nine degrees of freedom.

each t distribution by the size of the sample, these are identified by the number of *degrees of freedom*, which equals the number of observations minus the number of parameters that must be estimated beforehand.[2] Here, we calculate s using n observations and one estimated parameter (the sample mean); therefore, there are $n - 1$ degrees of freedom. In Figure 6.6, there are $10 - 1 = 9$ degrees of freedom.

6.5 Confidence Intervals Using the *t* Distribution

The interpretation of a confidence interval is not affected by our use of an estimated standard deviation. There is a slight adjustment in the actual calculation, in that we replace the known value of σ with its estimate s, and we replace the Z value with a somewhat larger t value.

Statisticians call the estimated standard deviation the estimator's *standard error*.[3]

$$\text{true standard deviation of } \overline{X} = \frac{\sigma}{\sqrt{n}}$$

$$\text{standard error of } \overline{X} = \frac{s}{\sqrt{n}}$$

[2] Another way to think about degrees of freedom is more closely related to the name itself. We calculate s from n deviations about the sample mean. We know from Chapter 3 that the sum of the deviations about the sample mean is always 0. Therefore, if we know the values of $n - 1$ of these deviations, we know the value of the last deviation, too. Only $n - 1$ deviations are freely determined by the sample.

[3] Sometimes the true standard deviation of the sample mean is called the *standard error* and the estimated standard deviation of the sample mean is called the *estimated standard error*.

In comparison with Equation 6.5, our confidence interval now uses the estimate s in place of σ and a t value t^* that depends on the sample size:

$$\text{confidence interval for } \mu = \overline{X} \pm t^* \frac{s}{\sqrt{n}} \tag{6.8}$$

In our poker example, here are the relevant statistics:

Sample size	203
Sample mean	25.5258
Sample standard deviation	8.5306
t value with 203 − 1 = 202 degrees of freedom	1.9716

Using Equation 6.8, a 95 percent confidence interval is

$$\overline{X} \pm t^*\left(\frac{s}{\sqrt{n}}\right) = 25.5258 \pm 1.9718\left(\frac{8.5306}{\sqrt{203}}\right) = 25.53 \pm 1.18$$

The calculations use four decimals places, but we round off the final answer to two decimal places.

It is spurious precision to report a large number of decimal places, far beyond the precision of the data. For example, it is spurious precision to report that the average income in a state is $40,326.88523612; a rounded-off value of $40,327 is more realistic. Spurious precision is especially problematic when making predictions. Remember Abraham Lincoln's prediction that the U.S. population in 1930 would be 251,669,914? The exact population at any point in time is never known with certainty and it is surely unreasonable to think that anyone can predict the exact population 70 years into the future.

Lincoln observed that the U.S. population had increased by 3 percent a year over the preceding 70 years and assumed that it would continue increasing by 3 percent a year over the next 70 years. As it turned out, the population grew by only 2 percent a year after 1860, and the 1930 population turned out to be only 123 million. (This is an example of the miracle of compound interest: over a 70-year period, the difference between a 2 percent and 3 percent growth rate is enormous.)

Table 6.1 shows how the sample size affects the width of 95 percent confidence intervals. For a reasonably sized sample, the value of the t statistic is pretty close to the 1.96 value that would be used if the standard deviation were known. Nonetheless, Gosset's t distribution is enormously valuable because it gives us a credible way to calculate confidence intervals. Before the t distribution was derived, statisticians simply used 1.96 and made the awkward assumption that the standard deviation was known.

Gosset derived the t distribution by assuming that the sample data are taken from a normal distribution. Subsequent research has shown that, because of the power of the central limit

Table 6.1: How the Sample Size Affects the Width of the Confidence Interval

Sample Size n	Degrees of Freedom $n-1$	95 Percent Confidence Interval
2	1	$\overline{X} \pm 12.71 \frac{s}{\sqrt{n}}$
10	9	$\overline{X} \pm 2.26 \frac{s}{\sqrt{n}}$
20	19	$\overline{X} \pm 2.09 \frac{s}{\sqrt{n}}$
30	29	$\overline{X} \pm 2.05 \frac{s}{\sqrt{n}}$
Infinite	Infinite	$\overline{X} \pm 1.96 \frac{s}{\sqrt{n}}$

theorem, confidence intervals based on the t distribution are remarkably accurate, even if the underlying data are not normally distributed, as long as we have at least 15 observations from a roughly symmetrical distribution or at least 30 observations from a clearly asymmetrical distribution. (A histogram can be used for a rough symmetry check.) Researchers armed with the t distribution can proceed confidently even when they do not know the probability distribution of X.

Do Not Forget the Square Root of the Sample Size

An easy mistake is to forget to divide s by the square root of n when calculating the standard error, since we already divided by $n-1$ when s was calculated. Remember that s is an estimate of the standard deviation of an individual observation and our confidence interval depends on the standard error of the sample mean, which we calculate by dividing s by the square root of the sample size.

Researchers at a pharmaceutical company once tested a new enzyme that they hoped would increase the yield from a manufacturing process. The data were calculated as the ratio of the yield with the new enzyme to the yield with the old process. An observed value of 110.2 meant that the yield was 10.2 percent higher with the enzyme than with the old process. If the enzyme had no effect on the yield, the measured yields would average about 100.

As it turned out, a sample of 41 observations found an average yield of 125.2 with a standard deviation of 20.1. A member of the company's research department was unimpressed:

> *[I]n the 41 batches studied, the standard deviation was 20.1. In my opinion, 20 points represents a large fraction of the difference between 125 and 100. There is no real evidence that the enzyme increases yield [3].*

This researcher forgot to divide the standard deviation by the square root of the sample size. If the standard deviation of the individual observations X is 20.1, then the standard error of the sample mean is only 3.14:

$$\text{standard error of } \overline{X} = \frac{s}{\sqrt{n}} = \frac{20.1}{\sqrt{41}} = 3.14$$

A 95 percent confidencc interval for μ is far from 100:

$$\overline{X} \pm t^*\left(\frac{s}{\sqrt{n}}\right) = 125.2 \pm 2.021\left(\frac{20.1}{\sqrt{41}}\right) = 125.2 \pm 6.3$$

Choosing a Confidence Level

Ninety-five percent confidence levels are standard, but there is no compelling reason why we cannot use others. If we want to be more confident—for example, 99 percent certain—that our interval will contain μ, then we use a wider interval. If we are satisfied with a 50 percent chance of encompassing μ, a narrower interval will do.

To illustrate, go back to our poker example. Table 6.2 shows 50 percent, 95 percent, and 99 percent confidence intervals. A 50 percent confidence interval is relatively narrow because we require only a 50 percent chance that it will include the population mean. If we want to be 95 or 99 percent confident that our interval will encompass the population mean, we make the interval progressively wider.

Choosing a Sample Size

We used a random sample of 203 high-stakes Internet poker players to determine this 95 percent confidence interval for the average looseness coefficient in the population:

$$\overline{X} \pm t^*\left(\frac{s}{\sqrt{n}}\right) = 25.5258 \pm 1.9718\left(\frac{8.5306}{\sqrt{203}}\right) = 25.53 \pm 1.18$$

Table 6.2: A Comparison of 50, 95, and 99 Percent Confidence Intervals

Confidence Level	*t* Value	Confidence Interval
50	0.6757	25.53 ± 0.40
95	1.9718	25.53 ± 1.18
99	2.6004	25.53 ± 1.56

Table 6.3: Quadrupling the Sample Halves the Width of a Confidence Interval, Using $s = 8.53$

Sample Size	Confidence Interval
25	25.53 ± 3.52
100	25.53 ± 1.70
400	25.53 ± 0.84
1,600	25.53 ± 0.42
3,200	25.53 ± 0.21

How could we obtain a more precise estimate, that is, one with a smaller margin for sampling error? The sample standard deviation s is what it is; there is no way of knowing whether a different sample would yield a larger or smaller value. The t value t^* depends on the sample size, but Table 6.1 shows that it changes little once we get past 20 or 30 observations. The only factor that is important and that we can control is the sample size. Because a confidence interval involves the square root of the sample size, Table 6.3 shows that a quadrupling of the sample size halves the width of a confidence interval.[4] (The effect is slightly larger for smaller samples, since the t value decreases a bit.)

We can use trial calculations like those shown in Table 6.3 to decide in advance how large a random sample we want. If we are happy with ± 1.70, then 100 observations are sufficient; if we want ± 0.42, then 1,600 observations are needed. The benefit of a larger sample is a more precise estimate; the cost is the expense involved in obtaining a larger sample.

Sampling from Finite Populations

One very interesting characteristic of a confidence interval is that it does not depend on the size of the population. Many people think that a sample of size 200 is less credible if it is from a population of 1 billion than from a population of 1 million. After all, 200 out of a billion is only 1 out of every 5 million. How can we obtain a reliable estimate if we look at such a small fraction of the population?

The solution to this puzzle is that the variation in sample means from one sample to the next depends on how much variation there is among the data, not the size of the population. In fact, the formula for a confidence interval implicitly assumes that the population is infinitely large, so that the probability distribution for X doesn't change as the random sample is selected.

The sample has to be fairly large relative to the population (say, more than 10 percent) for sampling to affect the probability distribution significantly. Furthermore, if the random sample does include a significant fraction of the population, our estimates are *more* reliable

[4] The sample standard deviation is held constant in Table 6.3 so that we can focus on the sample size.

than implied by Equation 6.8, because there is then less chance of obtaining an unlucky sample of observations that are consistently above or below the population mean. By assuming an infinite population, the confidence interval in Equation 6.8 actually *understates* the reliability of our estimates.

Exercises

6.1 Identify each of the following errors in a public opinion survey as either sampling error or systematic error:
  a. The survey misses many people who work during the day.
  b. The survey interviews only randomly selected people.
  c. The survey is conducted at McDonald's.
  d. The survey is conducted in a large apartment building.

6.2 Identify each of the following errors as either sampling error or systematic error:
  a. A study of income misses people whose income is not reported to the government.
  b. People tend to overreport their income in surveys.
  c. A study of after-tax income looks at take-home pay without considering tax refunds.
  d. A random sample might not select the person with the highest income.

6.3 Explain why the following procedure is likely to produce either biased or unbiased estimates of the amount of money people spend at restaurants each year. A random number between 1 and 7 is selected; this turns out to be 6. The survey then begins on January 6 and is conducted every 7 days for a full year: January 6, January 13, January 20, and so on.

6.4 There were 562,157 live births in California in 2006; information about each of these births was arranged in alphabetical order using the mother's name. State whether each of the following sampling procedures is likely to produced biased estimates of the average age of these mothers.
  a. A random number generator was used to select 100 numbers between 1 and 562,157.
  b. A random number generator was used to select 200 numbers between 1 and 562,157; every other number was then discarded, leaving 100 numbers.
  c. A random number generator was used to select 200 numbers between 1 and 562,157; among the 200 mothers selected, the 50 youngest mothers and the 50 oldest mothers were discarded.

6.5 There were 233,467 deaths in California in 2007; the names of these decedents were arranged in alphabetical order. A random number generator was used to select 100 numbers between 1 and 233,467. State whether each of the following sampling procedures is likely to produced biased estimates of the average age at which these people died.

a. The research assistant wrote down only the first 50 random numbers.
b. The research assistant put the random numbers in numerical order; for example, the numbers 7, 3, 9 were changed to 3, 7, 9.
c. The research assistant put the ages of the 100 randomly selected people in numerical order; for example, the people with ages 85, 72, 91 were put in this order: 72, 85, 91.

6.6 Compare the sampling distributions of two random samples that will be used to estimate the average age of a woman giving birth in the United States in 2010. One sample has 100 observations and the other has 200.
a. Which sampling distribution has the higher mean?
b. Which sampling distribution has the higher standard deviation?

6.7 Do you think that the probability distribution of household income in the United States is skewed left, skewed right, or symmetrical? If a random sample of 100 households is taken, do you think that the sampling distribution of the sample mean is skewed left, skewed right, or symmetrical?

6.8 A statistics class with 500 students obtained a database of the annual salaries of every government employee in the state of California. Each student took a random sample of 30 employees and calculated the average salary of these 30 people. Do you think that a histogram of these 500 average salaries is likely to be skewed left, skewed right, or symmetrical?

6.9 Which of these histograms do you think will look more like a normal distribution? Explain your reasoning.
a. One six-sided die is rolled 1 billion times and these 1 billion numbers are used to make a histogram.
b. Five six-sided dice are rolled and the average number is calculated. This experiment is repeated 1 million times and these 1 million averages are used to make a histogram.

6.10 Which of these experiments do you think has the higher average? Explain your reasoning.
a. One coin is flipped 1 billion times and the average number of heads is calculated (total number of heads divided by 1 billion).
b. Ten coins are flipped and the average number of heads is calculated (total number of heads divided by 10). This experiment is repeated 1 billion times and the average of the averages is calculated.

6.11 Each of the following three experiments will be repeated 1 million times and the variance of the 1 million results will be calculated. Which of these experiments do you think will have the largest variance? The smallest variance? Explain your reasoning.
a. A six-sided die is rolled and the outcome is recorded.

b. Ten six-sided dice are rolled and the sum of the ten numbers is recorded.
c. Ten six-sided dice are rolled and the average (the sum of the numbers, divided by 10) is recorded.

6.12 Each of the following three experiments will be repeated 1 million times and the variance of the 1 million results will be calculated. Which of these experiments do you think will have the largest variance? The smallest variance? Explain your reasoning.
a. A coin is flipped and a 1 is recorded if it lands heads, 0 if it lands tails.
b. A coin is flipped eight times and the number of heads is recorded.
c. A coin is flipped eight times and the number of heads, divided by 8, is recorded.

6.13 One grade of Washington navel oranges has a weight that is normally distributed with a mean of 12 ounces and a standard deviation of 2 ounces.
a. What is the probability that a randomly selected orange will weigh less than 10 ounces?
b. What is the probability that a bag containing 25 randomly selected oranges will have a net weight of less than 250 ounces?

6.14 The average life of a Rolling Rock tire is normally distributed with a mean of 40,000 miles and a standard deviation of 5,000 miles. Think about the four probabilities that follow. You do *not* have to calculate these probabilities. Instead, simply rank these four probabilities from largest to smallest; for example, if you think that (a) has the highest probability, (b) the next highest, (c) the next highest, and (d) the smallest probability, your answer would be a, b, c, d.
a. A tire will last more than 40,000 miles.
b. A tire will last more than 50,000 miles.
c. The average life of a random sample of four tires will be more than 40,000 miles.
d. The average life of a random sample of four tires will be more than 50,000 miles.

6.15 An investor believes that the returns on each of two investments over the next year can be described by a normal distribution with a mean of 10 percent and a standard deviation of 20 percent. This investor also believes that the returns on these two investments are independent. Investment Strategy A is to invest $100,000 in one of these investments, selected at random; Strategy B is to invest $50,000 in each investment. Explain your reasoning when you answer each of the following questions:
a. Which strategy has a larger probability of a return greater than 10 percent?
b. Which strategy has a larger probability of a return greater than 20 percent?
c. Which strategy has a larger probability of a negative return?

6.16 An economics professor told his students that, instead of spending hundreds of dollars for a very accurate watch, they should wear 20 cheap watches and calculate the average time shown on these 20 watches. Suppose that the reported time on each cheap watch is normally distributed with a mean equal to the actual time and a standard deviation of

10 seconds and the errors in the reported times on all the cheap watches are independent of each other. What are the mean and standard deviation of the average time on these 20 watches?

6.17 In a lawsuit concerning the discharge of industrial wastes into Lake Superior, the judge observed that a court-appointed witness had found the fiber concentration to be "0.0626 fibers per cc, with a 95 percent confidence interval of from 0.0350 to 0.900 fibers per cc" [4]. Explain the apparent statistical or typographical error in this statement.

6.18 An outside consultant for a small suburban private school with a reputation for academic excellence analyzed the test scores in Table 6.4 for 13 students (of the 20 students initially admitted) who stayed at the school from first grade through eighth grade. The scores are percentiles relative to students at suburban public schools. For example, on the first-grade reading comprehension test, Student A's score was higher than the scores of 79 percent of the students at suburban public schools. Calculate the difference X between the eighth-grade and first-grade reading comprehension scores for these 13 students. Assuming these are a random sample from a hypothetical population of students attending this school over the years, calculate a 95 percent confidence interval for a population mean of X.

6.19 Redo Exercise 6.18, this time using the mathematics scores.

6.20 Think of the database of 203 high-stakes poker players as a population, from which a random sample of 9 players is selected with these looseness coefficients:

25.75	15.14	22.88	31.62	26.56	17.71	47.89	23.52	23.32

Table 6.4: Exercise 6.18

	Reading Comprehension		Mathematics	
Student	First Grade	Eighth Grade	First Grade	Eighth Grade
A	79	33	81	57
B	91	38	93	51
C	98	92	97	98
D	99	75	99	74
E	94	4	96	34
F	98	75	96	91
G	99	68	98	95
H	97	33	93	78
I	99	68	98	90
J	99	99	96	67
K	99	9	97	53
L	97	86	97	67
M	91	86	96	83

Calculate a 95 percent confidence interval for the population mean. Does your confidence interval include the actual value of the population mean, $\mu = 25.53$?

6.21 Continuing the preceding exercise, another random sample of 16 players is selected. Does a 95 percent confidence interval using these 16 observations include the actual value of the population mean, $\mu = 25.53$?

20.60	51.13	27.89	18.39	28.42	21.44	20.59	22.24
28.38	21.61	24.51	17.64	24.42	21.26	21.73	28.36

6.22 Data were obtained for 16,405 women who immigrated to the United States and their adult daughters at matching times of their lives [5]. The median income in the mothers' and daughters' ZIP codes were used to measure economic status. Consider this database to be a population, from which a random sample of 25 mothers was selected with these ZIP-code incomes. Calculate a 95 percent confidence interval for the population mean. Does your confidence interval include the actual value of the population mean, $\mu = 37{,}408$?

32,644	39,225	33,445	45,267	36,769	26,505	29,872	24,207	28,651
33,656	26,505	22,091	47,806	30,174	41,621	46,342	23,498	39,906
49,686	39,747	22,151	36,566	39,225	46,468	53,881		

6.23 The following data are the median ZIP-code incomes of the daughters of the random sample of 25 mothers in the preceding exercise. Calculate a 95 percent confidence interval for the population mean. Does your confidence interval include the actual value of the population mean, $\mu = 40{,}064$?

40,018	32,496	25,860	55,123	36,769	26,505	50,543	61,800	30,375
33,656	25,593	20,275	35,846	46,012	39,225	46,175	23,538	23,841
46,892	39,747	38,010	56,416	64,947	40,057	20,034		

6.24 Using the data in Exercises 6.22 and 6.23, the following data are the differences between each daughter's ZIP-code income and her mother's ZIP-code income. Calculate a 95 percent confidence interval for the population mean of these differences. Does your confidence interval include zero?

7,374	−6,729	−7,585	9,856	0	0	20,671	37,593	1,724
0	−912	−1,816	−11,960	15,838	−2,396	−167	40	−16,065
−2,794	0	15,859	19,850	25,722	−6,411	−33,847		

6.25 A histogram of ZIP-code income is skewed right. Why do you suppose this is? If ZIP-code income is skewed right, can it also be normally distributed? If ZIP-code income is not normally distributed, is this an important problem for the calculations made in Exercises 6.22 and 6.23?

6.26 Forty randomly selected college students were asked how many hours they sleep each night. Use these data to estimate a 95 percent confidence interval for the mean of the population from which this sample was taken:

5	6	6	8	6	6	6.5	4
4	5	5	7	6	6	5	5
6.5	6.5	6.5	7	5	5.5	5	6
5	5	4	8	5	6.5	6	5
5	8	5	7	6	8	7	6

6.27 The first American to win the Nobel Prize in physics was Albert Michelson (1852–1931), who was given the award in 1907 for developing and using optical precision instruments. Between October 12, 1882, and November 14, 1882, he made 23 measurements [6] of the speed of light (in kilometers per second):

299,883	299,796	299,611	299,781	299,774	299,696
299,748	299,809	299,816	299,682	299,599	299,578
299,820	299,573	299,797	299,723	299,778	299,711
300,051	299,796	299,772	299,748	299,851	

Assuming that these measurements are a random sample, does a 99 percent confidence interval include the value 299,710.50 that is now accepted as the speed of light?

6.28 The population of Spain is approximately four times the population of Portugal. Suppose that a random sample is taken of 1,500 Spaniards and 1,500 Portuguese. Explain why you believe that the standard error of the sample mean is: (a) four times larger for the Spanish poll; (b) four times larger for the Portuguese poll; (c) twice as large for the Spanish poll; (d) twice as large for the Portuguese poll; or (e) the same for both polls.

6.29 Exercise 3.43 shows the percentage errors when 24 college students were asked to estimate how long it took them to read an article. Use these data to estimate a 95 percent confidence interval for the population mean.

6.30 In 1868, a German physician, Carl Wunderlich, reported his analysis of over 1 million temperature readings from 25,000 patients [7]. He concluded that 98.6° Fahrenheit, (which is 37.0° Celsius) is the average temperature of healthy adults. Wunderlich used the armpit for measuring body temperatures and the thermometers he used required 15 to 20 minutes to give a stable reading. In 1992, three researchers at the University of Maryland School of Medicine reported the results of 700 readings of 148 healthy persons using modern thermometers [8]. A random sample of their data follows. Use these data to calculate a 95 percent confidence interval for the population mean.

98.2	98.8	98.4	98.7	98.2	97.7	97.4	98.6	97.2	97.6
98.0	99.2	99.4	98.8	98.8	97.1	98.0	98.5	98.2	97.5
96.9	98.8	98.4	97.7	98.0	98.4	98.2	99.9	98.6	99.5
98.7	98.6	98.3	98.3	98.2	97.0	98.4	98.6	97.4	98.8
97.8	98.3	98.8	98.8	97.9	97.5	98.2	98.9	97.0	97.5

6.31 Explain why you either agree or disagree with each of these interpretations of the confidence interval calculated in the preceding exercise.
a. 95 percent of healthy adults have temperatures in this interval.
b. Healthy adults have temperatures in this interval 95 percent of the time.
c. If your temperature is in this interval, there is a 95 percent chance you are healthy.

6.32 An advertising agency is looking for magazines in which to advertise inexpensive imitations of designer clothing. Their target audience is households earning less than $30,000 a year. A study estimates that a 95 percent confidence interval for the average household income of readers of one magazine is $32,000 ± $1,200. Is the agency correct in concluding that the people in their target audience do not read this magazine?

6.33 A *Wall Street Journal* poll asked 35 economic forecasters to predict the interest rate on 3-month Treasury bills 12 months later [9]. These 35 forecasts had a mean of 6.19 and a variance of 0.47. Assuming these to be a random sample, give a 95 percent confidence interval for the mean prediction of all economic forecasters and explain why each of these interpretations is or is not correct:
a. There is a 0.95 probability that the actual Treasury bill rate is in this interval.
b. Approximately 95 percent of the predictions of all economic forecasters are in this interval.

6.34 Based on 11 separate studies, the Environmental Protection Agency (EPA) estimated that nonsmoking women who live with smokers have, on average, a 19 percent higher risk of lung cancer than similar women living in a smoke-free home. The EPA reported a 90 percent confidence interval for this estimate as 4 percent to 34 percent. After lawyers for tobacco companies argued that the EPA should have used a standard 95 percent confidence interval, a *Wall Street Journal* article reported "Although such a calculation wasn't made, it might show, for instance, that passive smokers' risk of lung cancer ranges from, say, 15% *lower* to 160% higher than the risk run by those in a smoke-free environment" [10].
a. Explain why, even without consulting a probability table, we know that a standard 95 percent confidence interval for this estimate does not range from −15 percent to +160 percent.
b. In calculating a 90 percent confidence interval, the EPA used a t distribution with 10 degrees of freedom. Use this same distribution to calculate a 95 percent confidence interval.

6.35 A statistics textbook [11] gives this example:

Suppose a downtown department store questions forty-nine downtown shoppers concerning their age.... The sample mean and standard deviation are found to be 40.1 and 8.6, respectively. The store could then estimate μ, *the mean age of all downtown shoppers, via a 95% confidence interval as follows:*

$$\overline{X} \pm 1.96\frac{s}{\sqrt{n}} = 40.1 \pm 1.96\frac{8.6}{\sqrt{49}}$$
$$= 40.1 \pm 2.4$$

Thus the department store should gear its sales to the segment of consumers with average age between 37.7 and 42.5.

Explain why you either agree or disagree with this interpretation: 95 percent of downtown shoppers are between the age of 37.7 and 42.5.

6.36 Statisticians sometimes report 50 percent confidence intervals, with the margin for sampling error called the *probable error*. For example, an estimate $\overline{X}$ of the average useful life of a television is said to have a probable error of 3 years if there is a 0.50 probability that the interval $\overline{X} \pm 3$ will include the value of the population mean. Calculate the probable error if a random sample of 25 televisions has an average useful life of 8.2 years with a standard deviation of 2.5 years.

6.37 An article about computer forecasting software explained how to gauge the uncertainty in a prediction: "[Calculate] the standard deviation and mean (average) of the data. As a good first guess, your predicted value ... is the mean value, plus or minus one standard deviation—in other words, the uncertainty is about one standard deviation" [12]. What did they overlook?

6.38 To investigate how infants first comprehend and produce words, six British children were observed in their homes every 2 weeks from age 6 to 18 months [13]. Among the data collected were the age (in days) at which each child first understood spoken English and first spoke a word of English, shown in Table 6.5. Assuming these six children to be a random sample of all British children, determine a 95 percent

Table 6.5: Exercise 6.38

	Age (days)		
	Understood	**Spoke**	**Difference**
Andrew	253	368	115
Ben	238	252	14
George	257	408	151
Katherine	220	327	107
Katy	249	268	19
Sebastian	254	326	72

confidence interval for the average difference (in days) between when a child first understands and speaks English. Does your estimate seem fairly precise? Why is it not more precise? Why might this be a biased sample?

6.39 The American Coinage Act of 1792 specified that a gold $10 eagle coin contain 247.5 grains of pure gold. In 1837, Congress passed a law stating that there would be an annual test of 1 coin and 1,000 coins, with the Mint failing the test if the weight of the single coin was more than 0.25 grains from 247.5, either above or below, or if the average weight of 1,000 eagle coins weighed as a group was more than 0.048 grains from 247.5, either above or below. If the weight of a single coin is normally distributed with a mean of 247.5 grains and a standard deviation σ, is the Mint more likely to fail the test of a single coin or the test of the average weight of 1,000 coins?

6.40 Do you think that a 95 percent confidence interval is wider if a researcher uses the normal distribution or t distribution? Explain your reasoning.

6.41 Other things being equal, explain why you think each of the following changes would increase, decrease, or have no effect on the width of a 95 percent confidence interval for the population mean:

a. The sample mean is smaller.
b. The population mean is smaller.
c. The sample standard deviation is smaller.
d. The sample size is smaller.

6.42 Explain why you either agree or disagree with this reasoning: "[T]he household [unemployment] survey is hardly flawless. Its 60,000 families constitute less than 0.1 percent of the work force" [14].

6.43 *Newsweek* once wrote [15] that the width of a confidence interval is inversely related to the sample size; for example, if a sample of 500 gives a confidence interval of plus or minus 5, then a sample of 2,500 would give a confidence interval of plus or minus 1. Explain the error in this argument.

6.44 Your assignment is to estimate the average number of hours a day that people spend doing e-mail. You think the standard deviation is probably around 2 hours. If so, how large a sample do you need for a 95 percent confidence interval to be approximately ± 10 minutes?

6.45 One researcher took a random sample of size 10 and estimated a 95 percent confidence interval for the average home price in an area to be $180,000 ± $40,000; another researcher took a random sample of size 20 and obtained a 95 percent confidence interval of $200,000 ± $30,000.

a. How would you use both of these studies to obtain a 95 percent confidence interval of $a \pm b$? (You do not need to do any calculations; simply explain how you would proceed.)

b. Explain why you think that the value of a is larger than, smaller than, or equal to \$190,000.
c. Explain why you think that the value of b is larger than, smaller than, or equal to \$30,000.

6.46 You believe that household income in a country has a mean of around \$40,000 with a standard deviation of about \$30,000. How large a sample should you take if you want the width of a 95 percent confidence interval for the mean to be approximately \$2,000?

6.47 Which of the three density functions shown in Figure 6.5 has the largest area under the curve? The smallest area?

6.48 The text reports a 95 percent confidence interval for the looseness coefficient for high-stakes Internet poker players to be 25.53 ± 1.18. Explain why you either agree or disagree with these interpretations of that number:
a. 95 percent of the players in this population have looseness coefficients in this interval.
b. If a poker player is randomly selected from this population, there is a 0.95 probability that his or her looseness coefficient is in this interval.
c. 95 percent of the confidence intervals estimated in this way include the average looseness coefficient of the players in the population.

6.49 A researcher calculated a 95 percent confidence interval using a random sample of size 10. When a colleague suggested that the sample was pretty small, the researcher doubled the size of the sample by assuming that he had an additional ten observations identical to his first ten observations. What do you think happened to his sample mean? The width of his confidence interval? If the width of his confidence interval decreased, would this be a good argument for following his procedure?

6.50 A government agency wants to estimate the number of fish in a lake. People from the agency catch 100 fish, tag them, and put them back in the lake. A short while later, they catch 100 fish and find that 10 have been tagged. Think of this second batch as a random sample of all of the fish in the lake. Based on this sample, what is your estimate of the fraction of the fish in the lake that are tagged? Because you know that 100 fish are tagged, what is your estimate of the total number of fish in the lake?

CHAPTER 7

Hypothesis Testing

Chapter Outline

A friend of mine once remarked to me that if some people asserted that the earth rotated from east to west and others that it rotated from west to east, there would always be a few well-meaning citizens to suggest that perhaps there was something to be said for both sides, and that maybe it did a little of one and a little of the other; or that the truth probably lay between the extremes and perhaps it did not rotate at all.

—Maurice G. Kendall

In Chapter 6, we saw how sample data can be used to estimate the value of a population mean and calculate a confidence interval that gauges the precision of this estimate. In this chapter, we see how sample data can be used to support or refute theories about the population mean, for example, whether experienced poker players tend to change their style of play after a big loss and whether, on average, the daughters of immigrant mothers are upwardly mobile. We use these principles in this chapter and later chapters to test other theories, including whether the average temperature of healthy adults is 98.6° Fahrenheit, the average student gains 15 pounds during the first year at college, interest rates affect stock prices, and unemployment affects presidential elections.

Essential Statistics, Regression, and Econometrics

You will see how to do these statistical tests and interpret the results. You will learn not only how to do tests correctly, but also how to recognize tests that are misleading.

7.1 Proof by Statistical Contradiction

Statistical tests are based on empirical data—ideally a random sample. These tests can never be as definitive as a mathematical or logical proof because a random sample is, well, random and therefore subject to sampling error. One random sample will yield one set of data and another sample will yield different data. There is always a chance, however small, that a new sample reverses a conclusion based on an earlier sample. We cannot be absolutely certain of the average looseness coefficient of all players unless we look at all players.

Most theories are vulnerable to reassessment because there never is a final tabulation of all possible data. New experiments and fresh observations continually provide new evidence—data that generally reaffirm previous studies but occasionally create doubt or even reverse conclusions that were once thought firmly established. Theories are especially fragile in the humanities and social sciences because it is difficult to control for extraneous influences. In the 1930s, John Maynard Keynes theorized that there was a very simple relationship between household income and consumer spending. This theory was seemingly confirmed by data for the 1930s and early 1940s but found inadequate when data for the 1950s became available. In the 1960s, economists believed that there was a simple inverse relationship between a nation's unemployment rate and its rate of inflation: When unemployment goes up, inflation goes down. In the 1970s, unemployment and inflation both went up, and economists decided that they had overlooked other important influences, including inflation expectations.

Even in the physical sciences, where laboratory experiments can generate mountains of data, theories are vulnerable to reassessment. For centuries before Copernicus, people were certain that the sun revolved around the earth. Until Galileo's experiments, the conventional wisdom was that heavy objects fall faster than lighter ones. Before Einstein, Newtonian physics was supreme. When scientists think of new ways to test old theories or fresh ways to interpret old data, the weakness of accepted theories may become exposed.

If we can never be absolutely sure that a theory is true or false, the next best thing would be to make probability statements, such as, "Based on the available data, there is a 0.90 probability that this theory is true." Bayesians are willing to make such subjective assessments, but other statisticians insist that a theory is either true or it is not. A theory is not true 90 percent of the time, with its truth or falsity determined by a playful Nature's random number generator.

Hypothesis tests have followed another route. Instead of estimating the probability that a theory is true, based on observed data, statisticians calculate the reverse probability—the probability that we would observe such data if the theory were true. An appreciation of the difference between these two probability statements is crucial to an understanding of the meaning and limitations of hypothesis tests.

Hypothesis tests are a proof by statistical contradiction. We calculate the probability that the sample data would look the way they do if the theory were true. If this probability is low, then the data are not consistent with the theory and we reject the theory, what Thomas Huxley called "the great tragedy of science—the slaying of a beautiful hypothesis by an ugly fact." This is proof by statistical contradiction: We find that our data would be unlikely to occur if the theory were true, and we consequently reject the theory. Of course, we can never be 100 percent certain, because unlikely events sometimes happen. Our theory might be correct but an unlucky random sample yielded improbable data.

Notice that our decision rule is to reject a theory if the data are not consistent with the theory. If the data are consistent with the theory, we do not reject the theory. But this does not prove that the theory is true, only that the data are consistent with the theory. This is a relatively weak conclusion because the data may be consistent with many other theories. For example, a 95 percent confidence interval for the population mean of 25.53 ± 1.18 is consistent with the theory that the population mean is 25 and is also consistent with the theory that the population mean is 26.

7.2 The Null Hypothesis

A specific example can make these general ideas more concrete. A substantial body of evidence indicates that decisions are shaped by a variety of cognitive biases. We will use the poker data described in Chapter 6 to investigate whether experienced poker players change their style of play after winning or losing a big pot. Poker is a very attractive source of data because it avoids many criticisms of artificial experiments, for example, that the participants are inexperienced or that they are inattentive because the payoffs are small.

For consistency, we look at six-player tables with blinds of $25/$50, which are considered high-stakes tables and attract experienced poker players. We consider a hand where a player wins or loses $1,000 to be a big win or loss. After a big win or loss, we monitor the player's behavior during the next 12 hands—two cycles around a six-player table. We follow two cycles because experienced players often make no voluntary bets, and 12 hands are reasonably close to the big win or loss. We restrict our attention to individuals who had enough big wins and losses to have played at least 50 hands in the 12-hand windows following big wins and at least 50 hands in the 12-hand windows following big losses.

The generally accepted measure of looseness is the percentage of hands in which a player voluntarily puts money into the pot. This can include a call or a raise, but does not include blind bets since these are involuntary. After a hand is dealt, everyone other than the player who put in the big blind must either bet or fold before they see the three-card flop. Thus, looseness generally measures how often a person puts money into the pot in order to see the flop cards. Tight players fold when the two cards they hold are not strong; loose players stay in, hoping that a lucky flop will strengthen their hand. At six-player tables, people are considered to be very tight players if their looseness is below 20 percent and to be extremely loose players if their looseness is above 50 percent.

We look at two *research questions*:

1. Do these players tend to have a loose or tight style of play?
2. Is their style of play different after a big loss than after a big win?

To answer the first research question, we make an assumption, called the *null hypothesis* (or H_0), about the population from which the sample is drawn. Typically, the null hypothesis is a "straw assumption" that we anticipate rejecting. To demonstrate, for example, that a medicine is beneficial, we see whether the sample data reject the hypothesis that there is no (*null*) effect.

Here, the null hypothesis might be that the average looseness coefficient is 35, which is halfway between 20 (tight) and 50 (loose). Although we expect experienced high-stakes players to have a relatively tight style of play, the way to demonstrate this by statistical contradiction is to see if the evidence rejects the straw hypothesis that the average looseness coefficient is 35.

The alternative hypothesis (usually written as H_1) describes the population if the null hypothesis is not true. Here, the natural alternative hypothesis is that the population mean is not equal to 35:

$$H_0\colon \mu = 35$$
$$H_1\colon \mu \neq 35$$

The alternative hypothesis is usually *two sided*, because even though we might have a hunch about how the study will turn out, we are reluctant to rule out beforehand the possibility that the population mean may be either lower or higher than the value specified by the null hypothesis. If, before seeing the data, we could rule out one of these possibilities, the alternative hypothesis would be *one sided*. If, for example, we were convinced beforehand that the average looseness coefficient of experienced players could not possibly be higher than 35, the one-sided alternative hypothesis would be $H_1\colon \mu < 35$.

7.3 *P* Values

Once we have specified the null and alternative hypotheses, we analyze our sample data. Because we want to test a null hypothesis about the population mean, we naturally look at the sample mean—because the sample mean is what we use to estimate the value of the population mean. The estimator used to test the null hypothesis is called the *test statistic*.

For our sample:

Sample size	203
Sample mean	25.53
Sample standard deviation	8.53

This sample is large enough to justify our assumption that the sample mean comes from a normal distribution. Equation 6.4, repeated here, gives the distribution of the sample mean:

$$\overline{X} \sim N\left[\mu, \frac{\sigma}{\sqrt{n}}\right] \tag{7.1}$$

We initially assume that the population standard deviation is known to be 8.53. If so, what values of the sample mean are likely if the null hypothesis is true and the population mean is 35? In this case, the distribution for the sample mean is

$$\overline{X} \sim N\left[35, \frac{8.53}{\sqrt{203}}\right]$$
$$\sim N[35, 0.60]$$

Our two-standard-deviations rule of thumb tells us that there is approximately a 0.95 probability that the sample mean will turn out to be within two standard deviations (or, more precisely, 1.96 standard deviations) of the mean. Here, that interval is $35 \pm 1.96(0.60) = 35 \pm 1.2$, or between 33.8 and 36.2.

Figure 7.1 shows a 0.025 probability that the sample mean will be less than 33.8, a 0.025 probability that the sample mean will be larger than 36.2, and a 0.95 probability that the sample mean will be between these two values.

The fact that our sample mean turns out to be 25.53 certainly casts doubt on the null hypothesis that this sample came from a population with a mean of 35. If the population mean really is 35, what are the chances that a random sample of 203 players would yield a sample mean as low as 25.53?

To calculate this probability, we convert to a standardized Z value:

$$Z = \frac{\overline{X} - \mu}{\sigma/\sqrt{n}} \tag{7.2}$$

For our data, with a population mean of 35, sample mean of 25.53, standard deviation of 8.53, and sample size of 203, the Z value is -15.82:

$$Z = \frac{25.53 - 35}{8.53/\sqrt{203}}$$
$$= -15.82$$

The sample mean is 15.82 standard deviations from 35.

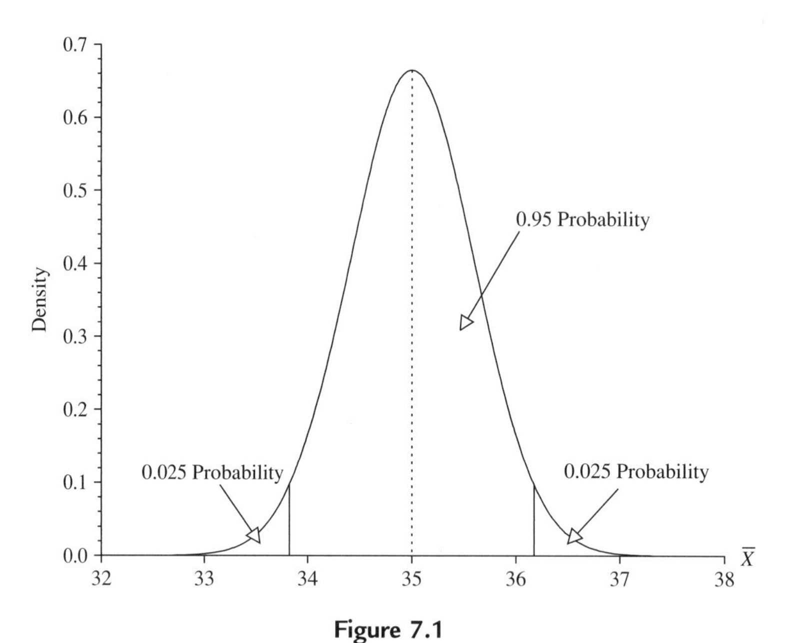

Figure 7.1
The population distribution of $\overline{X}$, if the null hypothesis is true.

We do not calculate the probability that Z would be *exactly* equal to −15.82. The probability that a normally distributed variable will equal any single number is 0. Instead, we calculate the probability that Z will be so far from 0, here, the probability that Z will be −15.82, or even lower:

$$P[Z \leq -15.82] = 1.13 \times 10^{-56}$$

In calculating the P value, we must take into account whether the alternative hypothesis is one sided or two sided. For a one-sided alternative hypothesis, the P value is the probability that the sample mean would be so far to one side of the population mean specified by the null hypothesis. For a two-sided alternative hypothesis, the P value is the probability that the test statistic would be so far, *in either direction*, from the population mean. In our poker example, we report the probability that a sample mean would be 15.82 standard deviations or more, in either direction, from the null hypothesis. Because the normal distribution is symmetrical, we double the 1.13×10^{-56} probability just calculated and report the *two-sided P value* as $2(1.13 \times 10^{-56}) = 2.26 \times 10^{-56}$.

Because this probability is minuscule, sampling error is an unconvincing explanation for why the average looseness coefficient in our sample is so far from 35. If the sample mean had been 34.5 or 35.5, we could have reasonably attributed this difference to the inevitable variation in random samples. However, a Z value of −15.82 is too improbable to be explained by the luck of the draw. We have shown, by statistical contradiction, that the average looseness coefficient in the population is not 35.

We can summarize our general procedure as follows. First, specify the null and alternative hypotheses:

$$H_0: \mu = 35$$
$$H_1: \mu \neq 35$$

Then, use the sample data to calculate the sample mean (25.53), which is our estimate of the value of μ. Use Equation 7.2 to calculate the *Z* value (−15,82), which measures how many standard deviations the estimate is from the null hypothesis. The *P* value (1.13×10^{-56}) is the probability, if the null hypothesis is true, that the sample mean would be this many standard deviations from the value of μ specified by the null hypothesis. If the alternative hypothesis is two sided, we double this probability to obtain the two-sided *P* value (2.26×10^{-56}).

Using an Estimated Standard Deviation

Our initial calculations assumed that the population standard deviation is known to be 8.53. However, this is an estimate based on sample data. The preceding chapter explains how the *t* distribution can be used in place of the normal distribution when the sample standard deviation *s* is used in place of the population standard deviation σ. Here, instead of the *Z* statistic in Equation 7.2, we use the *t* statistic:

$$t = \frac{\overline{X} - \mu}{s/\sqrt{n}} \tag{7.3}$$

The calculated *t* value is identical to the *Z* value we calculated previously:

$$t = \frac{25.53 - 35}{8.53/\sqrt{203}}$$
$$= -15.82$$

The difference is that, instead of using the normal distribution to calculate the *P* value, we use the *t* distribution with $n - 1 = 203 - 1 = 202$ degrees of freedom. The probability works out to be

$$P[t \leq -15.82] = 3.23 \times 10^{-37}$$

so that the two-sided *P* value is $2(3.23 \times 10^{-37}) = 6.46 \times 10^{-37}$.

A *P* value calculated from the *t* distribution is exactly correct if the data come from a normal distribution and, because of the power of the central limit theorem, is an excellent approximation if we have at least 15 observations from a generally symmetrical distribution or at least 30 observations from a very asymmetrical distribution. In our poker example, the

P value is minuscule and is strong evidence against the null hypothesis that the population mean is 35.

Significance Levels

A P value of 6.46×10^{-37} provides persuasive evidence against the null hypothesis. What about a P value of 0.01, or 0.10, or 0.25? Where do we draw a line and say that this P value is persuasive, but that one is not? In the 1920s, the great British statistician R. A. Fisher endorsed a 5 percent cutoff:

> *It is convenient to draw the line at about the level at which we can say: "Either there is something in this treatment, or a coincidence has occurred such as does not occur more than once in twenty trials...."*
>
> *If one in twenty does not seem high enough odds, we may, if we prefer, draw the line at one in fifty (the 2 percent point), or one in a hundred (the 1 percent point). Personally, the writer prefers to set a low standard of significance at the 5 percent point, and ignore entirely all results which fail to reach that level [1].*

Researchers today often report that their results either are or are not statistically significant at a specified level, such as 5 percent or 1 percent. What this cryptic phrase means is that the appropriate P value is less than this specified *significance level*. For example, because the P value of 6.46×10^{-37} is less than 0.05, we can report that we "found a statistically significant difference at the 5 percent level between the average looseness of a sample of experienced high-stakes poker players and 35." The reader is thereby told that there is less than a 0.05 probability of observing such a large difference between the sample mean and the value of the population mean given by the null hypothesis.

Fisher's endorsement of a 5 percent rule is so ingrained that some researchers simply say that their results are "statistically significant," with readers understanding that the P value is less than 0.05. Some say "statistically significant" if the P value is less than 0.05 and "highly significant" if the P value is less than 0.01. Simple rules are convenient and provide a common vocabulary, but we must not be blind to the fact that there is a continuum of P values. There is little difference between P values of 0.049 and 0.051. There is a big difference between P values of 0.049 and 6.46×10^{-37}. The most reasonable course is to report the P value and let readers judge for themselves. They will remember Fisher's rule of thumb, but they will also recognize that a P value of 0.051 is not quite significant at the 5 percent level, that 0.049 is barely significant, and that 6.46×10^{-37} is off the charts.

The reader's reaction may well be tempered by what is being tested, with some situations calling for more persuasive proof than others. The Food and Drug Administration may demand very strong evidence when testing a drug with modest benefits and potentially dangerous side effects. Similarly, in cases where the conventional wisdom is very strong, evidence to the contrary has to be very powerful to be convincing.

7.4 Confidence Intervals

Confidence intervals can be used for a statistical test when the alternative hypothesis is two sided. Specifically, if the two-sided *P* value is less than 0.05, then a 95 percent confidence interval does not include the value of the population mean specified by the null hypothesis. (And if the two-sided *P* value is less than 0.01, a 99 percent confidence interval does not include the null hypothesis.)

Consider our poker example. A test of the null hypothesis $\mu = 35$ has a *P* value less than 0.05 if the sample mean is more than (approximately) two standard deviations from 35. A 95 percent confidence interval for μ includes all values that are within two standard deviations of the sample mean. Thus, if the sample mean is more than two standard deviations from 35, the two-sided *P* value is less than 0.05 and a 95 percent confidence interval does not include 35. Therefore, a hypothesis test can be conducted by seeing whether a confidence interval includes the null hypothesis.

The nice thing about a confidence interval is that it can give us a sense of the practical importance of the difference between the sample mean and the null hypothesis. If we just report that the *P* value is 6.46×10^{-37} or that we "found a statistically significant difference at the 5 percent level," readers do not know the actual value of the estimator. In addition to the *P* value, we should report a confidence interval.

For our poker example, a 95 percent confidence interval was calculated in Chapter 6:

$$\overline{X} \pm t^* \frac{s}{\sqrt{n}} = 25.53 \pm 1.97 \frac{8.53}{\sqrt{203}}$$
$$= 25.53 \pm 1.18$$

We see that 35 is not inside this interval and that, as a practical matter, 25.53 ± 1.18 is far from 35.

The fact that confidence intervals can be used for hypothesis tests illustrates why not rejecting a null hypothesis is a relatively weak conclusion. When the data do not reject a null hypothesis, we have *not* proven that the null hypothesis is true. *Every* value inside a 95 percent confidence interval is consistent with the data. Therefore, we say "the data do not reject the null hypothesis," instead of "the data prove that the null hypothesis is true."

Two economists studied the effects of inflation on election outcomes [2]. They estimated that the inflation issue increased the Republican vote in one election by 7 percentage points, plus or minus 10 percentage points. Because 0 is inside this interval, they concluded that "in fact, and contrary to widely held views, inflation has no impact on voting behavior." That is not at all what their data show. The fact that they cannot rule out 0 does not prove that 0 is the correct value. Their 95 percent confidence interval does include 0, but it also includes everything from

−3 percent to +17 percent. Their best estimate is 7 percent, plus or minus 10 percent, and 7 percent is more than enough to swing most elections one way or another.

Here is another poker example. Our second research question is whether experienced poker players tend to play differently after big losses than they do after big wins. To answer this research question, we can compare each player's looseness coefficient in the 12 hands following big losses with his or her looseness coefficient in the 12 hands following big wins. This difference measures whether this person plays less cautiously after a big loss than after a big win. The natural null hypothesis is that the population mean is 0, H_0: $\mu = 0$. For our sample of 203 players, the mean is 2.0996 and the standard deviation is 5.5000. Using Equation 7.3, the t value is 5.439:

$$\begin{aligned} t &= \frac{\overline{X} - \mu}{s/\sqrt{n}} \\ &= \frac{2.0996 - 0}{5.5000/\sqrt{203}} \\ &= 5.439 \end{aligned}$$

With 203 − 1 = 202 degrees of freedom, the two-sided P value is 0.0000002.

Using Equation 7.7, the t value for a 95 percent confidence interval for the population mean is

$$\overline{X} \pm t^* \frac{s}{\sqrt{n}} = 2.0996 \pm 1.9716 \frac{5.5000}{\sqrt{203}} = 2.10 \pm 0.76$$

Poker players tend to play looser (less cautiously) after large losses, evidently attempting to recoup their losses.

7.5 Matched-Pair Data

It is sometimes thought that women pay more than men for comparable items—like simple haircuts, T-shirts, and jeans. To test this theory, data were gathered from 20 stores that do men's and women's haircuts. In each case, the adult had short hair and asked for the least expensive haircut. The prices are shown in Table 7.1.

One way to analyze such data is to consider the men's and women's data to be two independent samples and apply a difference-in-means test, which is covered in more encyclopedic textbooks. However, these data are not independent samples and we can use a more powerful test. These data are *matched pairs*, in that each pair of men's and women's prices is obtained from a single store.

If the data had been independent random samples, a statistical test would have had to take into account the possibility that, by the luck of the draw, the women went to stores that were, on average, more expensive (or less expensive) than the stores that the men visited.

Table 7.1: Prices of Inexpensive Haircuts in 20 Stores, in Dollars

	Women	Men	Difference (*X*)
	30	20	10
	15	15	0
	35	25	10
	45	45	0
	10	6	4
	25	20	5
	9	7	2
	55	40	15
	30	25	5
	15	12	3
	25	20	5
	50	45	5
	15	15	0
	30	20	10
	35	25	10
	15	15	0
	15	12	3
	45	40	5
	30	30	0
	35	25	10
Mean	28.20	23.10	5.10
Standard deviation	13.575	11.756	4.40

Therefore, the women's prices might have been higher, on average, not because women are charged more than men but because the women in this sample happened to go to more expensive stores.

Matched-pair data take care of that possibility in a simple but powerful way, by having the men and women go to the same stores. For a matched-pair test, we look at the difference between the matched prices, as given in the last column in Table 7.1. The natural null hypothesis is that the population mean is 0:

$$H_0\colon \mu = 0$$

For the data in Table 7.1, the t value is 5.184:

$$\begin{aligned} t &= \frac{\overline{X} - 0}{s/\sqrt{n}} \\ &= \frac{5.10}{4.400/\sqrt{20}} \\ &= 5.184 \end{aligned}$$

Therefore, the observed average difference of \$5.10 is more than five standard deviations from 0 and the two-sided P value is 0.00006, which is overwhelming statistical evidence against the null hypothesis. This observed difference is also substantial in that the average women's price is 22 percent higher than the average men's price ($5.10/$23.10 = 0.22).

Our poker study also used matched-pair data in that we compared each player's looseness coefficient after a big win and after a big loss. We will apply these principles to several other examples, which illustrate not only the practice of hypothesis testing but also the value of matched-pair data.

Immigrant Mothers and Their Adult Daughters

A study compared the economic status of 4,121 women who had immigrated to California from another country with the economic status of their adult daughters at comparable periods of their lives [3]. Economic status was measured by the median income in the ZIP codes in which the mothers and adult daughters lived, with ZIP-code income converted to percentiles to account for changes over time in income levels and in the number of ZIP codes. Thus, if the mother lived in a ZIP code that was in the 30th percentile and her adult daughter lived in a ZIP code that was in the 35th percentile, this was characterized as an increase in economic status.

These are matched-pair data because each daughter is matched with her mother and we calculate the difference between the daughter's ZIP-code percentile and her mother's ZIP-code percentile. This difference had a mean of 3.3378 and a standard deviation of 25.8856. The t value for testing the null hypothesis that the population mean is 0 is

$$\begin{aligned} t &= \frac{3.3378 - 0}{25.8856/\sqrt{4{,}121}} \\ &= 8.2775 \end{aligned}$$

and the two-sided P value is 1.68×10^{-16}. This observed difference is statistically persuasive and also substantial, representing a 3.34 percentage-point increase in economic status.

Fortune's *Most Admired Companies*

Each year, *Fortune* magazine compiles a list of America's most admired companies, with the ten most admired companies singled out for special praise. Do the stocks of these Top 10 companies do better than the average stock? To answer this question, a stock portfolio consisting of the Top 10 companies was compared to the overall U.S. stock market [4]. Each year, after the publication of the Top 10 list, the *Fortune* portfolio was adjusted to have an equal investment in each of the Top 10 stocks for that year.

There were 5,547 daily observations for the *Fortune* portfolio and for the S&P 500. These are matched-pair data in that each daily percentage return for the *Fortune* portfolio is matched with that day's percentage return on the market portfolio. We can consequently use the 5,547 differences between the daily returns to test the null hypothesis that the expected value of the difference is equal to 0.

The difference between the average daily percentage return on the *Fortune* portfolio and the market portfolio had a mean of 0.0213 and a standard deviation of 0.5832. The t value for testing the null hypothesis that the population mean is 0 is

$$t = \frac{0.0213}{0.5832/\sqrt{5547}}$$
$$= 2.71$$

and the two-sided P value is 0.0067. This observed difference is substantial. The average daily return for the *Fortune* portfolio was 0.0651 and the average daily return for the market portfolio was 0.0439. Over 250 trading days each year, these daily returns imply respective annual returns of 17.7 and 13.0.

Fateful Initials?

It has been claimed that people whose names have positive initials (such as ACE or VIP) live much longer than people with negative initials (such as PIG or DIE). This theory was tested using California mortality data [5]. The deceased were grouped by birth year, and for each birth year, the researchers calculated the difference between the average age at death for people with positive initials and people with negative initials. For males, there were 91 birth years and the 91 differences had a mean of –0.8254 with a standard deviation of 4.2874. The t value is -1.8365 and the two-sided P value is 0.0696:

$$t = \frac{-0.8254 - 0}{4.2874/\sqrt{91}}$$
$$= -1.84$$

On average, males with positive initials did not live quite as long as those with negative initials, though the observed difference is not statistically significant at the 5 percent level. For females, the sign is reversed but the observed differences are again not statistically significant.

7.6 Practical Importance versus Statistical Significance

It is easy to be confused by the distinction between statistical significance and practical importance. A statistically significant result may be of little practical importance. Conversely, a potentially important result may not be statistically significant. To illustrate what may seem paradoxical, consider our poker example and a test of the null hypothesis that the population mean of the looseness coefficient is 35.

Suppose the sample data look like this:

Sample size	10,000
Sample mean	35.05
Sample standard deviation	1.00

The t value is 5.0:

$$t = \frac{\overline{X} - \mu}{s/\sqrt{n}} = \frac{35.05 - 35}{1/\sqrt{10{,}000}} = 5.0$$

The two-sided P value is 0.0000006. Because of the small sample standard deviation and the large sample size, the sample mean of 35.05 is five standard deviations from 35. The evidence is overwhelming that the population mean is not *exactly* equal to 35, but the practical difference between 35 and 35.05 is inconsequential.

Now suppose the sample data look like this:

Sample size	9
Sample mean	17.50
Sample standard deviation	30.00

The t value is only -1.75:

$$t = \frac{\overline{X} - \mu}{s/\sqrt{n}} = \frac{17.5 - 35}{30/\sqrt{9}} = -1.75$$

The two-sided P value is 0.12. Because of the large sample standard deviation and the small sample size, these data are not statistically persuasive evidence against the null hypothesis, even though the sample mean is far from the null hypothesis. These results do not prove that the population mean is equal to 35. Instead, they suggest that the average looseness coefficient is much lower than that. We should collect additional data to see if this result holds up with a larger sample.

These examples are deliberate caricatures that are intended to demonstrate that P values measure only statistical significance, not practical importance. In the first case, there was statistical significance but no practical importance. In the second case, there was practical importance but no statistical significance. When you do research, you should report the P value and use common sense to judge the practical importance of your conclusions.

7.7 Data Grubbing

Academic journals are filled with statistical tests. The most useful and controversial tests find their way into the popular press. To be intelligent consumers or producers of such tests, so that we can separate useful information from misleading rubbish, we need to be able to recognize ways in which hypothesis tests can be misleading. Chapter 5 already alerted us to some of the pitfalls in collecting sample data, and "garbage in, garbage out" is a shorthand reminder of the fact that a beautifully executed statistical test can be ruined by biased data.

A selective reporting of results is also a serious misuse of hypothesis tests. Back when calculations were done by hand, researchers tended to think carefully about their theories *before* they did a statistical test. Today, with mountains of data, powerful computers, and incredible pressure to produce statistically significant results, untold numbers of worthless theories get tested. The harried researcher tries hundreds theories, writes up the result with the lowest P value, and forgets about the rest. Or hundreds of ambitious graduate students test different theories; some get statistically significant results, theses, and research grants. The problem for science and society is that we see only the tip of this statistical iceberg. We see the results that are statistically significant but not the hundreds of other tests that did not work out. If we knew that the published results were among hundreds of tests that were conducted, we would be much less impressed.

Three psychiatrists once set out to identify observed characteristics that distinguish schizophrenic persons from the nonschizophrenic [6]. They considered 77 characteristics, and 2 of their 77 tests turned up statistically significant differences at the 5 percent level. They emphasized the statistical significance of these two characteristics, overlooking the sobering thought that, by chance alone, about 4 out of every 77 independent tests should be statistically significant at the 5 percent level.

Not finding a statistically significant result can sometimes be as interesting as finding one. In 1887, Albert Michelson and Edward Morley conducted a famous experiment in which they measured the speed of light both parallel and perpendicular to the earth's motion, expecting to find a difference that would confirm a theory that was popular at that time. They did not find a statistically significant difference. Instead of throwing their study in the trash and looking elsewhere for something statistically significant, they reported their not statistically significant result. Their research laid the groundwork for Einstein's special theory of relativity and the development of quantum mechanics by others. Their "failed" study revolutionized physics.

In general, it is worth knowing that a well-designed study did not find a statistically significant result. However, a study [7] of psychology journals found that, of 294 articles using statistical tests, only 8 (3 percent) reported results that were not statistically significant at the 5 percent level [7]. Surely, far more than 3 percent of the statistical tests that were done had P values larger than 0.05. But researchers and editors generally believe that a test is not worth reporting if the P value is not less than 0.05.

This belief that only statistically significant results are worth publishing leads researchers to test many theories until the desired significance level is attained—a process known as *data grubbing* (or *fishing expeditions*). The eager researcher takes a body of data, tests countless theories, and reports the one with the lowest P value. The end result—the discovery of a test with a low P value—shows little more than the researcher's energy. We know beforehand that, even if only worthless theories are examined, a researcher will eventually stumble on a worthless theory with a low P value. On average, 1 out of every 20 tests of worthless theories will have a P value less than 0.05. We cannot tell whether a data grubbing expedition demonstrates the veracity of a useful theory or the perseverance of a determined researcher.

The Sherlock Holmes Inference

Hypothesis tests assume that the researcher puts forward a theory, gathers data to test this theory, and reports the results—whether statistically significant or not. However, some people work in the other direction: They study the data until they stumble upon a theory that is statistically significant. This is the Sherlock Holmes approach: examine the data and use inductive reasoning to identify theories that explain the data. As Sherlock Holmes put it: "It is a capital mistake to theorize before you have all the evidence" [8]. The Sherlock Holmes method can be very useful for solving crimes. Many important scientific theories, including Mendel's genetic theory, have been discovered by identifying theories that explain data. But the Sherlock Holmes approach has also been the source of thousands of quack theories.

How do we tell the difference between a good theory and quackery? There are two effective antidotes to unrestrained data grubbing: common sense and fresh data. Unfortunately, there are limited supplies of both. The first thing that we should think about when confronted with statistical evidence in support of a theory is, "Does it make sense?" If it is a ridiculous theory, we should not be persuaded by anything less than mountains of evidence, and even then be skeptical.

Many situations are not clear-cut, but we should still exercise common sense in evaluating statistical evidence. Unfortunately, common sense is an uncommon commodity, and many silly theories have been seriously tested by honest researchers. I include in this category the idea that people whose names have positive initials live years longer than people with negative initials. I will show you some other examples shortly.

The second antidote is fresh data. It is not a fair test of a theory to use the very data that were ransacked to discover the theory. For an impartial test, we should specify the theory before we see the data used to test the theory. If data have been used to concoct a theory, then look for fresh data that have not been contaminated by data grubbing to either confirm or contradict a suspect theory.

In chemistry, physics, and the other natural sciences, this is standard procedure. Whenever someone announces an important new theory, a dozen people rush to their laboratories to see if they can replicate the results. Unfortunately, most social scientists cannot use experiments to produce fresh data. Instead, they must record data as events occur. If they have a theory about presidential elections, the economy, or world peace, they may have to wait years or even decades to accumulate enough fresh data to test their theories (thus, the joke that, for some social scientists, *data* is the plural of *anecdote*).

When further testing is done with fresh data, the results are often disappointing, in that the original theory does not explain the new data nearly as well as it explained the data that were the inspiration for the theory. The usual excuse is that important changes necessitate modification of the original model. By mining the fresh data for a new theory, a provocative first paper can turn into a career. An alternative explanation of the disappointing results with fresh data is that, when a theory is uncovered by data grubbing, statistical tests based on these data typically exaggerate the success of the model. In the remainder of this chapter, we look at some theories that are statistically significant but unpersuasive.

The Super Bowl and the Stock Market

On Super Bowl Sunday in January 1983, both the business and sports sections of the *Los Angeles Times* carried articles on the Super Bowl Indicator [9]. The theory is that the stock market goes up if the National Football Conference (NFC) or a former NFL team now in the American Football Conference (AFC) wins the Super Bowl; the market goes down if the AFC wins. This theory had been correct for 15 out of 16 Super Bowls, and a stockbroker said "Market observers will be glued to their TV screens ... it will be hard to ignore an S&P indicator with an accuracy quotient that's greater than 94%." And, indeed, it was. An NFC team won, the market went up, and the Super Bowl system was back in news the next year, stronger than ever.

The accuracy of the Super Bowl Indicator is obviously just an amusing coincidence, since the stock market has nothing to do with the outcome of a football game. The Indicator exploits the fact that the stock market generally goes up and the NFC usually wins the Super Bowl. The correlation is made more impressive by the gimmick of including the Pittsburgh Steelers, an AFC team, in with the NFL. The excuse is that Pittsburgh once was in the NFL; the real reason is that Pittsburgh won the Super Bowl four times in years when the stock market went up.

The performance of the Super Bowl Indicator has been mediocre since its discovery—since there was nothing behind it but coincidence. What is genuinely surprising is that some people do not get the joke. The man who originated the Super Bowl Indicator intended it to be a humorous way of demonstrating that correlation does not imply causation [10]. He was flabbergasted when people started taking it seriously!

I once had a similar experience. Technical analysts try to predict whether stock prices are going up or down by looking for patterns in time series graphs of stock prices. I sent fake stock-price charts created from coin flips to a technical analyst at a large investment firm and asked him if he could find any patterns. My intention was to convince him that there are coincidental patterns even in random coin flips. Sure enough, he identified some "stocks" he wanted to buy. When I revealed that these were not real stocks, his response was that he had demonstrated that technical analysis could be used to predict coin flips!

What these examples really demonstrate is that some people have a hard time understanding that data grubbing will inevitably uncover statistical relationships that are nothing more than coincidence.

Extrasensory Perception

Professor J. B. Rhine and his associates at Duke University produced an enormous amount of evidence concerning extrasensory perception (ESP), including several million observations involving persons attempting to identify cards viewed by another person.

Even if ESP does not exist, about 1 out of every 20 people tested will make enough correct guesses to be statistically significant at the 5 percent level. With millions of tests, some incredible results are inevitable. This is a form of data grubbing—generate a large number of samples and report only the most remarkable results. Something with a one in a thousand chance of occurring is not really so remarkable if a thousand tests were conducted.

Our first antidote to data grubbing is common sense. Many people believe in ESP, but others do not. Perhaps the safest thing to say is that if ESP does exist, it does not seem to be of much practical importance. There is no public evidence that people can have long distance conversations without a telephone or win consistently in Las Vegas.

The second antidote to data grubbing is fresh data. People who score well should be retested to see whether their success was due to ESP or data grubbing. However, even if high scores are just luck, if enough people are tested and retested, some are bound to get lucky on more than one test—though most will not. Rhine in fact observed that high scorers who are retested almost always do worse than they did initially. His explanation is that "This fatigue is understandable... in view of the loss of original curiosity and initial enthusiasm" [11]. An alternative explanation is that the high scores in the early rounds were just lucky guesses.

The large number of persons tested is not the only difficulty in assessing the statistical significance of Rhine's results. He also looked for parts of a test where a subject did well, which is again data grubbing. A remarkable performance by 1 of 20 subjects or by one subject on 1/20th of a test is not really surprising. Rhine also looked for either "forward displacement" or "backward displacement," where a person's choices did not match the contemporaneous cards, but did match the next card, the previous card, two cards hence, or two cards previous. This multiplicity of potential matches increases the chances of finding coincidental matches. Rhine also considered it remarkable when a person got an unusually low score ("negative ESP"). He noted that a person

> *may begin his run on one side of the mean and swing to the other side just as far by the time he ends a run; or he may go below in the middle of the run and above it at both ends. The two trends of deviations may cancel each other out and the series as a whole average close to "chance" [12].*

With so many subjects and so many possibilities, it should be easy to find patterns, even in random guesses.

ESP research is interesting and provocative. But the understandable enthusiasm of the researchers can lead to a variety of forms of data grubbing that undermine their tests. They would be less controversial and more convincing—one way or the other—if there were more uses and fewer abuses of statistical tests.

How (Not) to Win the Lottery

Multimillion dollar Lotto jackpots offer people a chance to change their lives completely—and the popularity of Lotto games indicates that many are eager to do so. The only catch is that the chances of winning are minuscule. This chasm between dreams and reality gives entrepreneurs an opportunity to sell gimmicks that purportedly enhance one's chances of winning. One mail order company sold a Millionaire Maker for $19.95 using this pitch: "Lotto players face a dilemma each time they buy a ticket. What numbers to pick? Studies have shown that most Lotto *winners* don't use any sort of special system to select their numbers. Instead, they tap the power of *random selection*" [13]. Their $19.95 product is a battery powered sphere filled with numbered balls that "mixes the balls thoroughly" and chooses a "perfectly random set of numbers."

Gail Howard sold a very different system, a report on how to handicap lottery numbers, just like handicapping horses [14]. She reportedly appeared on the *Good Morning America* television show, wrote an article in *Family Circle* magazine, and published a monthly report (*Lottery Buster*). She says "You don't have to worry about the state lottery games being crooked or rigged… the winning numbers are selected through a completely random process." Nonetheless, she offered several tips for "greatly improving your chances of winning." For instance, do not bet on six consecutive numbers or six numbers that have won before, because these are highly unlikely to win. Well, yes, any six numbers are unlikely to win.

Just as a careful study can uncover worthless patterns in coin flips or dice rolls, so tedious scrutiny can discern worthless patterns in lottery numbers. The most persuasive reason for skepticism about lottery systems is that anyone who really had a system that worked would become rich buying lottery tickets rather than peddling books and battery powered gadgets.

Exercises

7.1 What would be a natural null hypothesis for the research question of whether or not the average U.S. passenger car gets more than 30 miles per gallon? What alternative hypothesis would you use?

7.2 What would be a natural null hypothesis for the research question of whether or not average home prices fell last year? What alternative hypothesis would you use?

7.3 Researchers used a random sample to test the null hypothesis that the average age of people visiting an Internet site is 25. Which of the following results has the higher t value for a test of this null hypothesis? Do not do any calculations; just explain your reasoning.
a. $\overline{X} = 26.43, s = 7.17, n = 50$
b. $\overline{X} = 32.14, s = 7.17, n = 50$

7.4 Which of the following results has the higher t value for a test of the null hypothesis that the average speed on a 40 miles-per-hour street is 40 miles per hour? Do not do any calculations; just explain your reasoning.
a. $\overline{X} = 43.71, s = 14.88, n = 60$
b. $\overline{X} = 43.71, s = 8.06, n = 60$

7.5 Which of the following results has the higher t value for a test of the null hypothesis that the average U.S. passenger car gets 30 miles per gallon? Do not do any calculations; just explain your reasoning.
a. $\overline{X} = 31.52, s = 8.33, n = 30$
b. $\overline{X} = 31.52, s = 8.33, n = 50$

7.6 Exercise 6.30 shows 50 body temperature readings for healthy persons using modern thermometers. Use these data to calculate the two-sided P value for a test of the null hypothesis that the population mean is 98.6° Fahrenheit.

7.7 Which of the following assumptions were used in calculating the P value in the preceding exercise?
a. Temperatures for healthy adults are normally distributed.
b. These 50 observations are a random sample.
c. The mean temperature for unhealthy adults is higher than the mean temperature for healthy adults.
d. On average, healthy adults have a temperature of 98.6° Fahrenheit.

7.8 Explain why you either agree or disagree with this reasoning: "I used a random sample to test the null hypothesis that the average body temperature of healthy adults is 98.6. I calculated a one-sided P value because the sample mean was less than 98.6."

7.9 Use the reading comprehension data in Exercise 6.18 to test the null hypothesis that the population mean of X is 0.
a. What is the two-sided P value?
b. Does the average value of X seem substantially different from 0?

7.10 Redo the preceding exercise, this time using the mathematics scores.

7.11 The data for deceased Major League Baseball (MLB) players in Table 7.2 were used [15] to investigate the research hypothesis that players with positive initials (like ACE) live longer, on average, than players with neutral initials (like GHR). For each birth year, the average age at death (AAD) was calculated for players with positive initials and for players with neutral initials (the control group). Calculate the difference between the positive and control AADs for each birth year and use these seven differences to test the null hypothesis that this population mean difference is equal to 0. (We implicitly assume that these data are a random sample.)
a. What is the two-sided P value?
b. Does the average value of the difference seem substantially different from 0?

7.12 Analogous to the preceding exercise, the data in Table 7.3 for deceased MLB players were used to investigate the research hypothesis that players with negative initials

Table 7.2: Exercise 7.11

Birth Year	Positive AAD	Control AAD
1853	75.00	65.09
1876	77.00	62.12
1884	55.00	68.09
1885	73.00	69.39
1897	80.50	70.95
1903	80.00	71.60
1913	93.00	73.68

Table 7.3: Exercise 7.12

Birth Year	Negative AAD	Control AAD
1858	57.00	62.94
1864	70.00	61.38
1874	85.00	65.28
1881	93.00	67.28
1886	57.00	70.61
1893	57.50	68.36
1916	41.00	72.53
1918	85.00	72.21

(like PIG) do not live as long, on average, as players with neutral initials (like GHR). Calculate the difference between the negative and control AADs for each birth year and use these eight differences to test the null hypothesis that the population mean difference is equal to zero.

a. What is the two-sided P value?

b. Does the average value of the difference seem substantially different from 0?

7.13 Critically evaluate this news article [16]:

BRIDGEPORT, Conn.—Christina and Timothy Heald beat "incredible" odds yesterday by having their third Independence Day baby.

Mrs. Heald, 31, of Milford delivered a healthy 8-pound 3-ounce boy at 11:09 a.m. in Park City Hospital where her first two children, Jennifer and Brian, were born on Independence Day in 1978 and 1980, respectively.

Mrs. Heald's mother, Eva Cassidy of Trumbull, said a neighbor who is an accountant figures the odds are 1-in-484 million against one couple having three children born on the same date in three different years.

7.14 A 1981 retrospective study published in the prestigious *New England Journal of Medicine* looked at the use of cigars, pipes, cigarettes, alcohol, tea, and coffee by patients with pancreatic cancer and concluded that there was "a strong association between coffee consumption and pancreatic cancer" [17]. This study was immediately criticized on several grounds, including the conduct of a fishing expedition in which multiple tests were conducted and the most statistically significant result reported. Subsequent studies failed to confirm an association between coffee drinking and pancreatic cancer [18]. Suppose that six independent tests are conducted, in each case involving a product that is, in fact, unrelated to pancreatic cancer. What is the probability that at least one of these tests will find an association that is statistically significant at the 5 percent level?

7.15 Researchers once reported that 1969–1990 California mortality data show that Chinese-Americans are particularly vulnerable to four diseases that Chinese astrology and traditional Chinese medicine associate with their birth years [19]. The two-sided P values were 0.022, 0.052, 0.090, and 0.010. However, a subsequent study [20] using data for 1960–1968 and 1991–2002 found the two-sided P values to be 0.573, 0.870, 0.235, and 0.981. How would you explain this discrepancy?

7.16 A researcher asked 100 college students, "How many hours did you sleep during the past 24 hours?" Identify several errors in this description of the statistical test: "The following equation is the basis for the Z value I obtained:

$$Z = \frac{x - \mu}{\sigma}$$

The P value is the area under the curve outside of two standard deviations of the mean."

7.17 A study of MLB players who were born between 1875 and 1930 reported that players whose first names began with the letter *D* died an average of 1.7 years younger than MLB players whose first names began with the letters *E* through *Z*, suggesting that "Ds die young" [21]. A subsequent study calculated the paired difference in the average age at death for each birth year [22]. For example, for MLB players born in 1900, the average age at death was 63.78 years for those with *D* first initials and 71.04 years for those with *E–Z* first initials, a difference of $X = -7.26$ years. This follow-up study looked at all MLB players, not just players born between 1875 and 1930. Overall, there were 51 birth years with at least five MLB players in each category and the 51 values of X had a mean of −0.57 and a standard deviation of 6.36. Calculate the two-sided P value for a test of the null hypothesis that the population mean is 0. How would you explain the fact that the two-sided P value was lower in the original study than in the subsequent study?

7.18 The follow-up study described in the previous exercise also looked at MLB players whose first or last initial was *D*. For MLB players born between 1875 and 1930, there were 53 birth years with the values of X having a mean of −0.81 and a standard deviation of 4.87. For all players, there were 73 birth years with the values of X having a mean of −0.26 and a standard deviation of 5.42. For each of these two data sets, calculate the two-sided P value for a test of the null hypothesis that the population mean is 0.

7.19 Medical researchers compared the measured calories in eight locally prepared health and diet foods purchased in Manhattan, New York, with the calories listed on the label [23]. The data in Table 7.4 are the percentage differences between the actual and labeled calories (a positive value means that the measured calories exceeded the amount shown on the product label). Determine the two-sided P value for a test of the null hypothesis that the population mean is 0.

7.20 The researchers in Exercise 7.19 also compared the actual and listed calories in 20 nationally advertised health and diet foods purchased in Manhattan (see Table 7.5). Determine the two-sided P value for a test of the null hypothesis that the population mean is 0.

7.21 Use the data in Exercise 6.24 to determine the two-sided P value for a test of the null hypothesis that the population mean of the differences between the daughter's income and her mother's income is equal to 0. Does the average difference seem substantial?

Table 7.4: Exercise 7.19

Chinese chicken	15	Florentine manicotti	6
Gyoza	60	Egg foo young	80
Jelly diet candy—red	250	Hummus with salad	95
Jelly diet candy—fruit	145	Baba ghanoush with salad	3

Table 7.5: Exercise 7.20

Noodles Alfredo	2	Imperial chicken	−4
Cheese curls	−28	Vegetable soup	−18
Green beans	−6	Cheese	10
Mixed fruits	8	Chocolate pudding	5
Cereal	6	Sausage biscuit	3
Fig bars	−1	Lasagna	−7
Oatmeal raisin cookie	10	Spread cheese	3
Crumb cake	13	Lentil soup	−0.5
Crackers	15	Pasta with shrimp	−10
Blue cheese dressing	−4	Chocolate mousse	6

7.22 In a letter to the prestigious *New England Journal of Medicine*, a doctor reported that 20 of his male patients with creases in their earlobes had many of the risk factors (such as high cholesterol levels, high blood pressure, and heavy cigarette usage) associated with heart disease [24]. For instance, the average cholesterol level for his patients with noticeable earlobe creases was 257 (mg per 100 mL), compared to an average of 215 with a standard deviation of 10 for healthy middle-aged men. If these 20 patients were a random sample from a population with a mean of 215 and a standard deviation of 10, what is the probability their average cholesterol level would be 257 or higher? Explain why these 20 patients may not be a random sample.

7.23 A researcher collected data on the number of people admitted to a mental health clinic's emergency room on 12 days when the moon was full [25]:

5	13	14	12	6	9	13	16	25	13	14	20

a. Calculate the two-sided P value for a test of the null hypothesis that these data are a random sample from a normal distribution with a population mean equal to 11.2, the average number of admissions on other days.

b. Explain why you either agree or disagree with this reasoning: "Because the sample mean was larger than 11.2, I calculated a one-sided P value."

7.24 The study of immigrant women and their adult daughters mentioned in the text also separated the data into 580 mother-daughter pairs where the adult daughter lived in the same ZIP code she was born in and 3,541 pairs where the daughters moved to a different ZIP code. For the 580 daughters who stayed, the difference in ZIP-code percentile had a mean of −2.5775 and a standard deviation of 8.1158. For the 3,541 daughters who moved, the difference in ZIP-code percentile had a mean of 4.3066 and a standard deviation of 27.6117. For each of these two data sets, calculate the two-sided P value for testing the null hypothesis that the mean is 0. Compare your results.

7.25 A builder contracted for the delivery of 4 million pounds of loam. The supplier delivered the loam in 1,000 truckloads that he claims average 4,000 pounds of loam. The builder weighed a random sample of 16 truckloads and obtained these data:

3,820	3,900	4,000	3,820	3,780	3,880	3,700	3,620
4,380	3,560	3,840	3,980	4,040	4,140	3,780	3,780

Using a one-tailed test, do these data reject the supplier's claim at the 5 percent level?

7.26 Chapter 6 discussed a test of a new enzyme in which a sample of 41 observations found an average yield of 125.2 with a standard deviation of 20.1. A member of the research department was unimpressed because the standard deviation of 20.1 was large relative to the difference between 125.2 and 100 (which would be the yield if the new process was ineffective). Use these data to test the null hypothesis that average yield is $\mu = 100$.

7.27 Use the data in Exercise 6.21 to calculate the two-sided P value for a test of the null hypothesis that the population mean is 25.53.

7.28 Use the data in Exercise 6.22 to calculate the two-sided P value for a test of the null hypothesis that the population mean is 37,408.

7.29 Use the data in Exercise 6.23 to calculate the two-sided P value for a test of the null hypothesis that the population mean is 40,064—which is, in fact, the actual value of the population mean. If this experiment were to be repeated a large number of times, what percentage of the time would you expect the two-sided P value to be less than 0.05?

7.30 Use the data in Exercise 6.26 to calculate the two-sided P value for a test of the null hypothesis that the population mean is 8.

7.31 Use the data in Exercise 6.27 to calculate the two-sided P value for a test of the null hypothesis that the population mean is 299,710.50.

7.32 While playing bridge recently, Ira Rate picked up his 13-card hand and found that it contained no spades. He promptly threw his cards down and declared that the deal had not been random. He read in a statistics textbook that (a) the probability of being dealt a hand with no spades is less than 5 percent and (b) this low a probability is sufficient evidence to show that the deal was not fair. Point out a flaw or two in his reasoning.

7.33 The following letter to the editor was printed in *Sports Illustrated* [26]:

In regard to the observation made by Sports Illustrated *correspondent Ted O'Leary concerning this year's bowl games (Scorecard, January 16), I believe I can offer an even greater constant. On Jan. 2 the team that won in all five games started the game going right-to-left on the television screen. I should know, I lost a total of 10 bets to my parents because I had left-to-right in our family wagers this year.*

If right to left makes no difference, the chances of losing all five bets is 0.0312. Are you convinced, then, that right to left does matter? What other explanation can you provide?

7.34 "As January goes, so goes the year" is an old stock market adage. The idea is that if the market goes up from January 1 to January 31, then it will go up for the entire year, January 1 to December 31. If the market drops in January, then it will be a bad year. *Newsweek* cited this theory and reported that "this rule of thumb has proved correct in 20 of the past 24 years" [27]. Explain why you are skeptical about this evidence.

7.35 In 1987, a brokerage firm observed that "since 1965, the Dow Jones Industrial Average declined 16 out of 21 times during the month of May. As of this writing, it appears as if another down May is in the offing. We offer no rational explanation for this 'May phenomenon.' We only observe that it often sets up attractive trading opportunities" [28]. Does this evidence persuade you that people should sell their stocks next April? Why or why not?

7.36 When she was in junior high school, Elizabeth Likens started listening to classical music after she was told that "studies show that students who listen to classical music get better grades." What would you advise?

7.37 Crime in downtown Houston increased by 40 percent during the first month after the Houston police launched an anticrime campaign, leading *Texas Monthly* to quip that "They should have launched a pro-crime campaign" [29]. What other logical explanation is there for these data?

7.38 Two Dutch researchers found that nondrinkers "are 50% more likely to get hurt on the [ski] slopes than skiers who imbibe heavily, and that people who sleep long and deeply are more prone to injury than those who are late-to-bed and early-to-rise" [30]. Provide an explanation other than that the route to safer skiing is to drink lots and sleep little?

7.39 Explain why you either agree or disagree with each of the following statements:

a. A test that is significant at the 1% level is also significant at the 5% level.

b. If the t value is 0, the P value is 0.

c. If the null hypothesis is $\mu = 0$, you can use a one-sided P value once you know whether your estimate is positive or negative.

7.40 A random sample of 800 first-year students at a large university included 558 students who had attended public high schools and 242 who had attended private schools. The average grade point average (GPA) was 3.52 for students from public schools and 3.26 for students from private schools. Does the observed difference seem substantial? The author calculated the P value to be 0.10 and concluded that although "this study proved the null hypothesis to be true, ... the sample size is minuscule in comparison to the population." Did she prove the null hypothesis to be true? Are her results invalidated by the small sample size?

7.41 A test of the null hypothesis that the average college student gains 15 pounds during the first year at college surveyed a random sample of 100 students and obtained a sample mean of 4.82 pounds with a standard deviation of 5.96 pounds. Explain why you either agree or disagree with each of these conclusions:
  a. "We assumed that the standard deviation of our sample equaled the population standard deviation."
  b. "We calculated the t valuc to be -17.08. The two-sided P value is 2.82×10^{-31}. According to Fisher's rule of thumb, our data are not statistically significant because our P value is considerably less than 0.05. This indicates that the probability of observing a large difference between the sample mean and the null hypothesis is greater than 0.05."
  c. "Our data strongly indicate that the Freshman 15 is just a myth. However, it must be recognized that we only took one sample of 100 students. Perhaps, if we took other samples, our results would be different."

7.42 Before 1996, the Environmental Protection Agency (EPA) concluded that secondhand smoke poses a cancer risk if a 95 percent confidence interval for the difference in the incidence of cancer in control and treatment groups excluded 0. In 1996, the EPA changed this rule to a 90 percent confidence interval [31]. If, in fact, secondhand smoke does not pose a cancer risk, does a switch from a 95 percent to a 90 percent confidence interval make it more or less likely that the EPA will conclude that secondhand smoke poses a cancer risk? Explain your reasoning.

7.43 A treatment group was given a cold vaccine, while the control group was given a placebo. Doctors then recorded the fraction of each group that caught a cold and calculated the two-sided P value to be 0.08. Explain why you either agree or disagree with each of these interpretations of these results:
  a. "There is an 8 percent probability that this cold vaccine works."
  b. "If a randomly selected person takes this vaccine, the chances of getting sick fall by about 8 percent."
  c. "These data do not show a statistically significant effect at the 5 percent level; therefore, we are 95 percent certain that this vaccine does not work."

7.44 A study of the relationship between socioeconomic status and juvenile delinquency tested 756 possible relationships and found 33 statistically significant at the 5 percent level [32]. What statistical reason is there for caution here?

7.45 In 1989, the *New York Times* reported that, if the Dow Jones Industrial Average increases between the end of November and the time of the Super Bowl, the football team whose city comes second alphabetically usually wins the Super Bowl [33]. How would you explain the success of this theory? Why did they choose the end of November for their starting date?

7.46 The results of hypothesis tests are often misinterpreted in the popular press. For instance, explain how this *Business Week* summary of a Data Resources Incorporated (DRI) study of insider trading in the stock market is misleading:

> *DRI also compared the action of a sample of takeover stocks in the month before the announcement with the action of those same stocks in other, less significant one-month intervals. The conclusion: There was only 1 chance in 20 that the strength of the takeover stocks was a fluke. In short, the odds are overwhelming that inside information is what made these stocks move [34].*

7.47 A *Wall Street Journal* editorial stated:

> *We wrote recently about a proposed Environmental Protection Agency ban on daminozide, a chemical used to improve the growth and shelf life of apples. EPA's independent scientific advisory panel last week ruled against the proposed prohibition. "None of the present studies are considered suitable" to warrant banning daminozide, the panel said. It also criticized the "technical soundness of the agency's quantitative risk assessments" for the chemical. EPA can still reject the panel's advice, but we see some evidence that the agency is taking such evaluations seriously. There is no better way to do that than to keep a firm grip on the scientific method, demanding plausible evidence when useful substances are claimed to be a danger to mankind [35].*

Does the *Journal*'s interpretation of the scientific method implicitly take the null hypothesis to be that daminozide is safe or that it is dangerous?

7.48 Researchers used a random sample to test the null hypothesis that the average change between 2000 and 2010 in the real hourly wages of construction workers was equal to 0. Their decision rule was to reject the null hypothesis if the two-sided P value is less than α. Use a rough sketch to answer the following questions. If the population mean is, in fact, equal to −\$0.75,

a. are they more likely to reject the null hypothesis if $\alpha = 0.05$ or if $\alpha = 0.01$?

b. are they more likely not to reject the null hypothesis if $\alpha = 0.05$ or if $\alpha = 0.01$?

7.49 Researchers used a random sample of size 100 to test the null hypothesis that the average difference between the household income of immigrant women and their adult daughters is equal to 0. Their decision rule was to reject the null hypothesis if the two-sided P value is less than α. Suppose that the population mean is, in fact, \$500 and the standard deviation is known to be \$5,000. Calculate the probability that they will

a. reject the null hypothesis if $\alpha = 0.05$. If $\alpha = 0.01$.

b. not reject the null hypothesis if $\alpha = 0.05$. If $\alpha = 0.01$.

7.50 In a 1982 racial discrimination lawsuit, the court accepted the defendant's argument that racial differences in hiring and promotion should be separated into eight job categories [36]. In hiring, it turned out that blacks were underrepresented by statistically significant amounts (at the 5 percent level) in four of the eight job categories. In the other four categories, whites were underrepresented in two cases and blacks were underrepresented in two cases, though the differences were not statistically significant at the 5 percent level. The court concluded that four of eight categories was not sufficient to establish a prima facie case of racial discrimination. Assume that the data for these eight job categories are independent random samples.

a. What is the null hypothcsis?
b. Explain why data that are divided into eight job categories might not show statistical significance in any of these job categories, even though there is statistical significance when the data are aggregated.
c. Explain why data that are divided into eight job categories might show statistical significance in each of the eight categories, even though there is no statistical significance when the data are aggregated.

CHAPTER 8

Simple Regression

Chapter Outline

> *"The cause of lightning," Alice said very decidedly, for she felt quite sure about this, "is the thunder—no, no!" she hastily corrected herself. "I meant it the other way."*
> *"It's too late to correct it," said the Red Queen. "When you've once said a thing, that fixes it, and you must take the consequences."*
>
> —**Lewis Carroll, Alice in Wonderland**

Regression models are used in the humanities, social sciences, and natural sciences to explain complex phenomena. Linguistic models use aptitude and motivation to explain language proficiency. Economic models use income and price to explain demand. Political models use incumbency and the state of the economy to explain presidential elections. Agricultural models use weather and soil nutrients to explain crop yields.

In this chapter, we analyze the simplest regression model, which has only one explanatory variable. We look at the assumptions that underlie this model and see how to estimate and use the model. Chapter 10 looks at more complex models that have more than one explanatory variable.

8.1 The Regression Model

In Chapter 1, we noted that one of the main components of Keynes' explanation of business cycles was a "consumption function" that relates household spending to income. When people lose their jobs and income, they cut back on spending—which jeopardizes other

Essential Statistics, Regression, and Econometrics

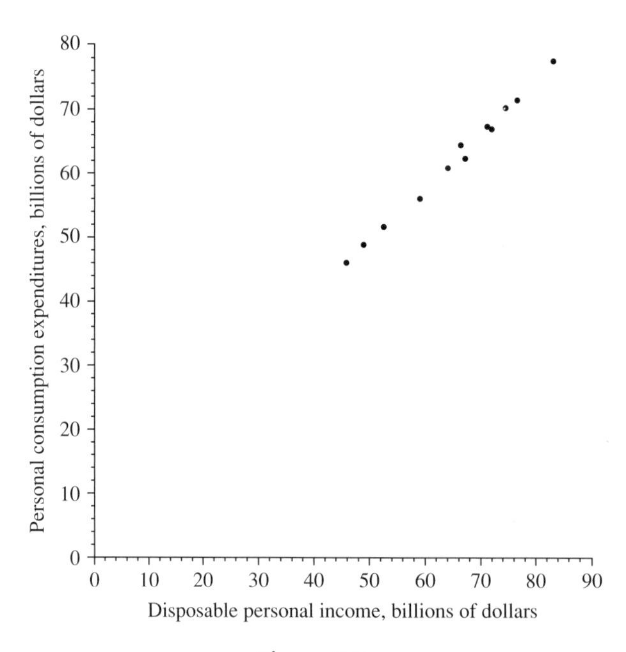

Figure 8.1
Household income and spending, 1929–1940.

peoples' jobs. Figure 8.1 is a scatterplot of the annual U.S. household income and spending data in Table 1.1.

There appears to be a positive statistical relationship, just as Keynes hypothesized. (Remember, he stated his theory before there were any data to test his theory.) To estimate this relationship, we need an explicit model. The linear (or straight-line) model is

$$Y = \alpha + \beta X$$

where Y is the dependent variable (spending), X is the explanatory variable (income), and α and β are two parameters. This model is shown in Figure 8.2.

The parameter α is the value of Y when $X = 0$. The parameter β is the slope of the line: the change in Y when X increases by 1. For instance, $\beta = 0.8$ means that a \$1.00 increase in household income causes spending to increase by \$0.80.

Regression models are often intended to depict a causal relationship, in that the researcher believes that changes in the explanatory variable cause changes in the dependent variable. The sign of the parameter β tells us whether this effect is positive or negative. In our example, a positive value for β means that an increase in income causes an increase in spending.

Sometimes, we use a regression model to investigate the statistical relationship between two variables without making any assumptions about causality. For example, we might look at

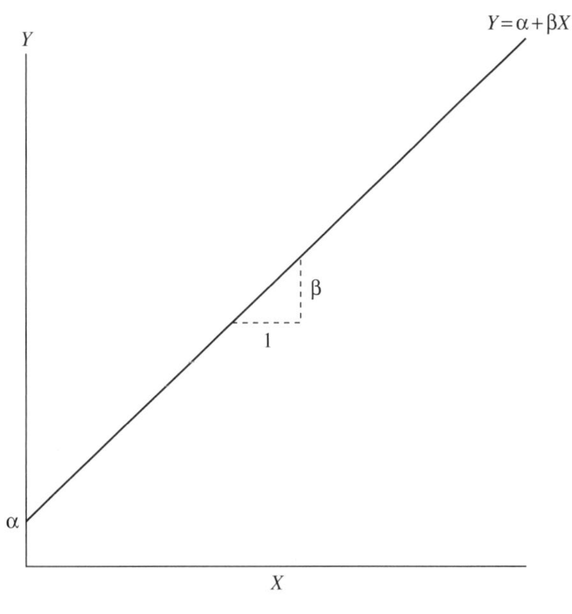

Figure 8.2
The linear model.

the statistical relationship between midterm and final examination scores. We do not think that midterm scores "cause" final examination scores. Instead, we believe that both scores are related to a student's knowledge of the subject, and we want to see how closely these test scores are correlated.

The data in Figure 8.1 do not lie exactly on a straight line, because household spending is affected by other things besides income (perhaps stock prices, interest rates, and inflation). In recognition of these other factors, the simple linear regression model includes an *error term* ε that encompasses the effects on Y of factors other than X:

$$Y = \alpha + \beta X + \varepsilon \tag{8.1}$$

Some of these other factors are intentionally neglected because we do not think they are important; some may be omitted because we simply lack any useful data for measuring them.

To reinforce the fact that the variables Y, X, and ε take on different values, while α and β are fixed parameters, the regression model can be written with i subscripts:

$$Y_i = \alpha + \beta X_i + \varepsilon_i$$

For notational simplicity, we typically omit these subscripts.

The omitted influences encompassed by the error term ε imply that, for any given value of X, the value of Y may be somewhat above or below the line $\alpha + \beta X$, depending on whether the value of ε is positive or negative. Because the error term represents the cumulative influence of a variety of omitted factors, the central limit theorem suggests that the values of the error term can be described by a normal distribution. In addition, we assume

1. The population mean of the error term is 0.
2. The standard deviation of the error term is constant.
3. The values of the error term are independent of each other.
4. The values of the error term are independent of X.

The first assumption is innocuous, since the value of the intercept is set so that the population mean of ε is 0. Because the expected value of the error term is 0, we can write the expected value of Y as

$$E[Y] = \alpha + \beta X$$

The second assumption is often reasonable, but sometimes violated, for example, if the size of the omitted influences depends on the size of the population. One way to guard against such effects is to scale our data, for example, by using per capita values.

The third assumption states that the values of the error term are independent of each other, in our consumption example, that the value of the error term in 1939 does not depend on the value of the error term in 1938. However, the error term involves omitted variables whose values may not be independent. The error term in a consumption function may include the rate of inflation, and 1939 inflation may not be independent of 1938 inflation. Advanced courses show how correlations among errors can be detected and how our statistical procedures can be modified to take this additional information into account. Violations of the second and third assumptions do not bias our estimates of α and β; however, the standard deviations of the estimates are unnecessarily large and underestimated, leading us to overestimate the statistical significance of our results.

The fourth assumption is probably the most crucial. It states that the error term is independent of X. Suppose that spending depends on both income and stock prices and that stock prices are positively correlated with income. If income is the explanatory variable and stock prices are part of the error term, then X and ε are correlated and the fourth assumption is violated. Here is why this is a problem. When income rises, stock prices tend to rise, too, and both factors increase spending. If we compare spending and income, with stock prices in the error term, we overestimate the effect of income on spending because we don't take into account the role of stock prices. The best way to deal with this problem is to use a multiple regression model with both income and stock prices as explanatory variables. In this chapter, we use the simple model to introduce the principles of regression analysis, and we assume that all four assumptions are correct.

The effects of the error term on the observed values of Y are depicted in Figure 8.3. For any value of X, the expected value of Y is given by the line $\alpha + \beta X$. If the value of ε happens to be positive, the observed value of Y is above the line; if the value of ε happens to be negative, the observed value of Y is below the line. In Figure 8.3, Y_1 and Y_3 are below the line, and Y_2

is above the line. Of course, we have no way of knowing which values of ε are positive and which are negative. All we see are the observed values of X and Y, as in Figure 8.4, and we use these observed values to estimate the slope and intercept of the unseen line. The next section does this.

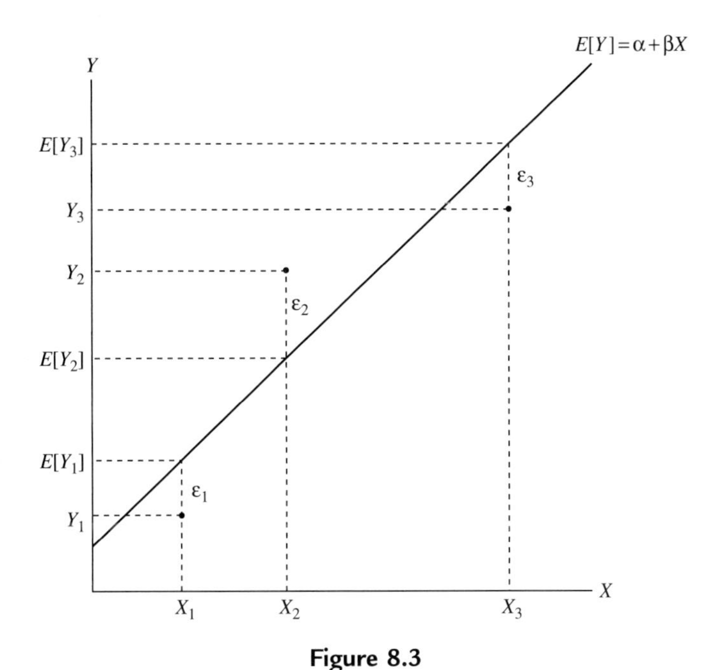

Figure 8.3
The error term causes Y to be above or below the line $E[Y] = \alpha + \beta X$.

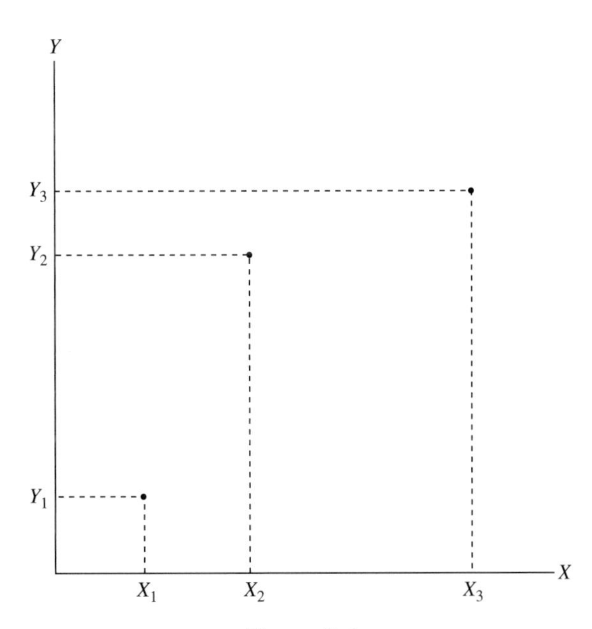

Figure 8.4
We see X and Y, but not the error term.

8.2 Least Squares Estimation

In 1765, J. H. Lambert, a Swiss-German mathematician and scientist, used scatter diagrams to show relationships between variables, for example, the relationship between the temperature of water and how rapidly it evaporates. He fit a line to these points simply by drawing a line that seemed to fit the points best: "It must therefore be drawn in such a way that it comes as near as possible to its true position and goes, as it were, through the middle of the given points" [1].

Two hundred years later, in the 1970s, the chief economic forecaster at one of the largest U.S. banks fit lines to scatter diagrams with a pencil and transparent ruler. Fitting lines by hand is hopelessly amateurish. It is often hard to know exactly where to draw the lines and even harder to identify the intercepts and calculate the slopes. In addition, we cannot fit lines by hand to multiple regression models with several explanatory variables. So, this bank forecaster had to work with models that had only one explanatory variable.

It is much better to have an explicit mathematical rule for determining the line that best fits the data. To do so, we need to specify what we mean by *best fit*. If our regression model will be used to predict the value of Y, then it makes sense to find the line that makes the most accurate predictions. The fitted line has an intercept a and a slope b, and the predicted value of Y for a given value of X is

$$\hat{Y} = a + bX \tag{8.2}$$

We use the symbols a and b for the fitted line to distinguish these estimates of the slope and intercept from the unknown parameters α and β.

For a given value of X, the prediction error is equal to the difference between the actual value of the dependent variable and the predicted value, $Y - \hat{Y}$. If we have n observations, then we might be tempted to calculate the sum of the prediction errors:

$$\sum_{i=1}^{n}(Y_i - \hat{Y}_i)$$

The problem is that positive and negative errors offset each other. If one prediction is 1 million too high and another prediction is 1 million too low, the sum of the prediction errors is 0, even though we are not making very accurate predictions. To keep positive and negative errors from offsetting each other, we can square the prediction errors:

$$\sum_{i=1}^{n}(Y_i - \hat{Y}_i)^2$$

It does not matter whether a prediction error is positive or negative, only how large it is. It is reasonable to say that the line that has the smallest sum of squared prediction errors is the line that best fits the data.

The equations for the slope b and intercept a of the line that minimizes the sum of squared prediction errors are the *least squares estimates* of α and β. The formula for the least squares estimator of β is

$$b = \frac{\sum_{i=1}^{n}(X_i - \overline{X})(Y_i - \overline{Y})}{\sum_{i=1}^{n}(X_i - \overline{X})^2} \tag{8.3}$$

The denominator is the sum of squared deviations of X about its mean and is always positive. The numerator is the sum of the products of the deviation of X from its mean times the deviation of Y from its mean. If Y tends to be above its mean when X is above its mean and Y tends to be below its mean when X is below its mean, then b is positive and there is a positive statistical relationship between X and Y. If, on the other hand, Y tends to be below its mean when X is above its mean and Y tends to be above its mean when X is below its mean, then there is a negative statistical relationship between X and Y.

After the slope has been calculated, the intercept is easily computed from the mathematical fact that the least squares line runs through the mean values of X and Y: $\overline{Y} = a + b\overline{X}$. Rearranging,

$$a = \overline{Y} - b\overline{X} \tag{8.4}$$

We do not need to do these calculations by hand. After we enter the values of X and Y in a statistical software program, the program almost immediately displays the least squares estimates, along with a wealth of other information. Figure 8.5 shows the least squares line

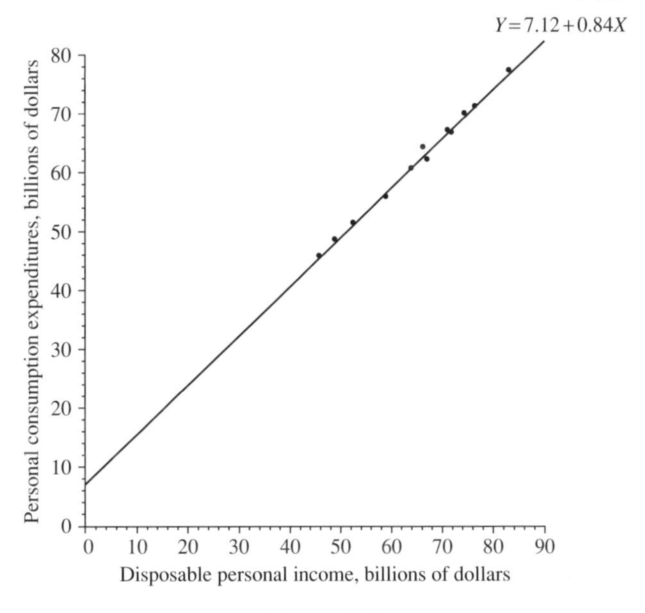

Figure 8.5
The least squares line for household income and spending, 1929–1940.

for the scatterplot of data on household income and spending graphed in Figure 8.1. No other line has a smaller sum of squared prediction errors. The least squares estimate of the slope is 0.84 and the least squares estimate of the intercept is 7.12. In the next two sections, confidence intervals and hypothesis tests are used to gauge the precision of these estimates and make inferences about α and β.

8.3 Confidence Intervals

When we use a sample mean to estimate a population mean, we recognize that the sample is but one of many samples that might have been selected, and we use a confidence interval to gauge the uncertainty in our estimate. We can do the same with a regression model.

Look back at Figure 8.3, which shows how the observed values of Y depend on the values of ε. As drawn, a line fit to these three points is fairly close to the true relationship. However, if the error term for the first observation had been positive instead of negative, Y_1 would be above the true line, rather than below it, and the least squares line would be flatter than the true relationship, giving an estimated slope that is too low and an estimated intercept that is too high. On the other hand, if the error term for the third observation had been positive, Y_3 would be above the true line and the least squares line would be steeper than the true relationship, giving an estimate of the slope that is too high and an estimate of the intercept that is too low.

Because we see only the observed values of X and Y and not the error term, we do not know whether our estimates of α and β are too high or too low, just as we do not know whether a sample mean is higher or lower than the population mean. However, we can gauge the reliability of our estimates by calculating standard errors and confidence intervals. If the error term is normally distributed and conforms to the four assumptions stated earlier, then our estimators are normally distributed with population means equal to the values of the parameters they are estimating:

$$a \sim N[\alpha, \text{standard deviation of } a]$$
$$b \sim N[\beta, \text{standard deviation of } b]$$

Figure 8.6 shows the probability distribution for our least squares estimator b.

Both a and b are unbiased estimators, in that the population means are equal to α and β. The formulas for the standard deviations are too complicated to present here, but we do not need the formulas since the calculations will be done by statistical software. There are four intuitively reasonable properties that are worth noting:

1. The standard deviations of the estimators are smaller if the standard deviation of the error term is small. If the values of ε are small, then the fitted line will be very close to the true relationship. If, however, the standard deviation of the error term is large, then

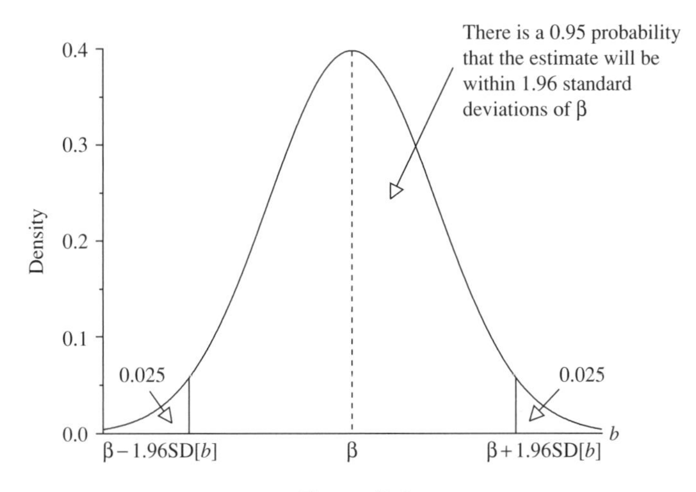

Figure 8.6
The sampling distribution of b.

the observations will be scattered widely about the true relationship, creating considerable uncertainty about the value of the slope and intercept.

2. The standard deviations of the estimators are smaller if there are a large number of observations. With lots of data, the observations above and below the true relationship are more likely to balance each other out, giving accurate estimates of both the slope and intercept. With only a handful of data, almost anything can happen.
3. The standard deviations of the estimators are smaller if there is a large amount of variation in X. If X moves around a lot, we can get a good idea of how changes in X affect Y. If, on the other hand, X barely moves, we cannot tell how changes in X affect Y.
4. The standard deviation of our estimate of the intercept is larger if the values of X are far from 0. We will know little about the value of Y when X is equal to 0 if we do not observe X near 0.

The *standard error of estimate* (SEE) is our estimate of the standard deviation of the error term ε:

$$\text{SEE} = \sqrt{\frac{\sum(Y_i - \hat{Y}_i)^2}{n-2}} \tag{8.5}$$

The SEE is the square root of the average prediction error (but we divide by $n - 2$ rather than n, because the estimation of α and β uses up two degrees of freedom). As in earlier chapters, the term *standard error* is used to distinguish a sample estimate from the population value. So, the estimated standard deviations of a and b are called *standard errors*.

The standard errors of a and b can be used to calculate confidence intervals for the slope and intercept. After choosing a confidence level, such as 95 percent, we use the t distribution with $n - 2$ degrees of freedom to determine the value t^* that corresponds to this probability.

The confidence intervals are equal to our estimates, plus or minus the appropriate number of standard errors:

$$a \pm t^*(\text{standard error of } a)$$
$$b \pm t^*(\text{standard error of } b)$$

For our consumption function, statistical software tells us that the standard errors are

$$\text{standard error of } a = 1.3396$$
$$\text{standard error of } b = 0.0202$$

With 12 observations, we have $12 - 2 = 10$ degrees of freedom. Table A.2 gives $t^* = 2.228$ for a 95 percent confidence interval with 10 degrees of freedom, so that a 95 percent confidence interval for the intercept is

$$95\% \text{ confidence interval for } \alpha = 7.12 \pm 2.228(1.3396)$$
$$= 7.12 \pm 2.98$$

and a 95 percent confidence interval for the slope is

$$95\% \text{ confidence interval for } \beta = 0.84 \pm 2.228(0.0202)$$
$$= 0.84 \pm 0.05$$

We are usually keenly interested in the slope because this tells us how changes in X affect Y. The intercept is typically less interesting, unless we are planning to predict the value of Y when X equals 0.

8.4 Hypothesis Tests

It is one short step from confidence intervals to hypothesis tests. The most common null hypothesis is that the value of the slope is 0:

$$H_0\text{: } \beta = 0$$
$$H_1\text{: } \beta \neq 0$$

If the slope is 0, there is no systematic linear relationship between the explanatory variable and the dependent variable. Of course, we chose our model because we think that Y is related to X. The null hypothesis $\beta = 0$ is a straw hypothesis that we hope to reject. The alternative hypothesis is two sided unless we can rule out positive or negative values for β *before* we look at the data.

The null hypothesis is tested with a t statistic that measures how many standard errors the estimate of β is from the value of β under the null hypothesis:

$$t = \frac{\text{estimated value of } \beta - \text{null hypothesis value of } \beta}{\text{standard error of the estimate of } \beta} \tag{8.6}$$

If the null hypothesis value of β is 0, then the t statistic is simply the ratio of the estimated slope to its standard error:

$$t = \frac{b}{\text{standard error of } b}$$

The null hypothesis is rejected if the t value is larger than the appropriate cutoff, using a t distribution with $n - 2$ degrees of freedom. For a test at the 5 percent level with a large number of observations, the cutoff is approximately 2, so that the null hypothesis $\beta = 0$ is typically rejected at the 5 percent level if the estimate of β is more than two standard errors from 0.

In our consumption example, the estimate of β is $b = 0.84$ and the standard error of b is 0.02, giving a t value of

$$t = \frac{0.84}{0.02}$$
$$= 41.42$$

The estimated slope is 41.42 standard errors from 0, far larger than the 2.228 cutoff for a two-tail test at the 5 percent level with $12 - 2 = 10$ degrees of freedom, and so rejects the null hypothesis decisively. Statistical software shows the two-sided P value to be 8.0×10^{-13}.

Some researchers report their estimated coefficients along with the standard errors:

> *Using annual U.S. household income and spending data for 1929–1940, a least squares regression yields this estimated consumption function:*

$$\underset{(1.34)}{Y = 7.12} + \underset{(0.02)}{0.84X}$$

( ): standard error

Most readers look at the ratio of the estimated slope to its standard error (here, 0.84/0.02) to see whether this ratio is larger than 2. This is such an ingrained habit that researchers often report t values along with their parameter estimates, to save readers the trouble of doing the arithmetic. Because they have little interest in the intercept, some do not bother to report its t value:

> *Using annual U.S. household income and spending data for 1929–1940, a least squares regression yields this estimated consumption function:*

$$Y = 7.12 + \underset{[41.4]}{0.84X}$$

[ ]: t value

Another reporting format is to show the two-sided P value.

The reported t value and P value are for the null hypothesis $\beta = 0$, since our foremost concern is generally whether there is a statistically significant linear relationship between the explanatory variable and the dependent variable. Sometimes, we are interested in a different null hypothesis. An automaker may want to know not only whether a 1 percent increase in prices will reduce the number of cars sold, but whether it will reduce sales by more than 1 percent. To test hypotheses other than $\beta = 0$, we substitute the desired value into Equation 8.6.

In addition to considering whether the estimated relationship between X and Y is statistically significant, we should consider if the estimated relationship is substantial and plausible. Here, we anticipated that an increase in income would increase spending and the estimated slope is indeed positive. If the estimate had turned out to be negative and statistically significant, the most likely explanation would be that an important variable had been omitted from our analysis and played havoc with our estimates. For example, it might have coincidentally turned out that the stock market fell in those years when income increased. A scatter diagram of spending and income would then suggest that higher income reduces spending, when it fact the positive effects of higher income were overwhelmed by falling stock prices. If we are able, on reflection, to identify such a confounding variable, then we can use the multiple regression techniques discussed in Chapter 10.

In addition to the right sign, we should consider whether the estimated value of the slope is substantial and plausible. We cannot determine this merely by saying whether the slope (here 0.84) is a big number or a small number. We have to provide some context by considering the magnitudes of the variables X and Y. The slope tells us the effect on Y of a one unit increase in X. (We say "one unit" because X might be measured in hundredths of an inch or trillions of dollars.) Here, X and Y are income and spending, each in billions of dollars. The 0.84 slope means that a $1 billion increase in income is predicted to increase spending by $0.84 billion. This is a substantial and plausible number.

8.5 R^2

One measure of the success of a regression model is whether the estimated slope has a plausible magnitude and is statistically significant. Another is the size of the prediction errors—how close the data are to the least squares line. Large prediction errors suggest that the model is of little use and may be fundamentally flawed. Small prediction errors indicate that the model may embody important insights. However, we need a scale to put things into perspective. An error of $1,000 in predicting the value of a million-dollar variable may be more impressive than a $1 error in predicting a $2 variable. We should also consider whether Y fluctuates a lot or is essentially constant; it is easy to predict the value of Y if Y does not change much.

An appealing benchmark that takes these considerations into account is the *coefficient of determination*, R^2, which compares the model's sum of the squared prediction errors to the sum of the squared deviations of Y about its mean:

$$R^2 = 1 - \frac{\sum_{i=1}^{n} (Y_i - \hat{Y}_i)^2}{\sum_{i=1}^{n} (Y_i - \overline{Y})^2} \tag{8.7}$$

The numerator is the model's sum of squared prediction errors; the denominator is the sum of squared prediction errors if we ignore X and simply use the average value of Y to predict Y.

The value of R^2 cannot be less than 0 nor greater than 1. An R^2 close to 1 indicates that the model's prediction errors are very small in relation to the variation in Y about its mean. An R^2 close to 0 indicates that the regression model is not an improvement over ignoring X and simply using the average value of Y to predict Y.

Mathematically, the sum of the squared deviations of Y about its mean can be separated into the sum of the squared deviations of the model's predictions about the mean and the sum of the squared prediction errors:

$$\sum_{i=1}^{n}(Y_i - \overline{Y})^2 = \sum_{i=1}^{n}(\hat{Y}_i - \overline{Y})^2 + \sum_{i=1}^{n}(Y_i - \hat{Y}_i)^2$$

$$\begin{matrix}\text{total} \\ \text{sum of squares}\end{matrix} = \begin{matrix}\text{explained} \\ \text{sum of squares}\end{matrix} + \begin{matrix}\text{unexplained} \\ \text{sum of squares}\end{matrix}$$

The variation in the predictions about the mean is the "explained" sum of squares, because this is the variation in Y attributed to the model. The prediction errors are the unexplained variation in Y. Dividing through by the total sum of squares, we have

$$1 = \frac{\text{explained}}{\text{total}} + \frac{\text{unexplained}}{\text{total}}$$

Because R^2 is equal to 1 minus the ratio of the sum of squared prediction errors to the sum of squared deviations about the mean, we have

$$\begin{aligned} R^2 &= 1 - \frac{\text{unexplained}}{\text{total}} \\ &= \frac{\text{explained}}{\text{total}} \end{aligned}$$

Thus R^2 can be interpreted as the fraction of the variation in the dependent variable explained by the regression model. The consumption function in Figure 8.5 has an R^2 of 0.994, showing that 99.4 percent of the variation in consumption during this period can be explained by the observed changes in income.

The correlation coefficient, R, is equal to the square root of R^2 and can also be calculated directly from this formula (Equation 3.7 in Chapter 3):

$$R = \frac{(\text{covariance between } X \text{ and } Y)}{(\text{standard deviation of } X)(\text{standard deviation of } Y)} = \frac{s_{XY}}{s_X s_Y} \tag{8.8}$$

The correlation coefficient is positive if the least squares line has a positive slope and negative if the least squares line has a negative slope. The correlation coefficient equals 1 or −1 if all of the points lie on a straight line and equals 0 if a scatter diagram shows no linear relationship between X and Y. (If there is no variation at all in X, so that the points lie on a vertical line, or no variation at all in Y, so that the points lie on a horizontal line, the correlation coefficient is defined as 0.)

Figure 8.7 gives four examples, with correlation coefficients of 0.90, −0.75, 0.50, and 0.00.

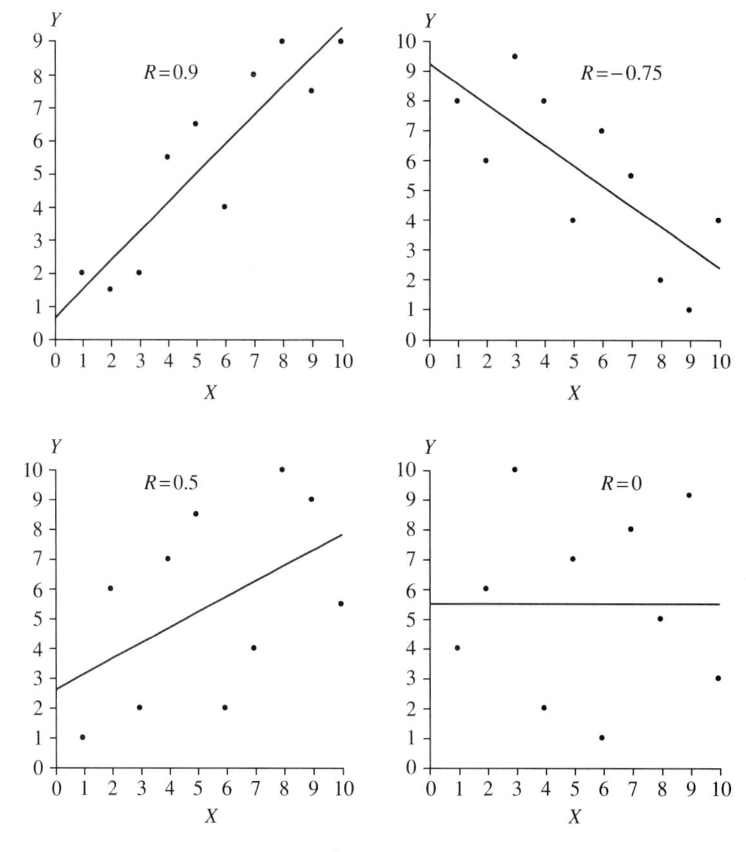

Figure 8.7
Four correlation coefficients.

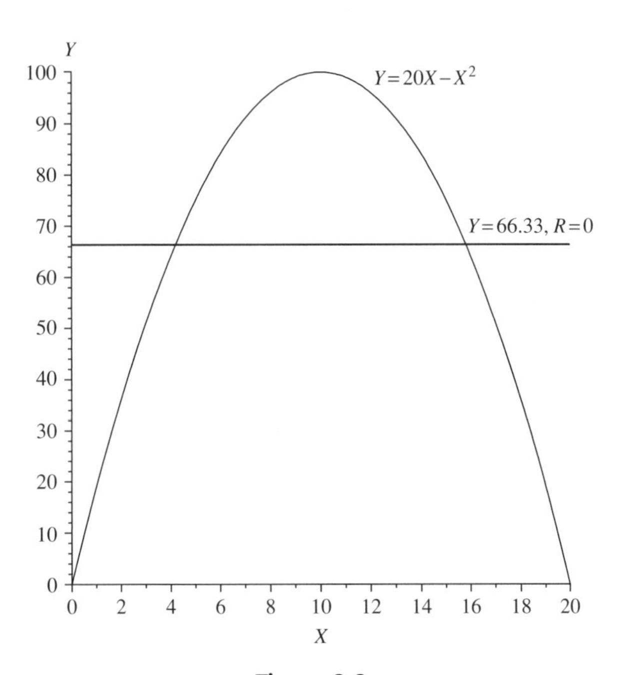

Figure 8.8
A zero correlation coefficient does not mean that X and Y are unrelated.

Figure 8.8 shows how a zero correlation coefficient does not rule out a perfect nonlinear relationship between two variables. In Figure 8.8, Y and X are exactly related by the equation $Y = 20X - X^2$. But the least squares line has a slope of 0 and a correlation coefficient of 0. Despite a perfect (nonlinear) relationship between X and Y, there is no linear relationship, in that the best linear predictor of Y is to ignore the value of X and use the average value of Y, 66.33. This example also demonstrates the value of looking at a scatterplot of the data.

Unlike the slope and intercept of a least squares line, the value of the correlation coefficient does not depend on which variable is on the vertical axis and which is on the horizontal axis. Therefore, it does not depend on which variable is considered the dependent variable or indeed whether there is any causal relationship at all. For this reason, correlation coefficients are often used to give a quantitative measure of the direction and degree of association between two variables that are related statistically but not necessarily causally. For instance, Exercise 8.47 concerns the correlation between the IQ scores of 34 identical twins who were raised apart. We do not believe that either twin causes the other to have a high or low IQ score but rather that both scores are influenced by common hereditary factors; and we want a statistical measure of the degree to which their scores are related statistically. The correlation coefficient does that.

The correlation coefficient does not depend on the units in which the variables are measured—inches or centimeters, pounds or kilograms, thousands or millions. For instance, the correlation between height and weight is not affected by converting heights from inches to centimeters, weights from pounds to kilograms, or both.

A statistical test of the null hypothesis that the population correlation coefficient is 0 is computationally equivalent to a *t* test of the null hypothesis that $\beta = 0$. Back when calculations were done by hand, Equation 8.8 provided a simple, direct way to calculate the correlation coefficient. Nowadays, it is easy to use statistical software to estimate a regression equation and thereby calculate the correlation coefficient, along with the *t* value and the *P* value for a statistical test of the equivalent null hypotheses that the slope and correlation are equal to 0.

8.6 Using Regression Analysis

So far, we have used a Keynesian consumption function to illustrate the mechanics of regression analysis. We now apply these lessons to three other examples.

Okun's Law Revisited

In Chapter 1, we noted that Arthur Okun, one of John F. Kennedy's economic advisers, estimated the relationship between the percent change in real gross domestic product (GDP) and the change in the unemployment rate. Using data for 1948–1960, his estimate, now known as *Okun's law*, was that real output would be about 3 percent higher if the unemployment rate were 1 percentage point lower [2]. We can see if this relationship still holds, using data through 2008.

Figure 8.9 shows a simple scatterplot. The line drawn through these points is the least squares line:

$$\begin{aligned} U = 1.301 &- 0.376Y, \quad R^2 = 0.73 \\ [10.28] &\ \ [12.31] \end{aligned}$$

where U is the change in the unemployment rate, Y is the percent change in real GDP, and the *t* values are in brackets. The relationship is highly statistically significant, with the 12.31 *t* value implying a two-sided *P* value of 1.1×10^{-17}. The slope is negative as expected. If a 1 percent change in real GDP is predicted to reduce the unemployment rate by 0.376 percentage points, then a 1 percentage-point change in the unemployment rate implies a 2.66 percent change in real GDP:

$$\frac{1}{0.376} = 2.66$$

Despite all the changes in the U.S. economy over the past 50 years, Okun's law is a remarkably robust statistical relationship.

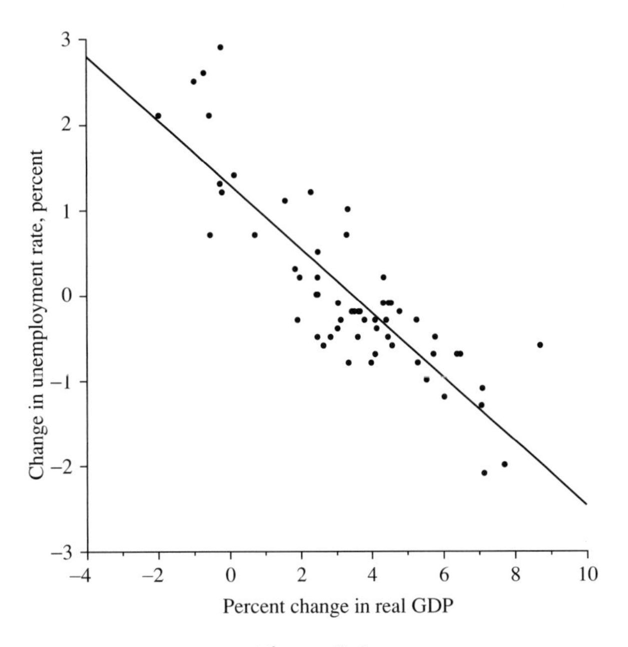

Figure 8.9
Okun's law, 1948–2008.

The Fair Value of Stocks

Chapter 2 introduced the Federal Reserve's stock valuation model, in which the "fair value" of stock prices is such that the earnings/price ratio (*E*/*P*) for the S&P 500 is equal to the interest rate *R* on 10-year Treasury bonds:

$$\frac{E}{P} = R \tag{8.9}$$

Figure 2.23 is a time series graph showing that *E*/*P* and *R* do tend to move up and down together.

Now, we are ready to examine the statistical relationship between these variables. Figure 8.10 is a scatter diagram using quarterly data. The line drawn through this scatterplot is the least squares line:

$$\frac{E}{P} = \underset{[2.90]}{1.156} + \underset{[14.75]}{0.796}R, \quad R^2 = 0.535$$

The t values are in brackets.

We noted in Chapter 2's discussion of the Fed's "fair value" model that we do not expect the earnings/price ratio to exactly equal the interest rate on 10-year Treasury bonds, though

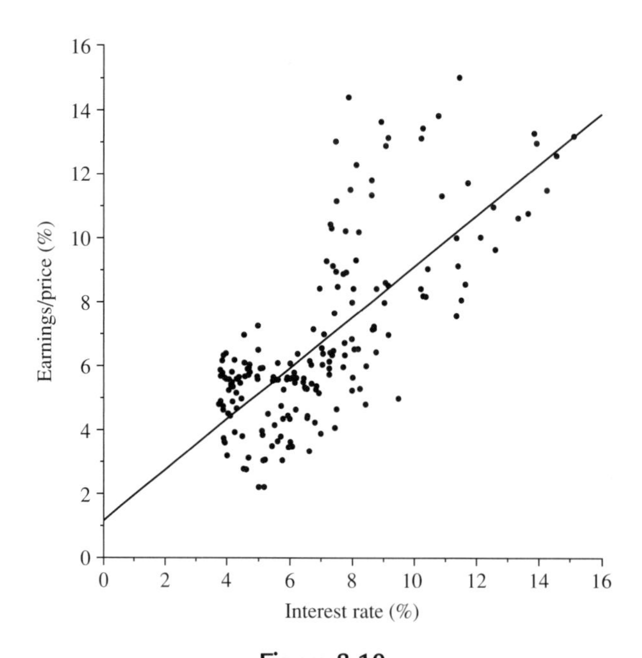

Figure 8.10
The S&P 500 Earnings/Price Ratio and the interest rate on 10-year Treasury bonds.

we might expect them to be positively correlated. The estimated equation confirms this. The 14.75 t value means that the two-sided P value for a test of the null hypothesis that E/P and R are uncorrelated is 2.9×10^{-33}. The 0.535 R^2 means that there is a 0.73 correlation ($\sqrt{0.535} = 0.732$) between E/P and R and that 53.5 percent of the variation in the earnings/price ratio can be explained by changes in the 10-year interest rate. For something as volatile and notoriously hard to predict as the stock market, that is very impressive.

The Nifty 50

In the early 1970s, many investors were infatuated by the Nifty 50, which were 50 growth stocks considered so appealing that they should be bought and never sold, regardless of price. Among these select few were Avon, Disney, McDonald's, Polaroid, and Xerox. Each was a leader in its field with a strong balance sheet, high profit rates, and double-digit growth rates.

Is such a company's stock worth any price, no matter how high? In late 1972, Xerox traded for 49 times earnings, Avon for 65 times earnings, and Polaroid for 91 times earnings. Then the stock market crashed. From their 1972–1973 highs to their 1974 lows, Xerox fell 71 percent, Avon 86 percent, and Polaroid 91 percent.

How well did these stocks do in the long run? And, is there any relationship between their 1972 price/earnings (P/E) ratio and subsequent performance? Figure 8.11 is a scatterplot of

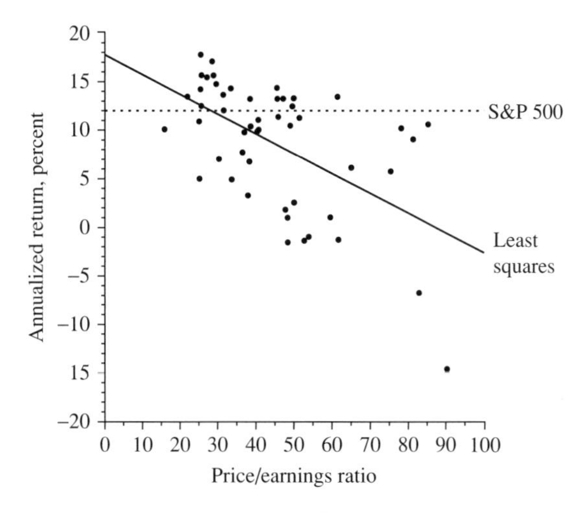

Figure 8.11
Annualized return versus P/E ratio for Nifty 50 stocks.

the price/earnings ratios of the Nifty 50 stocks in December 1972 and their annualized returns from December 31, 1972, through December 31, 2001. The horizontal line is the 12.01 percent annual return for the S&P 500 over this period.

The least squares line is

$$R = 17.735 - 0.2036\frac{P}{E}, \quad R^2 = 0.31$$
$$(2.13) \quad (0.0439)$$

with the standard errors in parentheses. The t value for testing the null hypothesis that the slope equals 0 is −4.64:

$$\begin{aligned} t &= \frac{b - 0}{\text{standard error of } b} \\ &= \frac{-0.2036 - 0}{0.0439} \\ &= -4.64 \end{aligned}$$

There are 50 observations and $50 - 2 = 48$ degrees of freedom. The two-sided P value is 0.000026, decisively rejecting the null hypothesis.

With 48 degrees of freedom, the t value for a 95 percent confidence interval is $t^* = 2.011$ and a 95 percent confidence interval for the slope is

$$\begin{aligned} b \pm t^* \text{standard error of } b &= -0.204 \pm 2.011(0.0439) \\ &= -0.204 \pm 0.088 \end{aligned}$$

It is estimated that a 1-point increase in the *P/E*, say from 50 to 51, reduces the annual rate of return by about 0.204 percent. A 10-point increase in the *P/E*, say from 50 to 60, is predicted to reduce the annual rate of return by about 2.04 percent, which is a lot.

8.7 Prediction Intervals (Optional)

Regression models are usually intended for predictions; for example, we might use a consumption function to predict what spending will be next year.

The predicted value of the dependent variable Y depends on our estimates of α and β, which depend on the values of the error term that happen to occur in the data used to estimate α and β. If different values of ε had occurred, we would have obtained somewhat different parameter estimates and therefore make somewhat different predictions.

All is not hopeless, however, because we can use an interval to quantify the precision of our prediction. For a selected value of the explanatory variable, X^*, the predicted value of Y is given by our least squares equation:

$$\hat{Y} = a + bX^* \tag{8.10}$$

After selecting a confidence level, such as 95 percent, we use a t distribution with $n - 2$ degrees of freedom to determine the appropriate value t^* and set our *prediction interval* for Y equal to the predicted value plus or minus the requisite number of standard errors of the prediction error, $Y - \hat{Y}$:

$$\hat{Y} \pm t^*(\text{standard error of } Y - \hat{Y}) \tag{8.11}$$

The interpretation of a prediction interval is analogous to a confidence interval—for a 95 percent prediction interval, there is a 0.95 probability that the interval will encompass the actual value of Y.

We can also predict the expected value of Y:

$$\text{estimate of } E[Y] = a + bX^* \tag{8.12}$$

and calculate a confidence interval for our estimate:

$$\hat{Y} \pm t^*[\text{standard error of } (a + bX^*)] \tag{8.13}$$

Our prediction of Y in Equation 8.10 and our estimate of $E[Y]$ in Equation 8.12 are identical. However, because of the error term, the expected value of Y is more certain than any single value of Y, thus a confidence interval for $E[Y]$ is narrower than a prediction interval for a single value of Y.

The formulas for the standard errors used in Equations 8.11 and 8.13 are very complicated, and we should let statistical software do the calculations for us. Unfortunately, some software packages show only prediction intervals for the observed values of X. If we want prediction intervals for other values, we must do the calculations ourselves using these formulas:

$$\text{standard error of } Y-\widehat{Y} = \text{SEE}\sqrt{1+\frac{1}{n}+(X^*-\overline{X})^2\left(\frac{\text{standard error of } b}{\text{SEE}}\right)^2}$$

$$\text{standard error of } a+bX^* = \text{SEE}\sqrt{\frac{1}{n}+(X^*-\overline{X})^2\left(\frac{\text{standard error of } b}{\text{SEE}}\right)^2}$$

For our consumption example, we will calculate a 95 percent prediction interval and confidence interval when X is equal to the sample mean, 65.35. The predicted value of Y is equal to 61.83:

$$\begin{aligned}\widehat{Y} &= a+b\overline{X}\\ &= 7.12+0.83(65.35)\\ &= 61.83\end{aligned}$$

Table A.2 gives $t^* = 2.228$ for a 95 percent interval with $12-2 = 10$ degrees of freedom. Statistical software gives the values of the standard errors:

$$\text{standard error of } Y-\widehat{Y} = 0.80394$$
$$\text{standard error of } a+bX^* = 0.22297$$

Therefore,

$$\text{95 percent prediction interval for } Y\text{: } 61.83 \pm 2.228(0.80394) = 61.83 \pm 1.79$$
$$\text{95 percent confidence interval for } E[Y]\text{: } 61.83 \pm 2.228(0.22297) = 61.83 \pm 0.50$$

As stated before, the error term makes the value of Y less certain than the expected value of Y. Figure 8.12 shows how the prediction interval widens as we move away from the middle

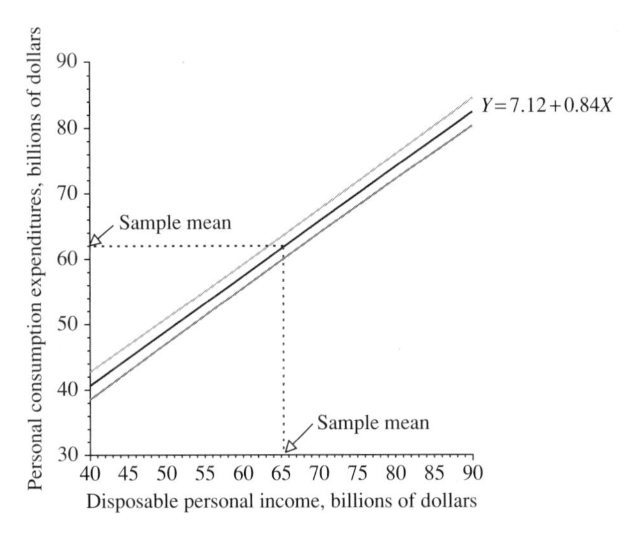

Figure 8.12
Consumption function 95 percent prediction interval.

of our data, showing that we can be more confident of predictions near the center of the data and should be more cautious on the fringes of the data.

Exercises

8.1 For each of the following pairs, explain why you believe that they are positively related, negatively related, or essentially unrelated. If they are related, identify the dependent variable.

a. The cost of constructing a house and its market price.
b. The winter price of corn and the number of acres of corn planted by Iowa farmers the following spring.
c. The price of steak and the amount purchased by a college dining hall.
d. The price of gasoline and sales of large cars.
e. The number of wins by the New York Yankees and the price of tea in China.

8.2 Figure 8.13 is a scatter diagram of the annual rates of return on Apple and Microsoft stock during the years 1998–2008:

a. What was the highest annual return for Apple?
b. What was the lowest annual return for Microsoft?
c. Was the average annual return higher for Apple or for Microsoft?
d. Was there a greater dispersion in Apple or Microsoft returns?
e. Were Apple and Microsoft stock returns positively related, negatively related, or unrelated?

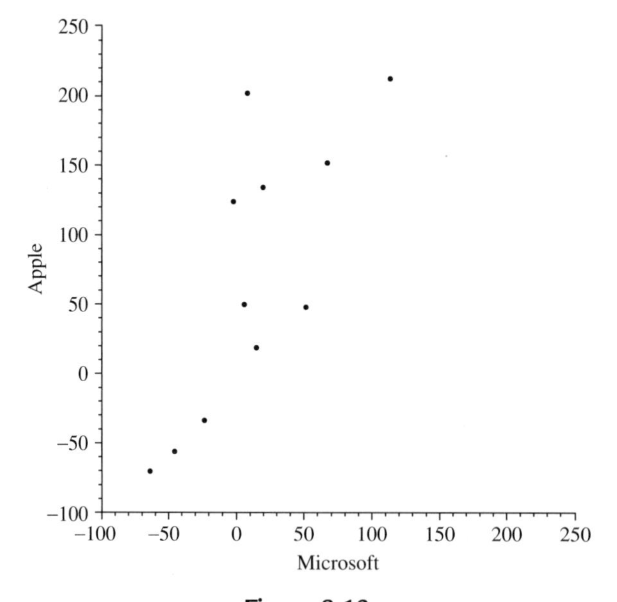

Figure 8.13
Exercise 8.2.

8.3 Use the data in Table 1.2 to calculate least squares estimates of the relationship between income and spending. What is the estimated slope?

8.4 Use the data in Table 1.6 to estimate a simple linear regression model where the dependent variable is 2003 prices and the explanatory variable is the hard drive size.

8.5 Use the data in Table 1.6 to estimate a simple linear regression model where the dependent variable is 2007 prices and the explanatory variable is the hard drive size.

8.6 Use the data in Table 1.6 to estimate a simple linear regression model where
   a. The dependent variable is the price of a 80 GB hard drive and the explanatory variable is the year.
   b. The dependent variable is the natural logarithm of the price of a 80 GB hard drive and the explanatory variable is the year.

   Which of these equations seems to describe the data better? Explain your reasoning.

8.7 Use the data in Table 1.6 to estimate a simple linear regression model where
   a. The dependent variable is the price of a 200 GB hard drive and the explanatory variable is the year.
   b. The dependent variable is the natural logarithm of the price of a 200 GB hard drive and the explanatory variable is the year.

   Which of these equations seems to describe the data better? Explain your reasoning.

8.8 Table 8.1 shows average monthly U.S. housing starts (in thousands). Estimate a simple linear regression model where the dependent variable is the number of housing starts and the explanatory variable is time, with $t = 0$ in 1960.
   a. Interpret the value of the estimated slope.
   b. Does the estimated slope have a two-sided P value less than 0.05?
   c. State the null hypothesis in words, not symbols.

Table 8.1: Exercise 8.8

1960	104.34	1970	118.47	1980	107.68	1990	98.38	2000	130.72
1961	109.42	1971	171.02	1981	90.34	1991	84.50	2001	133.58
1962	121.90	1972	196.38	1982	88.52	1992	98.97	2002	142.08
1963	133.59	1973	170.43	1983	141.92	1993	107.31	2003	153.98
1964	127.41	1974	111.48	1984	145.78	1994	121.42	2004	162.97
1965	122.74	1975	96.70	1985	145.14	1995	112.85	2005	172.34
1966	97.08	1976	128.11	1986	150.45	1996	123.08	2006	150.08
1967	107.63	1977	165.59	1987	135.05	1997	122.83	2007	112.92
1968	125.62	1978	168.36	1988	124.00	1998	134.75	2008	75.42
1969	122.23	1979	145.42	1989	114.68	1999	136.77	2009	46.17

8.9 A least squares regression was used to estimate the relationship between the dollar price P of 45 used 2001 Audi A4 2.8L sedans with manual transmission and the number of miles M the car had been driven:

$$P = 16{,}958 - \underset{(0.0233)}{0.0677}M, \quad R^2 = 0.70$$
$$(2{,}553) \qquad (0.0233)$$

The standard errors are in parentheses.

a. Does the value 16,958 seem reasonable?
b. Does the value −0.0677 seem reasonable?
c. Does the value 0.70 seem reasonable?
d. Is the estimated relationship between M and P statistically significant at the 5 percent level?
e. Should the variables be reversed, with M on the left-hand side and P on the right-hand side? Why or why not?

8.10 Use the government saving and GDP data in Exercise 2.29 to calculate the ratio of government saving to GDP each year from 1990 through 2009. Then use least squares to estimate the model $Y = \alpha + \beta X + \varepsilon$, where Y is the ratio of government saving to GDP and X is the year.

8.11 Plutonium has been produced in Hanford, Washington, since the 1940s, and some radioactive waste has leaked into the Columbia River, possibly contaminating the water supplies of Oregon residents living near the river. A 1965 study [3] of the statistical relationship between proximity to the Columbia River and cancer mortality compared an exposure index and the annual cancer mortality rate per 100,000 residents for nine Oregon counties near the Columbia River (Table 8.2).

a. Which variable would be the dependent variable in a simple linear regression model?
b. Does the estimated slope have a two-sided P value less than 0.05?
c. State the null hypothesis in words, not symbols.
d. What is the value of R^2?

8.12 Use the Vostok data in Exercise 2.28 to estimate the regression models $C = \alpha_1 + \beta_1 A + \varepsilon_1$ and $A = \alpha_2 + \beta_2 D + \varepsilon_2$. Write a brief essay summarizing your findings.

8.13 Many investors look at the beta coefficients provided by investment advisors based on least squares estimates of the model $R_i = \alpha + \beta R_M + \varepsilon$, where R_i is the rate of return on

Table 8.2: Exercise 8.11

Exposure index	11.64	8.34	6.41	3.83	3.41	2.57	2.49	1.62	1.25
Cancer mortality	207.5	210.3	177.9	162.3	129.9	130.1	147.1	137.5	113.5

a stock and R_M is the return on a market average, such as the S&P 500. Historical plots of the annual rates of return on two stocks and the market as a whole over a 10-year period are shown in Figure 8.14. During this period, which of these two stock returns had the higher

a. Mean?
b. Standard deviation?
c. Beta coefficient?
d. R^2?

8.14 Many investors look at the beta coefficients provided by investment advisors based on least squares estimates of the model $R_i = \alpha + \beta R_M + \varepsilon$, where R_i is the rate of return on a stock and R_M is the return on a market average, such as the S&P 500. Explain why you either agree or disagree with this advice:

a. "A stock with a low beta is safer than a high-beta stock, because the lower the beta is, the less the stock's price fluctuates."
b. "If the market was expected to rise, an investor might increase the R^2 of the portfolio. If the market was expected to decline, a portfolio with a low R^2 would be appropriate."

8.15 In a 1993 election in the second senatorial district in Philadelphia, which would determine which party controlled the Pennsylvania State Senate, the Republican candidate won based on the votes cast on Election Day, 19,691 to 19,127. However, the Democrat won the absentee ballots, 1,391 to 366, thereby winning the election by 461 votes, and Republicans charged that many absentee ballots had been illegally solicited or cast. A federal judge ruled that the Democrats had engaged in a "civil conspiracy" to win the election and declared the Republican the winner [4]. Among the evidence presented were the data in Figure 8.15 for 22 state senatorial elections in Philadelphia during the period 1982–1993. On the horizontal axis is the difference

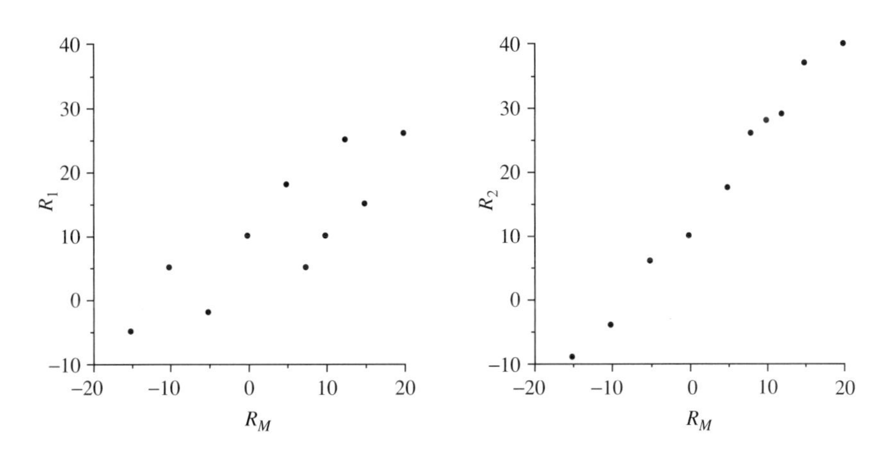

Figure 8.14
Exercise 8.13.

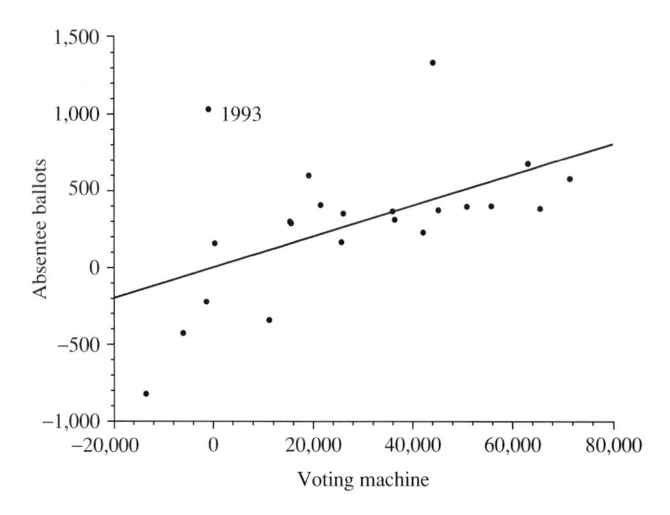

Figure 8.15
Exercise 8.15.

between the number of votes for the Democratic and Republican candidates cast on Election Day and recorded by voting machines; on the vertical axis is the difference between the number of votes for the Democratic and Republican candidates cast by absentee ballot.

a. What criterion do you think was used to draw a line through the points?
b. Explain why this line's positive or negative slope is either plausible or counterintuitive.
c. Was this evidence cited by the Republican or Democrat? Explain your reasoning.

8.16 The Australian Bureau of Meteorology calculates the Southern Oscillation Index (SOI) from the monthly air pressure difference between Tahiti and Darwin, Australia. Negative values of the SOI indicate an El Niño episode; positive values indicate a La Niña episode. Since 1989, Australia's National Climate Center has used the SOI to predict rainfall 3 months in advance. A model was estimated [5] by least squares using 43 years of annual data on the wheat yield relative to the long-run upward trend (positive values indicate unusually high yields, negative readings unusually low yields) and the average SOI reading for June, July, and August (Australia's winter). In these data, the average wheat yield is 0.000 and the average SOI reading is 0.267.

$$\hat{Y} = \underset{(0.0296)}{-0.0032} + \underset{(0.0036)}{0.0123}X, \quad R^2 = 0.216$$

a. Which variable do you think is the dependent variable? Explain your reasoning.
b. Is the estimated slope statistically significant at the 5 percent level?
c. Are El Niño or La Niña episodes most often associated with high wheat yields?

8.17 Table 8.3 shows decade averages of annual precipitation (mm) in the contiguous United States. Letting Y be precipitation and X be the decade midpoints, 1905, 1915, and so on, draw a scatter diagram and use least squares to estimate the equation $Y = \alpha + \beta X + \varepsilon$. Is there statistically persuasive evidence of a trend in precipitation?

8.18 The effective delivery and care of newborn babies can be aided by accurate estimates of birth weight. A study [6] attempted to predict the birth weight of babies for 11 women with abnormal pregnancies by using echo-planar imaging (EPI) up to 1 week before delivery to estimate each baby's fetal volume. Draw a scatter diagram and use least squares estimates of the regression model to see if there is a statistically significant relationship at the 1 percent level between the child's estimated fetal volume X (decimeters cubed) and the birth weight Y (kilograms), both of which are in Table 8.4. Summarize your findings, and be sure to interpret the estimated slope of your fitted line.

8.19 How do you suppose that least squares coefficient estimates, standard errors, t values, and R^2 would be affected by a doubling of the number of observations, with the new data exactly replicating the original data? Use the data in Exercise 8.18 test your theory.

8.20 Table 8.5 shows the prepregnancy weights of 25 mothers and the birth weights of their children, both in kilograms [7].

a. Which would you use as the dependent variable and which as the explanatory variable in the regression model $Y = \alpha + \beta X + \varepsilon$?

b. Draw a scatter diagram.

Table 8.3: Exercise 8.17

1900s	1910s	1920s	1930s	1940s	1950s	1960s	1970s	1980s
747.50	714.80	731.77	685.08	754.84	711.62	717.67	761.53	753.79

Table 8.4: Exercise 8.18

X	3.11	1.54	2.88	2.15	1.61	2.71	3.35	2.22	3.42	2.31	3.22
Y	3.30	1.72	3.29	2.30	1.62	3.00	3.54	2.64	3.52	2.40	3.40

Table 8.5: Exercise 8.20

Mother	Child	Mother	Child	Mother	Child	Mother	Child	Mother	Child
49.4	3.52	70.3	4.07	73.5	3.23	65.8	3.35	63.5	4.15
63.5	3.74	50.8	3.37	59.0	3.57	61.2	3.71	59.0	2.98
68.0	3.63	73.9	4.12	61.2	3.06	55.8	2.99	49.9	2.76
52.2	2.68	65.8	3.57	52.2	3.37	61.2	4.03	65.8	2.92
54.4	3.01	54.4	3.36	63.1	2.72	56.7	2.92	43.1	2.69

c. Use least squares to see if there is a positive or negative relationship.
d. Interpret your estimates of α and β.

8.21 The following regression equation was estimated using data on college applicants who were admitted both to Pomona College and another college ranked among the top 20 small liberal arts colleges by *U.S. News & World Report*:

$$\hat{Y} = 0.2935 + 0.0293X, \quad R^2 = 0.562$$
$$(0.0781) \quad (0.0065)$$

where

( ) = standard deviation.
Y = fraction of students who were admitted to both this college and Pomona College who enrolled at Pomona, average value = 0.602.
X = *U.S. News & World Report* ranking of this college.

a. Redraw Figure 8.16 and add the estimated regression line.
b. Use your graph to explain how the estimates 0.2935 and 0.0293 were obtained. (Do not show the formulas; explain in words the basis for the formulas.)
c. Explain why you are not surprised that the R^2 for this equation is not 1.0.

8.22 Use the estimated equation reported in the preceding exercise to answer these questions:
a. Does the estimated coefficient of X have a plausible value?
b. Is the estimated coefficient of X statistically significant at the 5 percent level?
c. What is the null hypothesis in question b?
d. What is the predicted value of Y for $X = 30$?
e. Why should we not take the prediction in question d seriously?

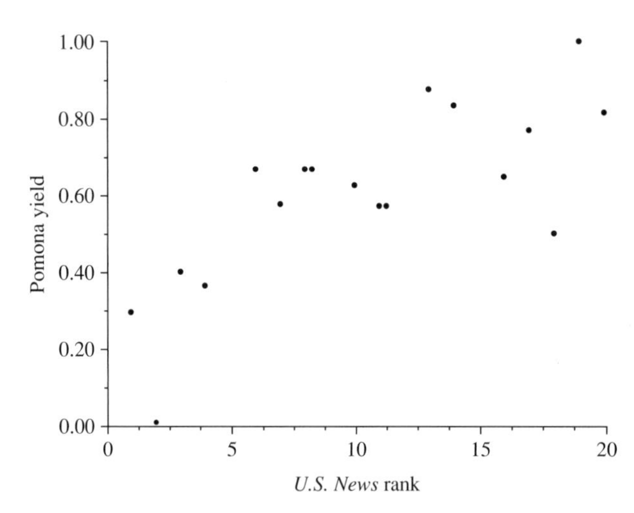

Figure 8.16
Exercise 8.21.

Table 8.6: Exercise 8.23

	Crawling Age	Temperature		Crawling Age	Temperature
January	29.84	66	July	33.64	33
February	30.52	73	August	32.82	30
March	29.70	72	September	33.83	33
April	31.84	63	October	33.35	37
May	28.58	52	November	33.38	48
June	31.44	39	December	32.32	57

8.23 A Denver study looked at the relationship between climate and the age at which infants begin crawling [8]. (It was suspected that crawling is delayed when infants wear heavy clothing.) Tables 8.6 shows the average age (in weeks) when children born in different months began crawling and the average Denver temperature for the sixth month after birth.

a. Draw a scatter diagram and the least squares line.
b. What is your dependent variable? Explain your reasoning.
c. Is the estimated slope statistically significant at the 5 percent level?
d. Is the sign of the estimated slope consistent with the researchers' theory?
e. Is the value of the estimated slope plausible?

8.24 A student used least squares to estimate the relationship between the number of games won by a football team (Y) and the number of serious injuries suffered by the players (X). He concluded that "Using a t distribution with 13 degrees of freedom, a t value of 1.58 is not sufficient to reject the null hypothesis $\beta = 0$ at a 95 percent confidence interval. Therefore, we can conclude that the number of injuries that a team suffers does affect its chances of winning games." Explain the error in his interpretation of his results.

8.25 In American college football, Rivals.com constructs a widely followed index for rating colleges based on the high school players they recruit and Jeff Sagarin constructs a highly regarded system for rating college football teams based on their performance. Table 8.7 shows the average 2005–2008 performance and recruiting ratings for the 20 colleges with the highest recruiting averages. Use these data to estimate a simple linear regression model with performance as the dependent variable and recruiting as the explanatory variable.

a. Draw a scatter diagram and the least squares line.
b. Is the relationship statistically significant at the 5 percent level?
c. Is the size of the estimated relationship substantial?
d. Which team overperformed the most, in that its performance was much higher than predicted by your estimated equation?
e. Which team underperformed the most, in that its performance was much lower than predicted by your estimated equation?

Table 8.7: Exercise 8.25

	Recruiting	Performance
Alabama	2,044.00	81.538
Auburn	1,773.00	81.918
California	1,444.00	83.048
Clemson	1,710.00	80.545
Florida	2,481.75	91.615
Florida State	2,232.50	80.308
Georgia	2,116.50	85.590
LSU	2,100.50	89.480
Miami-FL	1,920.00	76.290
Michigan	1,984.75	80.435
Nebraska	1,625.75	78.618
Notre Dame	1,881.50	77.013
Ohio State	1,928.00	90.390
Oklahoma	2,186.75	88.978
Penn State	1,349.50	86.678
South Carolina	1,514.50	78.425
Southern California	2,680.50	95.073
Tennessee	1,850.25	79.040
Texas	2,043.25	92.398
UCLA	1,355.75	77.525

Table 8.8: Exercise 8.26

Maternal Age	Births	Late Fetal Deaths	Early Neonatal Deaths
20–24	70,557	251	212
25–29	68,846	253	177
30–34	26,241	125	78
35–39	6,811	34	33
40–52	1,069	7	5

8.26 A study [9] of 173,524 births in Sweden between 1983 and 1987 used the data in Table 8.8 on late fetal deaths (stillbirth at a gestation of 28 weeks or longer) and early neonatal deaths (during the first 6 days of life). Convert the late fetal death data to rates per 1,000 births by dividing the total number of late fetal deaths by the number of births and multiplying the result by 1,000.

a. Draw a scatter diagram using your calculated death rates and the maternal-age midpoint (22.5, 27.5, 32.5, 37.5, and 46.5). Which is your dependent variable?
b. Use least squares to estimate the regression model $Y = \alpha + \beta X + \varepsilon$.
c. What is the two-sided P value for the slope?
d. What is a 95 percent confidence interval for the slope?
e. Explain why the estimated value of the slope does or does not seem substantial.

8.27 Answer the questions in Exercise 8.26, this time using early neonatal death rates per 1,000 births.

8.28 Table 8.9 shows the systolic and diastolic blood pressure readings (both in mmHG) for 26 healthy young adults [10]. Ignoring gender, use all 26 observations to estimate a linear regression model with systolic blood pressure as the dependent variable and diastolic blood pressure as the explanatory variable.
a. Is the estimated relationship statistically significant at the 5 percent level?
b. What is the correlation between systolic and diastolic blood pressure?

8.29 Table 8.10 shows cricket chirps per minute and temperature in degrees Fahrenheit for the striped ground cricket [11]. Which is the dependent variable? Draw a scatter diagram with the cricket chirps axis running from 0 to 20, and the temperature axis running from 60 to 100. Does the relationship seem linear? Estimate the equation $Y = \alpha + \beta X + \varepsilon$ by least squares and draw this fitted line in your scatter diagram. In ordinary English, interpret your estimate of the slope.

8.30 A student estimated a regression model that used the number of pages in best-selling hardcover books to predict the book's price. Explain why this conclusion is

Table 8.9: Exercise 8.28

Systolic	Diastolic	Gender	Systolic	Diastolic	Gender
108	62	Female	96	64	Female
134	74	Male	114	74	Male
100	64	Female	108	68	Male
108	68	Female	128	86	Male
112	72	Male	114	68	Male
112	64	Female	112	64	Male
112	68	Female	124	70	Female
122	70	Male	90	60	Female
116	70	Male	102	64	Female
116	70	Male	106	70	Male
120	72	Male	124	74	Male
108	70	Female	130	72	Male
108	70	Female	116	70	Female

Table 8.10: Exercise 8.29

Chirps	Temperature	Chirps	Temperature	Chirps	Temperature
20.0	88.6	15.5	75.2	17.2	82.6
16.0	71.6	14.7	69.7	16.0	80.6
19.8	93.3	15.4	69.4	17.0	83.5
18.4	84.3	16.2	83.3	17.1	82.0
17.1	80.6	15.0	79.6	14.4	76.3

Table 8.11: Exercise 8.31

Age	Bone Density	Age	Bone Density	Age	Bone Density
15	104.66	28	88.92	38	111.49
17	90.68	29	108.49	39	88.82
17	105.28	30	84.06	41	93.89
19	98.55	34	85.30	43	128.99
23	100.52	35	101.66	44	111.39
26	126.60	35	91.41	45	114.60
27	82.09	35	91.30	45	100.10
27	114.29	36	75.78		
28	95.24	37	110.46		

wrong: "The 0.75 R^2 is pretty large, when you consider that the price of a book is \$20; that is, \$15 of the price is explained by the number of pages."

8.31 Table 8.11 shows the age and bone density of 25 women between the ages of 15 and 45 who died between 1729 and 1852 [12]. For a sample of modern-day British women of similar ages, the relationship between bone density and age is statistically significant at the 1 percent level, with an estimated slope of −0.658. Draw a scatter diagram and estimate the model $Y = \alpha + \beta X + \varepsilon$ by least squares.
a. Which variable did you use as the dependent variable? Explain your reasoning.
b. Is the estimated slope statistically significant at the 1 percent level?
c. Compare your results to those reported for modern-day British women.

8.32 Soccer enthusiasts have long debated how to decide tournament games in which the score is a tie but a winner must be determined. If the game is allowed to continue until one team scores a deciding goal, this team may be too tired to play well in the next round; yet other criteria (such as a succession of penalty kicks) seem arbitrary. Some have argued that a team's dominance can be measured by the number of corner kicks it is awarded during the game. To test this theory, a student estimated this equation using data from 17 games played by a college men's soccer team. Write a paragraph summarizing his results.

$$\begin{aligned} Y = \underset{(0.594)}{0.376} + \underset{(0.125)}{0.200}X, \quad R^2 = 0.147 \end{aligned}$$

Y is equal to the number of goals by this team minus the number of goals by the opposing team (average value 0.59), X is equal to the number of corner kicks by this team minus the number of corner kicks by the opposing team (average value = 1.06), and the standard errors are in parentheses.

8.33 A random sample of 100 college students was asked to rate their courses on a scale of 1 to 10 (10 being best) and state their grade in the course on a scale of 1 to 12 (A = 12,

A− = 11, and so on). The average rating was 5.53 and the average grade 9.69. A least squares regression of rating (Y) on grade (X) gave these results:

$$\begin{aligned} Y = &-2.14 + 0.79X \\ &(2.91) \quad (0.296) \end{aligned}$$

The standard errors are in parentheses.

a. What does a least squares regression tell us that a comparison of the average values, 5.53 versus 9.69, does not?
b. How do you interpret the estimate −2.14?
c. How do you interpret the estimate 0.79?
d. Is the estimated relationship between X and Y statistically significant at the 1 percent level?
e. Do you think that grades influence ratings or that ratings influence grades?
f. If so, what difference does it make to a regression equation?

8.34 A 1991 survey of 47 college sophomores investigated the relationship between the amount of time spent studying and one's grades. A least squares regression gave these results:

$$\begin{aligned} Y = &2.945 + 0.2022X, \quad R^2 = 0.26 \\ &(0.100) \quad (0.006) \end{aligned}$$

where Y = grade point average on a 4-point scale, X = average hours per week spent studying, and the standard errors are in parentheses.

a. Does the estimated equation indicate that more studying leads to higher or lower grades?
b. Is the relationship statistically significant at the 5 percent level?
c. The average grade point average in this sample was 3.296; what is the average number of hours per week spent studying?
d. This researcher assumed that grade point average is the dependent variable and studying is the explanatory variable. Could it possibly be the other way around?

8.35 Data were obtained for 108 cars on their price (X) and the time (Y) in seconds for the car to accelerate from 0 to 60 miles per hour. The average value of X is $23,429 and the average value of Y is 8.9 seconds. Interpret and evaluate this least squares regression:

$$\begin{aligned} Y = &10.31 - 0.0001X, \quad R^2 = 0.32 \\ &[38.6] \qquad [7.0] \end{aligned}$$

[ ]: t values

8.36 In the 1960s, the data in Table 8.12 were collected for 21 countries on deaths from coronary heart disease per 100,000 persons of age 35–64, and annual per capita cigarette consumption [13]. Draw a scatter diagram and estimate a regression equation to see if there is a statistical relationship between these variables.

a. Which variable did you use as the dependent variable? Explain your reasoning.
b. Is the estimated slope statistically significant at the 5 percent level?
c. Does the estimated value of the slope seem plausible?

8.37 To investigate whether economic events influence presidential elections, draw a scatter diagram and estimate the equation $Y = \alpha + \beta X + \varepsilon$ using the Table 8.13 data on X, the change in the unemployment rate during a presidential election year, and Y, the percentage of the vote for major party presidential candidates received by the incumbent party.

a. Does the estimated coefficient of X have a plausible sign and magnitude?
b. Test at the 5 percent level the null hypothesis that $\beta = 0$.
c. Does the intercept have a plausible magnitude?
d. Test at the 5 percent level the null hypothesis that $\alpha = 50$.
e. How close is the predicted value of Y in 2008, when $X = 1.2$, to the actual value, $Y = 46.3$?

Table 8.12: Exercise 8.36

	Cigarettes	Deaths		Cigarettes	Deaths
United States	3,900	259.9	Greece	1,800	41.2
Canada	3,350	211.6	Austria	1,770	182.1
Australia	3,220	238.1	Belgium	1,700	118.1
New Zealand	3,220	211.8	Mexico	1,680	31.9
United Kingdom	2,790	194.1	Italy	1,510	114.3
Switzerland	2,780	124.5	Denmark	1,500	144.9
Ireland	2,770	187.3	France	1,410	144.9
Iceland	2,290	110.5	Sweden	1,270	126.9
Finland	2,160	233.1	Spain	1,200	43.9
West Germany	1,890	150.3	Norway	1,090	136.3
Netherlands	1,810	124.7			

Table 8.13: Exercise 8.37

	X	Y		X	Y		X	Y
1900	−1.5	53.2	1936	−3.2	62.5	1972	−0.3	61.8
1904	1.5	60.0	1940	−2.6	55.0	1976	−0.8	48.9
1908	5.2	54.5	1944	−0.7	53.8	1980	1.3	44.7
1912	−2.1	54.7	1948	−0.1	52.4	1984	−2.1	59.2
1916	−3.4	51.7	1952	−0.3	44.6	1988	−0.7	53.9
1920	3.8	36.1	1956	−0.3	57.8	1992	0.7	37.7
1924	2.6	54.4	1960	0.0	49.9	1996	−0.2	54.7
1928	0.9	58.8	1964	−0.5	61.3	2000	−0.3	50.3
1932	7.7	40.9	1968	−0.2	49.6	2004	−0.5	51.2

8.38 Data were collected on eruptions of the Old Faithful geyser [14]. A least squares regression using 99 observations on the time between eruptions Y (in minutes) and the duration X (in seconds) of the preceding eruption yielded this estimated equation: $Y = 34.549 + 0.2084X$. The value of R^2 is 0.87, and the estimated coefficient of X has a standard error of 0.0076 and a t value of 27.26.

a. Interpret the estimated coefficient of X.
b. Calculate a 95 percent confidence interval for the coefficient of X.
c. Is the relationship statistically significant at the 5 percent level?
d. What percentage of the variation in the time interval between Old Faithful's eruptions can be explained by the duration of the preceding eruption?

8.39 A least squares regression using 99 observations on the height Y (in feet) of Old Faithful eruptions and the duration X (in seconds) of an eruption yielded these results (R^2 = 0.02 and the standard errors are in parentheses):

$$\begin{array}{rl} Y = 141.51 & -\ 0.0213X \\ (3.15) & \quad (0.0149) \end{array}$$

a. Interpret the estimated coefficient of X.
b. Calculate a 95 percent confidence interval for the coefficient of X.
c. Is the relationship statistically significant at the 5 percent level?
d. What percentage of the variation in the height of Old Faithful's eruptions can be explained by the duration of an eruption?

8.40 Figure 8.15 shows a graph of state senatorial elections in Philadelphia from 1982 through 1993. Table 8.14 shows the data (Democrat minus Republican) for the 21 elections preceding the disputed 1993 election. Estimate the regression model $Y = \alpha + \beta X + \varepsilon$, where the dependent variable is the absentee ballots and the explanatory variable is the Election Day ballots. In the disputed 1993 election, the difference in the Election Day ballots was −564. Find a 95 percent prediction interval for the absentee ballots in 1993. Is 1,025, the actual value for the absentee ballots, inside this prediction interval?

Table 8.14: Exercise 8.40

Election Day	Absentee	Election Day	Absentee
26,427	346	65,862	378
15,904	282	−13,194	−829
42,448	223	56,100	394
19,444	593	700	151
71,797	572	11,529	−349
−1,017	−229	26,047	160
63,406	671	44,425	1,329
15,671	293	45,512	368
36,276	360	−5,700	−434
36,710	306	51,206	391
21,848	401		

Table 8.15: Exercise 8.41

Height	2	3	4	5	6	7	8	9	10	11	12	13	14
Weight	130	135	139	143	148	152	157	162	166	171	175	180	185

Table 8.16: Exercise 8.42

Height	−3	−2	−1	0	1	2	3	4	5	6	7	8
Weight	111	114	118	121	124	128	131	134	137	141	144	147

8.41 Table 8.15 shows the average weight (in pounds) for 18- to 24-year-old U.S. men of different heights (measured in inches above 5 feet). Use least squares to estimate the relationship between height (X) and weight (Y).

a. Interpret the estimated coefficient of X.
b. Calculate a 95 percent confidence interval for the coefficient of X.
c. Is the relationship statistically significant at the 5 percent level?
d. The American Heart Association says that the ideal male weight is $110 + 5.0X$. How does your fitted line compare to this guideline?

8.42 Table 8.16 shows the average weight (in pounds) for 18- to 24-year-old U.S. women of different heights (measured in inches above 5 feet). Use least squares to estimate the relationship between height (X) and weight (Y).

a. Interpret the estimated coefficient of X.
b. Calculate a 95 percent confidence interval for the coefficient of X.
c. Is the relationship statistically significant at the 5 percent level?
d. The American Heart Association says that the ideal female weight is $100 + 5.0X$. How does your fitted line compare to this guideline?

8.43 Use the data in the two preceding exercises to compare the least squares lines for men and women.

a. Which line has the larger slope?
b. Do the lines cross? Where?
c. Are the predicted values of Y for $X = -60$ sensible? If not, should we discard our results as implausible?
d. Which equation has a higher R^2?

8.44 Chapter 8 includes a discussion of a study comparing the economic status of 4,121 women who immigrated to California from another county with the economic status of their adult daughters. Economic status was measured by ZIP-code income percentile. For example, if the mother lived in a ZIP code with a 30th-percentile median income and her adult daughter lived in a ZIP code with a 35th-percentile median income, this was characterized as an increase in economic status. The equation $Y = \alpha + \beta X + \varepsilon$ was estimated using the natural logarithms of ZIP-code percentiles for each mother-daughter pair. The coefficient

β is the elasticity of Y with respect to X; for example, if $\beta = 0.5$, then a 1 percent increase in X is predicted to give an 0.5 percent increase in Y. A similar equation was estimated for 6,498 California-born mothers and their adult daughters. Here are the results (the standard errors are in parentheses):

$$\begin{aligned} \text{foreign-born: } Y &= \underset{(0.141)}{6.593} + \underset{(0.013)}{0.376}X \\ \text{California-born: } Y &= \underset{(0.116)}{5.987} + \underset{(0.010)}{0.436}X \end{aligned}$$

a. Is the dependent variable the logarithm of the mother's income percentile or the daughter's income percentile? Explain your reasoning.
b. Does the sign of the estimated slope make sense to you?
c. Are the estimated slopes statistically significant at the 5 percent level?
d. The estimated elasticity is used to measure intergenerational economic mobility. Does a larger elasticity indicate more or less mobility? Explain your reasoning.
e. Do these results suggest that there is more economic mobility for the daughters of foreign-born mothers or the daughters of California-born mothers?

8.45 The procedure used in Exercise 8.44 was also applied to mother-daughter pairs where the adult daughter lived in the same ZIP code she was born in and to mother-daughter pairs where the daughter moved to a different ZIP code. Write a paragraph identifying the most interesting findings. For daughters who move:

$$\begin{aligned} \text{foreign-born: } Y &= \underset{(0.153)}{7.409} + \underset{(0.015)}{0.300}X \\ \text{California-born: } Y &= \underset{(0.130)}{6.863} + \underset{(0.012)}{0.353}X \end{aligned}$$

For daughters who stay:

$$\begin{aligned} \text{foreign-born: } Y &= \underset{(0.179)}{0.105} + \underset{(0.017)}{0.985}X \\ \text{California-born: } Y &= \underset{(0.119)}{0.300} + \underset{(0.011)}{0.969}X \end{aligned}$$

8.46 A zero coupon bond (a "zero") pays a specified amount of money when it matures and nothing before the bond's maturity. Financial theory teaches that the volatility of the price of a zero is proportional to its maturity. Use the data in Table 8.17 to estimate a least squares regression equation.
a. Which variable did you use for the dependent variable? Explain your reasoning.
b. Is the sign of the estimated slope consistent with financial theory?

Table 8.17: Exercise 8.46

Maturity (years)	5	10	15	20
Standard deviation of price (percent)	10.8	20.7	28.8	38.1

c. What is the P value for a two-tailed test of the null hypothesis that the value of the slope is equal to 0?
d. What is the P value for a two-tailed test of the null hypothesis that the value of the slope is equal to 1?
e. Give a 95 percent confidence interval for the slope.

8.47 A classic study [15] of the IQ scores of 34 identical twins who were raised apart used the intelligence scores in Table 8.18. Draw a scatter diagram and calculate the R^2 for a regression of the intelligence score of the second-born on the intelligence score of the first-born. (This R^2 was used to measure the influence of genes on intelligence, with $1 - R^2$ measuring the influence of environment.) If 100 were added to all the first-born IQ scores or all the second-born IQ scores, how do you think R^2 would be affected? Check your expectations by adding 100 to all the IQs and recalculating R^2.

8.48 Southern California casinos were given "location values" based on how close they are to households of various income levels and the location of competing casinos. A casino has a higher location value if potential customers with money to spend are nearby and there are fewer competing casinos. Data were also collected on the number of gaming devices (primarily slot machines) installed at each casino. These data are in Table 8.19.

Table 8.18: Exercise 8.47

First	Second	First	Second	First	Second	First	Second	First	Second
22	12	30	26	30	34	6	10	41	41
36	34	29	35	27	24	23	21	19	9
13	10	26	20	32	18	38	27	40	38
30	25	28	22	27	28	33	26	12	9
32	28	21	27	22	23	16	28	13	22
26	17	13	4	15	9	27	25	29	30
20	24	32	33	24	33	4	2		

Table 8.19: Exercise 8.48

Gaming Devices	Location Value	Gaming Devices	Location Value	Gaming Devices	Location Value
1,996	120.75	1,956	71.91	349	0.07
2,000	110.28	2,045	158.13	2,000	404.02
2,268	128.99	751	60.07	230	0.01
1,049	70.89	2,000	79.66	302	25.24
1,261	92.62	961	54.80	2,000	132.10
1,599	75.28	1,039	58.27	2,139	169.47

a. Which variable would you consider to be the dependent variable?
b. Do you expect the relationship between these two variables to be positive, negative, or 0?
c. Estimate the linear relationship between these two variables, $Y = \alpha_1 + \beta_1 X + \varepsilon_1$.
d. Estimate the log-linear relationship between the natural logarithms of these two variables, $\ln[Y] = \alpha_2 + \beta_2 \ln[X] + \varepsilon_2$.
e. Which of these two estimated relationships seems to fit these data better?

8.49 The data in Table 8.20 are for 16 three-digit ZIP codes within 4 hours' driving time of a Southern California casino. The driving time to the casino is in hours and the gambling activity is the total dollars lost at the casino by residents of thc ZIP code divided by the median income, in thousands of dollars, in that ZIP code.
a. Which variable would you consider to be the dependent variable?
b. Do you expect the relationship between these two variables to be positive, negative, or 0?
c. Estimate the linear relationship between these two variables, $Y = \alpha_1 + \beta_1 X + \varepsilon_1$.
d. Estimate the log-linear relationship between the natural logarithms of these two variables, $\ln[Y] = \alpha_2 + \beta_2 \ln[X] + \varepsilon_2$.
e. Which of these two estimated relationships seems to fit these data better?

8.50 The high desert area of Southern California is 80 to 90 miles from Newport Beach, Disneyland, and the Staples Center, and perhaps more significantly, 35–45 miles from jobs in the San Bernardino area. A study of the high desert housing market looked at the annual data in Table 8.21 on building permits for single-family homes and the ratio of the median home price in the high desert to the median home price in San Bernardino.
a. Which variable is the dependent variable?
b. Do you expect the relationship between these two variables to be positive, negative, or 0?

Table 8.20: Exercise 8.49

Driving Time	Gambling Activity	Driving Time	Gambling Activity	Driving Time	Gambling Activity	Driving Time	Gambling Activity
0.35	1.2897	1.35	0.0650	1.52	0.0462	1.67	0.0104
0.67	0.3783	1.35	0.0405	1.53	0.0026	1.90	0.0094
0.82	0.2332	1.35	0.0033	1.57	0.0340	2.00	0.0031
0.98	0.1299	1.37	0.0943	1.58	0.0173	2.35	0.0182
1.07	0.1198	1.38	0.0213	1.58	0.0153	2.55	0.0048
1.22	0.0318	1.43	0.0277	1.58	0.0130	2.78	0.0001
1.22	0.0503	1.47	0.0233	1.63	0.0059	3.02	0.0010
1.27	0.0075	1.52	0.0351	1.63	0.0379	3.65	0.0020

Table 8.21: Exercise 8.50

Year	Price Ratio	Building Permits
1997	0.844	152
1998	0.833	200
1999	0.741	315
2000	0.791	419
2001	0.690	655
2002	0.645	986
2003	0.519	2,102
2004	0.635	2,699
2005	0.649	2,258
2006	0.548	3,078
2007	0.625	1,090

c. Estimate the linear relationship between these two variables, $Y = \alpha_1 + \beta_1 X + \varepsilon_1$.
d. Estimate the log-linear relationship between the natural logarithms of these two variables, $\ln[Y] = \alpha_2 + \beta_2 \ln[X] + \varepsilon_2$.
e. Which of these two estimated relationships seems to fit these data better?

CHAPTER 9

The Art of Regression Analysis

Chapter Outline

> *There are few statistical facts more interesting than regression to the mean for two reasons. First, people encounter it almost every day of their lives. Second, almost nobody understands it.*
>
> —***Anonymous journal referee***

Regression models are estimated by using software to calculate the least squares estimates, t values, P values, and R^2. But there is more to good regression analysis than entering data in a software program. The art of regression analysis involves

1. Specifying a plausible model.
2. Obtaining reliable and appropriate data.
3. Interpreting the output.

Throughout this book we have emphasized the importance of beginning with a believable research hypothesis, for example, that income affects spending or that interest rates affect stock prices. The first theme of this chapter is that problems can arise when there is no solid research hypothesis—when there is data without theory. A second theme is the importance of good data. As is often said: garbage in, garbage out. What that means in this context is that poor data yield unreliable estimates. This chapter's third theme is that it is not enough to specify a reasonable model, obtain good data, and plug the numbers into a software package. We should scrutinize the output for clues of possible errors and omissions.

9.1 Regression Pitfalls

Although regression is very elegant and powerful, it can also be misused and misinterpreted. We consider several pitfalls to be avoided.

Significant Is Not Necessarily Substantial

It is easy to confuse statistical significance with practical importance. The estimated linear relationship between two variables is statistically significant at the 5 percent level if the estimated slope is more than (approximately) two standard deviations from 0. Equivalently, a 95 percent confidence interval does not include 0.

This does not mean that the estimated slope is large enough to be of practical importance. Suppose that we use 100 observations to estimate the relationship between consumption C and household wealth W, each in billions of dollars:

$$\begin{aligned} C = {} & 15.20 + 0.000022W \\ & (2.43) \quad (0.000010) \end{aligned}$$

The standard errors are in parenthesis. The t value for testing the null hypothesis that the slope is 0 is 2.2:

$$\begin{aligned} t &= \frac{b - 0}{\text{standard error of } b} \\ &= \frac{0.000022}{0.000010} \\ &= 2.2 \end{aligned}$$

which gives a two-sided P value less than 0.05.

With $100 - 2 = 98$ degrees of freedom, $t^* = 1.9846$ and a 95 percent confidence interval excludes 0:

$$\begin{aligned} b \pm t^* \text{SE}[b] &= 0.000022 \pm 1.9846(0.000010) \\ &= 0.000022 \pm 0.000020 \end{aligned}$$

There is a statistically significant relationship between wealth and spending.

Is this relationship substantial? The estimated slope is 0.000022, which means that a \$1 billion increase in wealth is predicted to increase spending by \$0.000022 billion, which is \$22,000. Is this of any practical importance? Probably not. Practical importance is admittedly subjective. The point is that it is not the same as statistical significance.

It can also turn out that the estimated slope is large but not statistically significant, as in this example:

$$\begin{aligned} C = {} & 15.20 + 0.22W \\ & (2.43) \quad (0.12) \end{aligned}$$

Now the estimated slope is 0.22, which means that a $1 billion increase in wealth is predicted to increase spending by $0.22 billion, which is a substantial amount.

However, the t value is less than 2:

$$t = \frac{b - 0}{\text{standard error of } b} = \frac{0.22}{0.12} = 1.8$$

and a 95 percent confidence interval includes 0:

$$b \pm t^{*}\text{SE}[b] = 0.22 \pm 1.9846(0.12) = 0.22 \pm 0.24$$

In this second example, the estimated effect of an increase in wealth on spending is large but not statistically significant at the 5 percent level. That is the point. Statistical significance and practical importance are separate questions that need to be answered separately.

Correlation Is Not Causation

Regression models are sometimes used simply to quantify empirical relationships. We look at average life expectancies over time, not because we think that the mere passage of time causes life expectancies to change, but because we want to see if there has been a long-run trend in life expectancies. We look at the statistical relationship between midterm and final examination scores, not because we believe that midterm scores cause final scores, but because we want to see how well one test score predicts the other. In these cases, we say that the variables are correlated statistically, without suggesting a causal relationship.

In other cases, we use a regression model to quantify or test our belief that changes in the explanatory variable *cause* changes in the dependent variable. Keynes' consumption function was motivated by his belief that household spending is affected by income. The Fed's fair-value stock model was motivated by the belief that stock prices are affected by interest rates. If we attach a causal interpretation to our regression results, we should have this causal theory in mind *before* we estimate a regression model.

Researchers sometimes stumble upon an unexpected correlation then search for a plausible explanation. This is what Ed Leamer calls "Sherlock Holmes inference," since Sherlock Holmes was famous for "reasoning backwards" by constructing theories that fit the data. In statistics, reasoning backward sometimes leads to a new and useful model but often leads to useless models. Unfortunately, there are innumerable examples of statistical correlations without causation. Two safeguards are: (a) use common sense to judge a model's plausibility; and (b) retest the model using fresh data.

There are three reasons why a causal interpretation of an empirical correlation may be misleading: simple chance, reverse causation, and omitted factors. The first explanation, simple chance, refers to the fact that unrelated data may be coincidentally correlated. In the model $Y = \alpha + \beta X + \varepsilon$, X has a statistically significant effect on Y at the 5 percent level if there is less than a 5 percent chance that the estimate of β would be so far from 0, if β really is 0. What we must recognize is that, even if β really is 0, there is a 5 percent chance that the data will indicate a statistically significant relationship. Suppose, for example, that we use a random number generator to create 200 numbers and call the first 100 numbers X and the remaining 100 numbers Y. Even though the values of X and Y are randomly generated, if we run a least squares regression of Y on X, there is a 5 percent chance that the empirical relationship will be statistically significant at the 5 percent level. A hapless researcher who spends his entire life looking at unrelated variables will find statistical significance on about 1 out of every 20 tries. We can hope that no one is so misguided as to study only unrelated variables, but we should also recognize that statistical correlation without a plausible explanation may be a coincidence—particularly if the researcher estimates dozens (or hundreds) of equations and reports only those with the lowest P values.

To illustrate the dangers of data grubbing, an economist once ran a large number of regressions and found an astounding correlation between the stock market and the number of strikeouts by a professional baseball team [1]. British researchers found a very high correlation between the annual rate of inflation and the number of dysentery cases in Scotland the previous year [2]. We should not be impressed with a high, or even spectacular, correlation unless there is a logical explanation, and we should be particularly skeptical when there has been data grubbing.

The second reason why an observed correlation may be misleading is that perhaps Y determines X rather than the other way around. Okun's law originated with Arthur Okun's observation that when demand increases, firms meet this demand by hiring more workers, which reduces the unemployment rate. In Chapter 8, we estimated this equation:

$$\begin{aligned} U &= 1.301 - 0.376Y, \quad R^2 = 0.73 \\ &\quad\ [10.28] \quad [12.31] \end{aligned}$$

where U is the change in the unemployment rate and Y is the percent change in real gross domestic product (GDP). By this reasoning, the increase in demand causes firms to employ more workers, thereby reducing the unemployment rate.

The empirical relationship can also be expressed the other way around. If firms hire more workers, perhaps because the labor supply increases, unemployment will fall and output will increase. If we reverse the dependent variable and explanatory variable, our least squares estimated equation is

$$\begin{aligned} Y &= 3.449 - 1.932U, \quad R^2 = 0.73 \\ &\quad\ [21.44] \quad [12.31] \end{aligned}$$

The value of R^2 and the t value for the slope are not affected by the reversal of the dependent variable and explanatory variable. The estimated slope does change. In the first case, the estimated coefficients minimize the sum of squared errors in predicting unemployment and, in the second case, the estimated coefficients minimize the sum of squared errors in predicting output. These are different criteria and the results are seldom identical.

Here, the first equation predicts that a 1 percent increase in real GDP will reduce the unemployment rate by 0.376 percentage points. If so, a 1 percentage-point increase in the unemployment rate implies a 2.66 percent decrease in real GDP:

$$\frac{1}{-0.376} = -2.66$$

In contrast, the second equation predicts that a 1 percentage-point increase in the unemployment rate will reduce real GDP by 1.93 percent.

As stated earlier, the R^2 and the t value for the slope are not affected by the reversal of the dependent variable and the explanatory variable. Either way, these statistics tell us that there is a highly significant correlation between output and unemployment. What they cannot tell us is which variable is causing changes in the other. For that, we need logical reasoning, not t values. Sometimes, the causal direction is clear-cut. If we are looking at the relationship between exposure to radioactive waste and cancer mortality, it is obvious that exposure to radioactive waste may increase cancer mortality, but cancer mortality does not cause exposure to radioactive waste.

Other cases can be more subtle. Data from six large medical studies found that people with low cholesterol levels were more likely to die of colon cancer [3]; however, a later study indicated that the low cholesterol levels may have been caused by colon cancer that was in its early stages and therefore undetected [4]. For centuries, residents of New Hebrides believed that body lice made a person healthy. This folk wisdom was based on the observation that healthy people often had lice and unhealthy people usually did not. However, it was not the absence of lice that made people unhealthy, but the fact that unhealthy people often had fevers, which drove the lice away.

The third reason why impressive empirical correlations may be misleading is that an omitted factor may affect both X and Y. In the regression model $Y = \alpha + \beta X + \varepsilon$, we assume that the omitted factors collected in the error term ε are uncorrelated with the explanatory variable X. If they are correlated, we may underestimate or overestimate the effect of X on Y, or even identify an effect when there is none. Many Arizona residents die from bronchitis, emphysema, asthma, and other lung diseases, but this does not mean that Arizona's climate is especially hazardous to one's lungs. On the contrary, doctors recommend that patients with lung problems move to Arizona for its beneficial climate. Many do move to Arizona and benefit from the dry, clean air. Although many eventually die of lung disease in Arizona, it is not because of Arizona's climate.

A less morbid example involves a reported positive correlation between stork nests and human births in northwestern Europe. Few believe that storks bring babies. A more logical explanation is that storks like to build their nests on buildings. Where there are more people, there are usually more human births as well as more buildings for storks to build nests.

The multiple regression procedures introduced in the next chapter provide a way to deal with the confounding effects of omitted variables. For example, if we want to investigate the influence of a particular diet on death rates and know that death rates depend on the age of the population, we can estimate a multiple regression equation that includes several explanatory variables—diet, age, and anything else we believe to be important.

A particularly common source of coincidental correlations is that many variables are related to the size of the population and tend to increase over time as the population grows. If we pick two such variables at random, they may appear to be highly correlated, when in fact they are both affected by a common omitted factor—the growth of the population. With only a small amount of data grubbing, I uncovered the simple example in Table 9.1 using annual data on the number of U.S. golfers and the total number of missed workdays due to reported injury or illness (both in thousands).

A least squares regression gives

$$\begin{array}{ll} D = 304{,}525 + 12.63G, & R^2 = 0.91 \\ \quad\quad [14.19] \quad\quad [6.37] & \end{array}$$

with the t values in brackets. This relationship is highly statistically significant and very substantial—every additional golfer leads to another 12.6 missed days of work. It is semi-plausible that people may call in sick to play golf (or that playing golf may cause injuries). But, in fact, most workers are not golfers and most missed days are not spent playing golf or recovering from golf. The number of golfers and the number of missed days both increased over time, not because one was causing the other, but because both were growing with the population.

A simple way to correct for the coincidental correlation caused by a growing population is to deflate both sets of data by the size of the population, P, giving us per capita data. Table 9.2 shows per capita data on golfers and missed days.

Table 9.1: Golf Addiction

	Golfers (G)	Missed Days (D)
1960	4,400	370,000
1965	7,750	400,000
1970	9,700	417,000
1975	12,036	433,000
1980	13,000	485,000
1985	14,700	500,000

Table 9.2: Golf per Capita

	Golfers per Capita (G/P)	Missed Days per Capita (D/P)
1960	0.02435	2.0479
1965	0.03989	2.0586
1970	0.04731	2.0336
1975	0.05472	1.9684
1980	0.05708	2.1295
1985	0.06144	2.0896

There was a large increase in the number of golfers per capita, while missed days per capita has no discernible trend. A least squares regression gives

$$\frac{G}{P} = 2.01 + 0.76\frac{D}{P}, \quad R^2 = 0.04$$
$$[21.05] \quad [0.39]$$

Once we take into account the growing population by converting our data to per capita values, the coincidental correlation disappears and we find that there is virtually no statistical relationship between the number of golfers and the number of missed workdays.

Detrending Time Series Data

This kind of spurious correlation is especially likely to occur with time series data, where both X and Y trend upward over time because of long-run increases in population, income, prices, or other factors. When this happens, X and Y may appear to be closely related to each other when, in fact, both are growing independently of each other.

In Chapter 8 we estimated Keynes' consumption function using time series data for the years 1929–1940. Are these estimates tainted by the fact that income and spending both trend upward over time? Table 1.1, repeated here as Table 9.3, shows the time series data used to estimate this consumption function. Because of the Great Depression, income and spending were actually somewhat lower in 1940 than in 1929. It is probably not the case that our earlier estimates are tainted by common trends, but we will check.

There are a variety of ways to adjust time series data for trends. The simplest way is to work with changes or percentage changes. For each year, we calculate the percentage change in income and spending from the preceding year. For example, the percentage change in income in 1930 was

$$\begin{aligned} \text{percentage change} &= 100\left(\frac{\text{new} - \text{old}}{\text{old}}\right) \\ &= 100\left(\frac{74.7 - 83.4}{83.4}\right) \\ &= -10.43 \end{aligned}$$

Table 9.3: U.S. Disposable Personal Income and Consumer Spending, Billions of Dollars

	Income	Spending
1929	83.4	77.4
1930	74.7	70.1
1931	64.3	60.7
1932	49.2	48.7
1933	46.1	45.9
1934	52.8	51.5
1935	59.3	55.9
1936	67.4	62.2
1937	72.2	66.8
1938	66.6	64.3
1939	71.4	67.2
1940	76.8	71.3

Table 9.4: Percentage Changes in Income and Spending

	Income	Spending
1930	−10.43	−9.43
1931	−13.92	−13.41
1932	−23.48	−19.77
1933	−6.3	−5.75
1934	14.53	12.2
1935	12.31	8.54
1936	13.66	11.27
1937	7.12	7.4
1938	−7.76	−3.74
1939	7.21	4.51
1940	7.56	6.1

Table 9.4 shows the percentage changes in income and spending for each year. We then estimate the simple regression model using the percentage changes in spending and income instead of the levels of spending and income. Figure 9.1 shows the data and the least squares line.

Another way to detrend time series data is to fit a variable to a trend line and calculate the deviations from this trend line. For example, Figure 9.2 shows income along with the trend line obtained by estimating a linear regression model with income as the dependent variable and year as the explanatory variable.[1] Detrended income in each year is equal to the difference between observed income and trend income.

[1] If the percentage growth rate is more relevant, the trend lines can also be estimated from a regression model with the logarithm of the variable as the dependent variable.

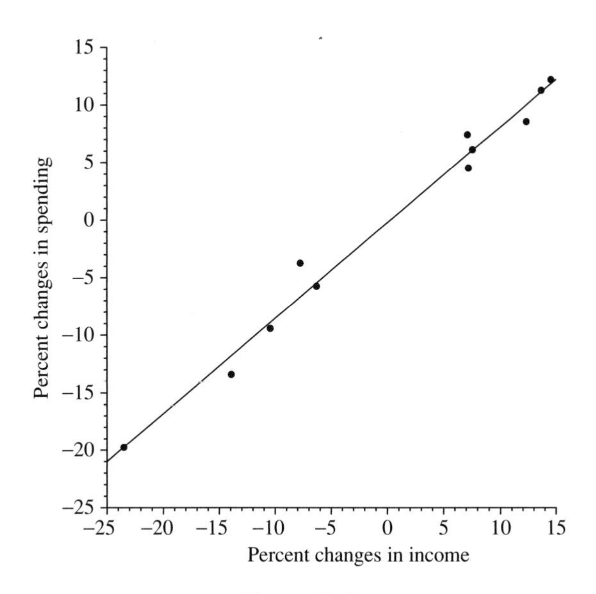

Figure 9.1
The relationship between percentage changes in income and spending.

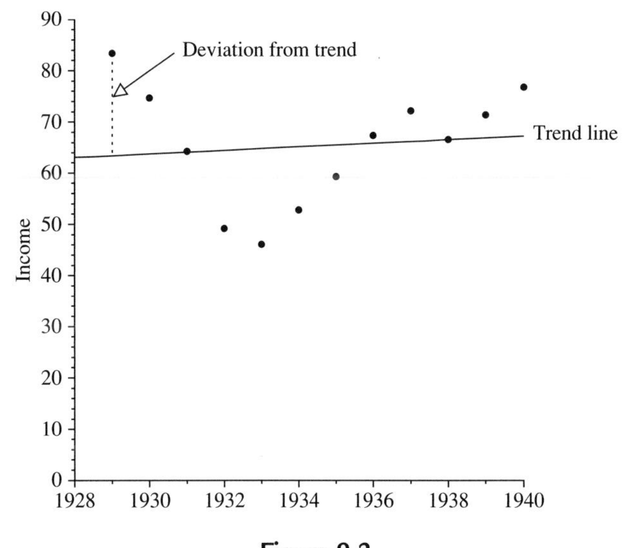

Figure 9.2
Fitting income to a trend line.

Consumption can be detrended in the same way. We then estimate a regression model using the deviations of spending and income from their trends. Figure 9.3 shows the data and the least squares line.

Table 9.5 summarizes the results of estimating a Keynesian consumption function using the levels of spending and income, percentage changes in spending and income, and the

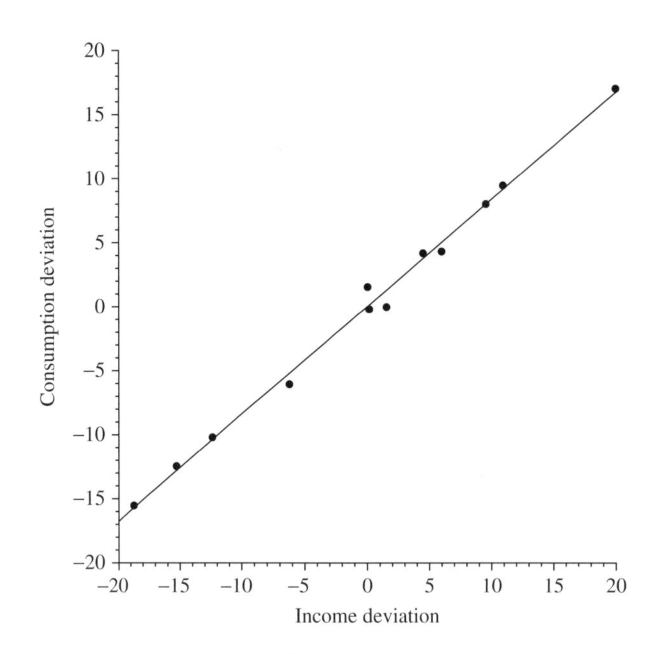

Figure 9.3
The relationship between the deviations of income and spending from trend.

Table 9.5: Consumption Function Estimates

Data	Slope	*t* Value	R^2
Levels	0.837	41.42	0.994
Percent changes	0.831	23.72	0.984
Deviations from trend	0.837	41.42	0.994

deviations of spending and income from their trend values. The results are not perfectly comparable since the variables are not identical. Nonetheless, it is clear that detrending the data has little effect on our earlier conclusion that there is a substantial and highly statistically significant relationship between income and spending. Our initial estimates of Keynes' consumption function were evidently not tainted by the use of time series data. Exercises 9.37 and 9.38 show a very different example, using hypothetical income and spending data where detrending has a very large effect on the results.

Incautious Extrapolation

When we use a regression model to make a prediction, it is a good idea to calculate a 95 percent or 99 percent prediction interval to gauge how much uncertainty there is in our prediction. Prediction intervals are narrowest at the average value of the explanatory variable and get wider as we move farther away from the mean, warning us that there is more uncertainty about predictions on the fringes of the data.

Even a very wide prediction interval may understate how much uncertainty there is about predictions far from the sample mean because the relationship between the two variables may not always be linear. Linear models are very convenient, and often very successful. However, a linear model assumes that, for all values of X, the effect on Y of a one-unit change in X is a constant, β. It is often the case that, as X increases, the effect of X on Y eventually either becomes stronger or weaker, which means that the relationship is no longer linear. Fertilizer increases crop yields, but eventually the effect gets smaller and then turns negative because too much fertilizer damages crops. College graduates, who go to school for 16 years, generally earn substantially higher incomes than do high school graduates. However, incomes do not keep increasing by a constant amount if a person goes to school for 20, 24, 28, or 32 years.

Figure 9.4 shows a consumption function that is a linear approximation to a hypothetical true relationship between income and spending. The linear approximation works fine for values of income between, say, 10 and 25, but fails miserably for values of income far outside this range. Before using a model to make predictions far from the data used to estimate the model, think about whether it is reasonable to assume that the effect of a change in X on Y is constant.

Sometimes, extrapolations fail because a major event disrupts a historical pattern or relationship. Figure 9.5 shows U.S. cigarette sales per capita increasing steadily between 1900 and 1963. A regression line fit to these data has an R^2 of 0.95, a t value of 34.47, and a two-sided P value of 6.7×10^{-43}. Surely, we can use this regression equation to project cigarette sales reliably for many years to come.

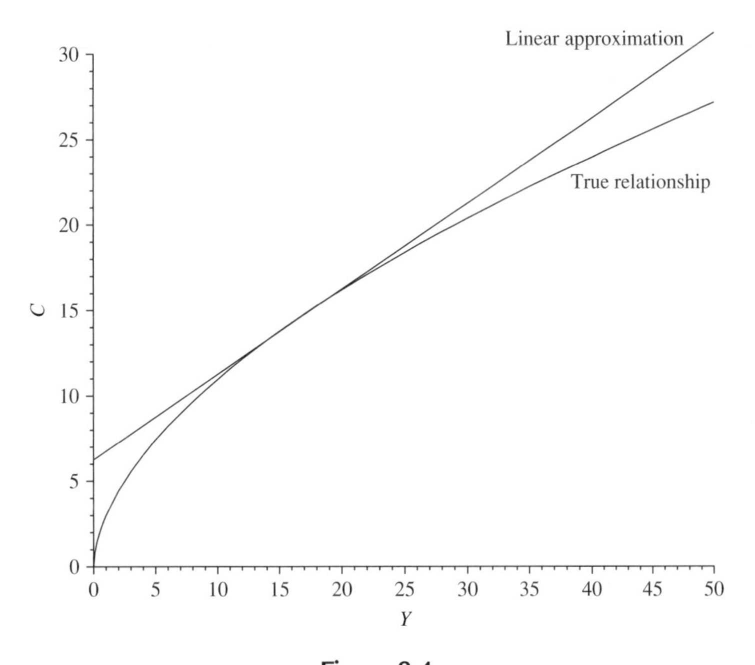

Figure 9.4
Be careful about extrapolating linear approximations.

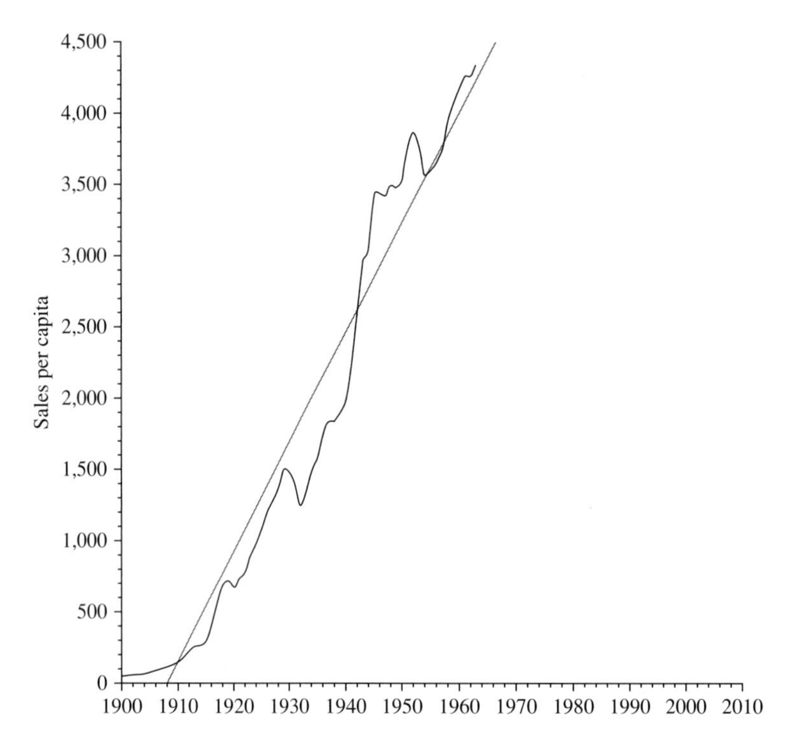

Figure 9.5
Cigarette sales increase relentlessly.

Not so fast! In 1964, the U.S. Surgeon General released a widely publicized report warning of the health risks associated with smoking. Figure 9.6 shows that U.S. cigarette sales per capita declined steadily after the publication of this report.

Regression Toward the Mean

Suppose that a college student named Cory takes a test of his soccer knowledge. We can think of Cory's ability as the expected value of his score—his average score on a large number of tests. Some students have an ability of 90, some 80, and so on. Cory's ability is 80, but his score on any single test depends on the questions asked. Imagine a test pool consisting of a very large number of possible questions, from which 100 questions are selected. Although Cory knows the correct answers to 80 percent of all questions, it can be shown that there is only about a 10 percent chance that he will know the answers to exactly 80 out of 100 randomly selected questions. There is a 17 percent chance that his score will be below 75 or above 85. He will average 80 percent correct on a large number of tests, but his scores vary substantially from test to test. His scores can be also affected by other factors, such as his health and the environment in which the test is given.

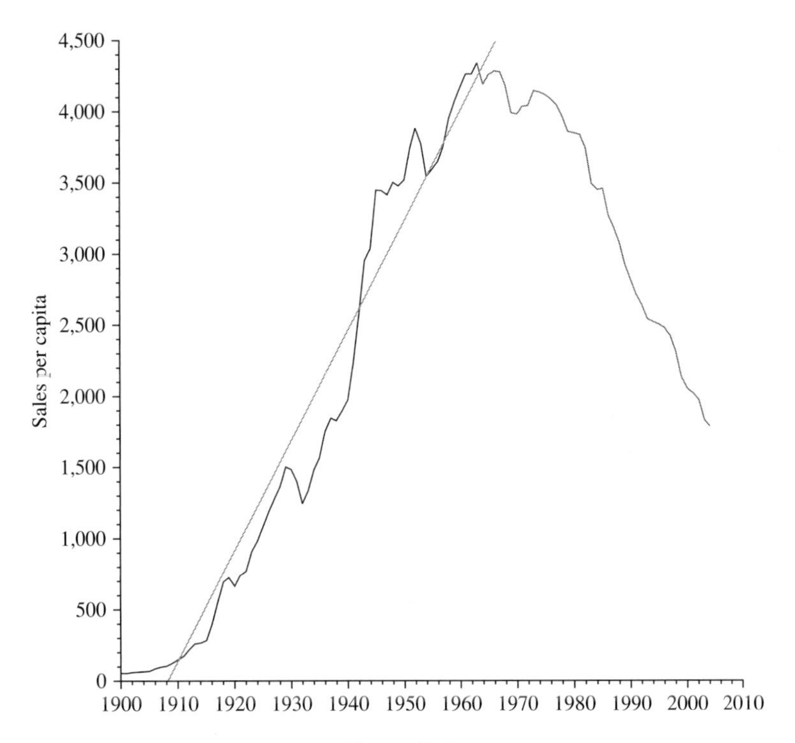

Figure 9.6
Per capita cigarette sales collapse after the 1964 report of the Surgeon General.

The challenge for researchers is that we observe Cory's test score, not his ability, and we don't know whether his score is above or below his ability. There is a clue with important implications. A student whose test score is high relative to other students probably scored above his own ability. For example, a student who scores in the 90th percentile is more likely to be someone of somewhat more modest ability who did unusually well than to be someone of higher ability who did poorly, because there are more people with abilities below 90 than above 90. When he takes another test—the next day, the next week, or 3 years later—he will probably not do as well. Students and teachers should anticipate this drop off and not blame themselves if it occurs. Similarly, students who score far below average are more likely to have had bad luck than good luck and can anticipate scoring somewhat higher on later tests. Their subsequently higher scores may be a more accurate reflection of their ability rather than an improvement in their ability.

This tendency of people who score far from the mean to score closer to the mean on a second test is an example of a statistical phenomenon known as *regression to the mean*.

There is well-established evidence that most people (even statisticians) are largely unaware of regression to the mean. For example, a study reported improved test scores for college students whose initial scores placed them in a remedial group that was given special tutoring. A skeptic noted that, because of regression to the mean, their scores could be expected to improve on a second test even if the instructor "had no more than passed his hand over the students' heads before he retested them" [5].

Many data regress toward the mean. Sir Francis Galton (1822–1911) observed regression to the mean in his study of the relationship between the heights of parents and their adult children [6]. Because the average male height is about 8 percent larger than the average female height, he multiplied the female heights by 1.08 to make them comparable to the male heights. The heights of each set of parents were then averaged to give a "mid-parent height." The mid-parent heights were divided into nine categories and he calculated the median height of the children of parents in each category. Figure 9.7 shows a least squares line fit to his data.

The slope of the least squares line is 0.69, which means that every inch that the parents' height is above the mean, the child's height is only 0.69 inches above the mean. And every inch that the parents' height is below the mean, the child's height is only 0.69 inches below the mean. This is regression to the mean.

An incorrect interpretation of this statistical phenomenon is that everyone will soon be the same height. That is analogous to saying that everyone will soon get the same score on a test of soccer knowledge. That is *not* what regression to the mean implies. Regression to the mean says that, because test scores are an imperfect measure of ability, those who score highest on any single test probably scored above their ability and will not do as well on a second test. Not that they will score below average, but that they will score closer to the mean. There will still be high scorers on a second test, including students who are above average in ability and happen to score above their ability. The observation that high scorers probably scored above their ability does not imply that abilities are converging.

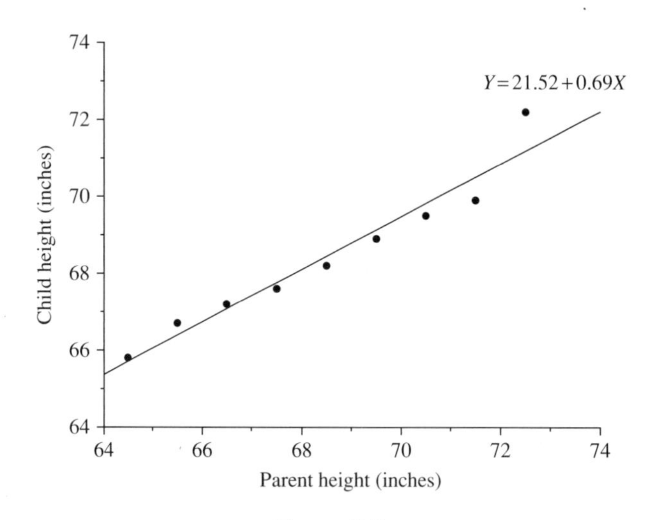

Figure 9.7
Height regression to the mean.

Similarly, athletic performances are an imperfect measure of skills and consequently regress. Among Major League Baseball players who have batting averages of 0.300 or higher in any season, 80 percent do worse the following season [7]. Of those MLB teams that win more than 100 out of 162 baseball games in a season, 90 percent do not do as well the next season. The regression toward the mean fallacy is to conclude that the skills of good players and teams deteriorate. The correct conclusion is that those with the best performance in any particular season usually are not as skillful as their lofty record suggests. Most have had more good luck than bad, so that this season's record is better than the season before or the season after.

An economic example is a book with the provocative title *The Triumph of Mediocrity in Business* [8], which was written by a statistics professor. This professor found that businesses with exceptional profits in any given year tend to have smaller profits the following year, while firms that do poorly usually do better the next year. From this evidence, he concluded that strong companies were getting weaker, and weak companies stronger, so that soon all will be mediocre. The president of the American Statistical Association wrote an enthusiastic review of this book, but another statistician pointed out that the author had been fooled by regression to the mean. In any given year, companies with very high (or low) profits probably experienced good (or bad) luck. While their subsequent performance will usually be closer to the mean, their places at the extremes will be taken by other companies experiencing fortune or misfortune.

An investments textbook written by a Nobel Prize winner makes this same error [9]. The author discusses a model of stock prices that assumes, "ultimately, economic forces will force the convergence of the profitability and growth rates of different firms." To support this assumption, he looked at the 20 percent of firms with the highest profit rates in 1966 and the 20 percent with the lowest profit rates. Fourteen years later, in 1980, the profit rates of both groups are more nearly average: "convergence toward an overall mean is apparent... the phenomenon is undoubtedly real." The phenomenon is regression toward the mean, and the explanation is statistical, not economic.

The Real Dogs of the Dow

The Dow Jones Industrial Average contains 30 blue chip stocks that represent the nation's most prominent companies. When companies falter, they are replaced by more successful companies. Regression to the mean suggests that companies taken out of the Dow may not be as bad as their current predicament indicates and the companies that replace them may not be as terrific as their current performance suggests. If investors are insufficiently aware of this statistical phenomenon, stock prices may be too low for the former and too high for the latter—mistakes that will be corrected when these companies regress to the mean. Thus, stocks taken out of the Dow may outperform the stocks that replace them. This hypothesis was tested [10] with the 50 substitutions made since the Dow expanded to 30 stocks in 1928.

Two portfolios were followed, one consisting of stocks that were deleted from the Dow and the other stocks that were added. Whenever a Dow substitution was made, the deleted stock was added to the deletion portfolio and the replacement stock was added to the addition portfolio. The 20,367 daily differences between the percentage returns on the deletion portfolio and the addition portfolio are matched-pair data that can be used to test the null hypothesis that the expected value of the difference is equal to 0.

These differences had a mean of 0.0155 percent and a standard deviation of 0.8344 percent. The t value for testing the null hypothesis that the population mean is 0 is

$$t = \frac{0.0155}{0.8344/\sqrt{20{,}367}} \\ = 2.65$$

The two-sided P value is 0.0080. This observed difference is substantial. The deletion portfolio had an annual return of 14.5 percent, the addition portfolio only 10.1 percent. Compounded over nearly 80 years, the deletion portfolio grew to 21 times the size of the addition portfolio.

9.2 Regression Diagnostics (Optional)

Regression analysis assumes that the model is correctly specified, for example, that the error term ε has a constant standard deviation and the values of the error term are independent of each other. Violations of these two assumptions do not bias the estimates of the slope and intercept, but their standard deviations are unnecessarily large and underestimated, causing us to overestimate the statistical significance of our results. Nobel laureate Paul Samuelson advised econometricians to "always study your residuals." A study of a regression model's residuals (the deviations of the scatter points from the fitted line) can reveal outliers—perhaps an atypical observation or a clerical error—or violations of the assumptions that the error term has a constant standard deviation and independent values. It is generally a good idea to use a scatter diagram to look at the residuals and the underlying data.

Looking for Outliers

In the early 1990s, researchers collected data for nine cigarette brands on each brand's percentage of total cigarette advertising and percentage of total cigarettes smoked by persons 12 to 18 years of age [11]. A least squares regression yielded a positive relationship with a t value of 5.83 and a two-sided P value of 0.001, leading to nationwide newspaper stories about the relationship between cigarette advertising and teen smoking. However the scatter diagram in Figure 9.8 shows that these results are due entirely to the Marlboro outlier. For the eight other brands, there is a negative relationship between advertising and smoking, although it is not statistically significant at the 5 percent level. (The two-sided P value is 0.15.)

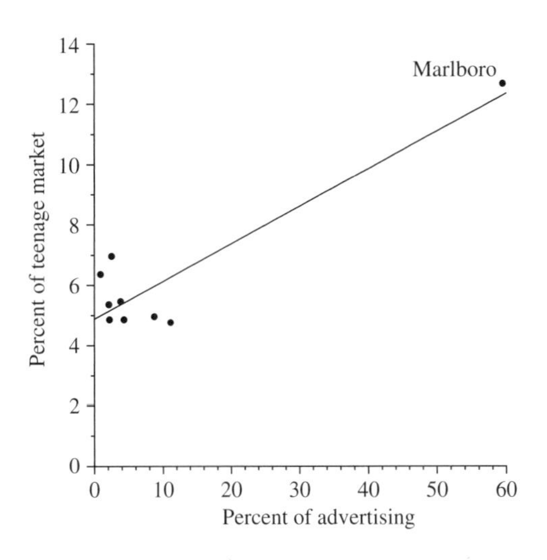

Figure 9.8
The Marlboro outlier.

The reader must decide whether Marlboro is decisive evidence or a misleading outlier. If we had not looked at a scatter diagram, we would not have known that this is the real issue.

Checking the Standard Deviation

Suppose that we are investigating the variation among cities in annual spending at restaurants. Believing that spending depends primarily on a city's aggregate income, we use the following data for ten cities:

Income X (millions)	620	340	7,400	3,000	1,050	2,100	4,300	660	2,300	1,200
Spending Y (millions)	96	46	1,044	420	131	261	522	95	283	156

Least squares estimation of the simple regression model yields

$$Y = -10.99 + 0.138X, \quad R^2 = 0.992$$
$$[0.8] \qquad [31.0]$$

[ ]: t values

The results seem satisfactory. The relationship between income and restaurant spending is, as expected, positive and, as hoped, statistically significant. The value of R^2 is nearly 1.

Following Samuelson's advice, the prediction errors are plotted in Figure 9.9 with income on the horizontal axis. The residuals appear to be larger for high-income cities. On reflection, it is plausible that the errors in predicting restaurant spending in large cities tend to be larger than the errors in predicting spending in small cities, violating our assumption that the standard deviation of the error term is constant.

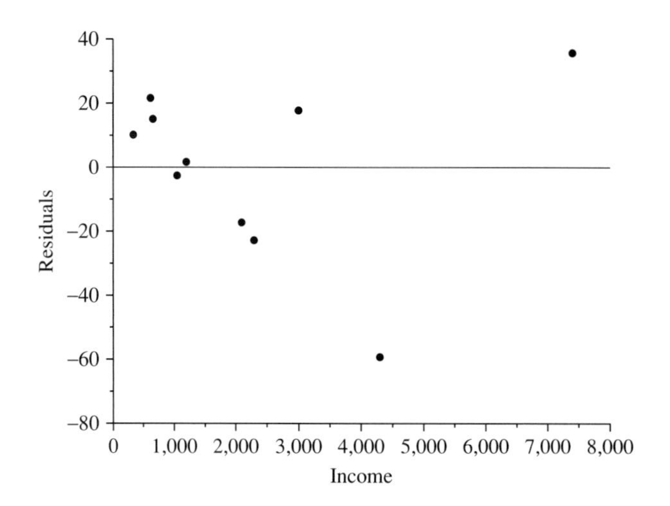

Figure 9.9
Spending-income residuals.

A variety of formal statistical procedures for testing this assumption are described in more advanced courses. Here, we discuss one possible solution. The error term encompasses the effects of omitted variables. With cross-section data, the scale of the omitted variables may be larger for some observations (like large cities); with time series data, the scale of the omitted variables may increase over time. One solution is to respecify the model, rescaling the variables so that the error term has a constant standard deviation. In our restaurant spending example, the problem is apparently caused by the fact that some cities are much larger than others. To give the data a comparable scale, we can respecify our model in terms of per capita income and per capita spending, using the following data on population *P*, in thousands:

P	40	20	400	150	50	100	200	30	100	50
X/P	15,500	17,000	18,500	20,000	21,000	21,000	21,500	22,000	23,000	24,000
Y/P	2,410	2,310	2,610	2,800	2,620	2,610	2,610	3,160	2,830	3,120

Least squares estimation of a regression equation using per capita data yields

$$\frac{Y}{P} = 1{,}020 + 0.083\frac{X}{P}, \quad R^2 = 0.641$$
$$[2.3] \qquad [3.8]$$

[ ]: *t* values

The plot of the residuals in Figure 9.10 is a big improvement. Evidently, the problem was caused by the use of aggregate data for cities of very different sizes and solved by a rescaling to per capita data.

Notice that this rescaling caused a substantial change in the estimated slope of the relationship. Our initial regression implies that a $1 increase in income increases restaurant spending by

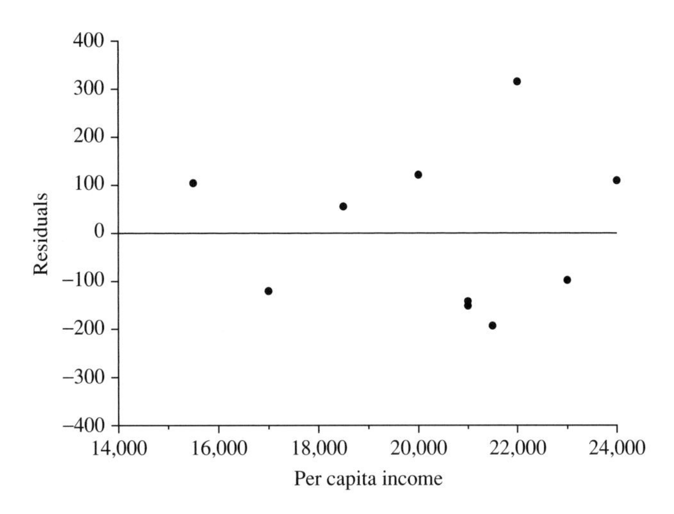

Figure 9.10
Per capita sales-income residuals.

\$0.138. The rescaled regression gives an estimate that is 40 percent lower: A \$1 increase in per capita income increases per capita spending by \$0.083. If our rescaled estimates are an improvement, then our initial estimates were misleading.

Some may be concerned by the fact that the rescaling caused a substantial decline in the t value and R^2, suggesting that the rescaled model is not an improvement. However, we cannot compare these statistical measures directly because the variables are different. The first R^2 measures the accuracy in predicting aggregate spending; the second R^2 measures the accuracy in predicting per capita spending. If, as here, aggregate spending varies more than per capita spending, the per capita model could be more accurate and yet have the lower R^2.

To compare the predictive accuracy of these two models, we need to focus on a single variable—either aggregate spending or per capita spending. For the unscaled model, predicted aggregate spending is given by the regression equation

$$\hat{Y} = -10.99 + 0.138X$$

For the per capita model, aggregate spending can be predicted by multiplying the city's population times predicted per capita spending:

$$\begin{aligned}\hat{Y} &= P\left(\frac{Y}{P}\right)\\ &= P\left(1{,}020 + 0.083\frac{X}{P}\right)\end{aligned}$$

Table 9.6 shows these predictions. The per capita model generally has smaller prediction errors, often substantially so. The standard error of estimate (the square root of the sum of the

Table 9.6: Aggregate Predictions of the Unscaled and per Capita Models

	Unscaled Model		Per Capita Model	
Actual Spending	Predicted ($\hat{Y}$)	Error	Predicted $P(Y/P)$	Error
96	74.4	22.0	92.2	4.2
46	35.8	10.4	48.6	−2.4
1,044	1,008.0	36.0	1,021.8	22.2
420	402.1	17.9	401.8	18.2
131	133.6	−2.6	138.1	−7.1
261	278.2	−17.2	276.2	−15.2
522	581.1	−59.1	560.6	−38.6
95	79.9	14.9	85.3	9.5
283	305.7	−22.7	292.7	−9.7
156	154.3	1.7	150.5	5.5

squared prediction errors, divided by $n-2$) is 29.0 for the aggregate model and 18.8 for the per capita model. Despite its lower t value and R^2, the per capita model fits the data better than the aggregate model!

This example illustrates not only a strategy for solving the problem of unequal standard deviations, but also the danger of overrelying on t values and R^2 to judge a model's success. A successful model should have plausible estimates, and residuals that are consistent with the model's assumptions.

Independent Error Terms

Sometimes, a scatter diagram of the residuals reveals that the residuals are not independent of each other. Figure 9.11 shows some consumption function residuals that have a definite pattern: negative residuals for low values of income, then positive, then negative again. This systematic pattern indicates that the residuals are not independent.

Figure 9.12 shows why the residuals have this particular pattern: There is a nonlinear relationship between income and spending. When a linear equation is fit to these nonlinear data, the residuals are negative for low values of income, positive for middle values, and negative for high values. The solution is to fit a nonlinear equation to these nonlinear data.

Figure 9.13 shows an example of nonindependent residuals using time series data. These data are from the Fed's stock valuation model in Figure 8.10. The residuals are plotted in Figure 9.13, not against the explanatory variable, but against time, to show the sequence in which the residuals appear. These residuals are positively correlated in that positive values tend to be followed by positive values and negative values by negative values. In this case, the cause might be a pattern in an omitted variable that is part of the error term. For example, Keynes wrote that investors' "animal spirits" sometimes cause stock prices to wander away from fundamental values for extended periods of time. If so, the Fed's fundamental value model might have positive residuals for several months (or even years) and negative residuals for other time periods.

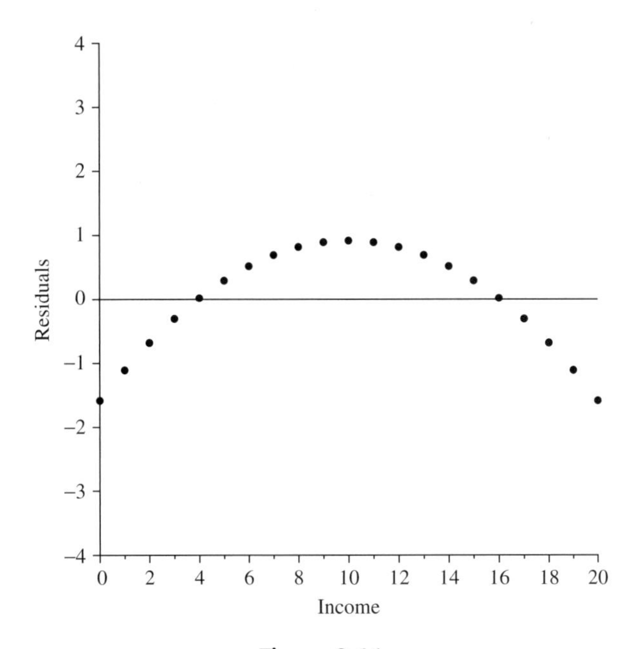

Figure 9.11
Suspicious residuals.

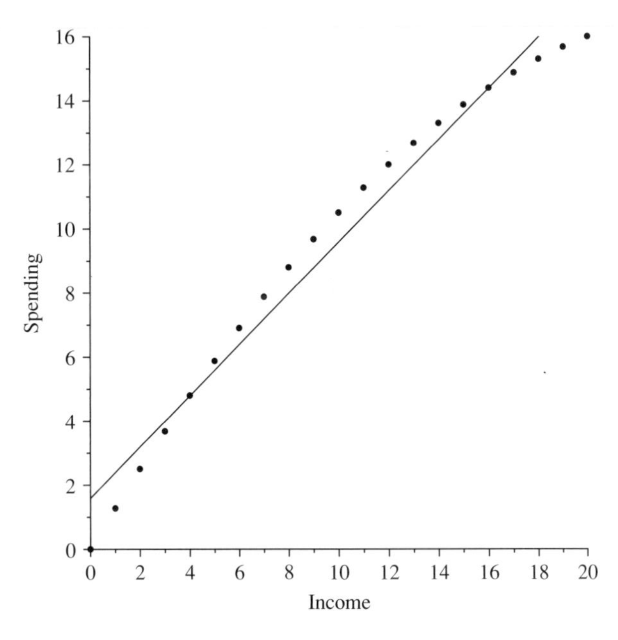

Figure 9.12
Mystery solved.

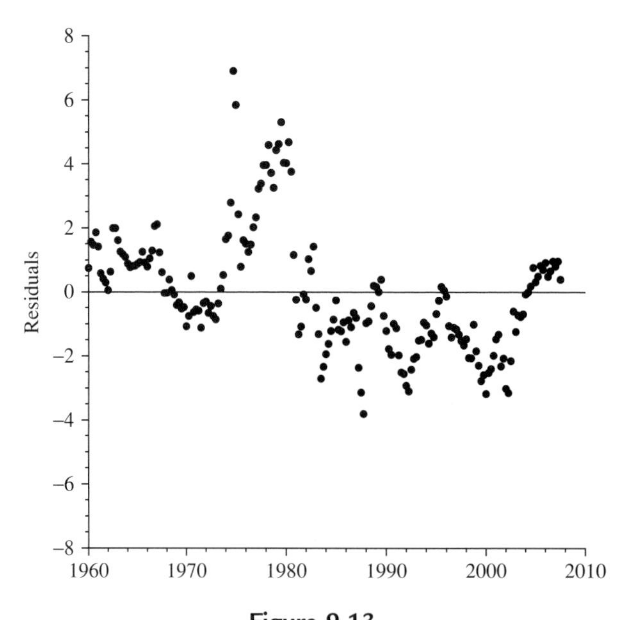

Figure 9.13
Residuals for the Fed's fair value model.

Negatively correlated residuals, in contrast, zigzag back and forth, alternating between positive and negative values. These are relatively rare; an example might be monthly spending for a household that sticks to an annual budget: If it spends too much one month, it cuts back the following month. More advanced courses discuss a number of tests and possible cures for both positively and negatively correlated residuals.

Exercises

9.1 Explain why you either agree or disagree with this interpretation of the results from estimating a regression model: "This study concludes that the data are statistically substantial because there are more than 5 degrees of freedom, and statistically significant because the P value is not below 0.05."

9.2 A study of 16,405 adult daughters of women who immigrated to California from other countries measured the economic status of the daughters by the median household income Y for the ZIP code they lived in [12]. A least squares regression used a variable D that equals 1 if the daughter was her mother's first-born child and 0 otherwise:

$$Y = 38{,}976 + 1{,}513D$$
$$[208.10] \quad [6.85]$$

The t values are in brackets. Is the estimated relationship between birth order and income

a. Statistically significant at the 5 percent level?

b. Substantial?

9.3 The study described in the preceding exercise also looked at 11,794 first-born daughters at the time they gave birth to their first child. A least squares regression was used to examine the relationship between income Y and the daughter's age A (in years) at the time she gave birth:

$$Y = 28{,}891 + 615A$$
$$[30.07] \quad [12.17]$$

The t values are in brackets. Is the estimated relationship between age and income
a. Statistically significant at the 5 percent level?
b. Substantial?

9.4 Table 9.7 shows the SAT scores (600–2400) and college GPAs (four-point scale) for 24 second-year college students at a highly selective college. Use a simple regression model to estimate the relationship between these variables. Explain your answers to these questions:
a. Which variable did you use as the dependent variable?
b. Is the relationship statistically significant at the 5 percent level?
c. Is the estimated relationship of any practical importance?
d. Do you think that your least squares equation can be used to make reliable GPA predictions for a student with a 1200 SAT score?

9.5 The Boston Snow (B.S.) indicator uses the presence or absence of snow in Boston on Christmas Eve to predict the stock market the following year: "the average gain following snow on the ground was about 80 percent greater than for years in which there was no snow" [13]. How would you explain the success of the B.S. indicator?

9.6 Explain the following observation, including the use of the word *sadly*: "There have always been a considerable number of pathetic people who busy themselves examining the last thousand numbers which have appeared on a roulette wheel, in search of some repeating pattern. Sadly enough, they have usually found it" [14].

9.7 Two psychologists found a strong positive correlation between family tension and the number of hours spent watching television [15]. Give a logical explanation other than television shows increase family tension.

Table 9.7: Exercise 9.4

SAT	GPA	SAT	GPA	SAT	GPA	SAT	GPA
1900	3.60	2060	3.00	2200	3.80	2260	3.30
1950	3.27	2080	3.80	2210	3.40	2280	3.60
1990	2.40	2090	3.00	2240	3.80	2300	3.80
2010	3.40	2100	3.73	2250	3.83	2300	3.80
2020	3.70	2100	3.60	2260	3.75	2340	3.83
2050	3.80	2190	3.50	2260	3.92	2350	3.75

9.8 Give a logical statistical explanation for the Cy Young jinx:

> *Of 71 winners of the Cy Young Award [for the best pitcher] since it was instituted in 1956, only three followed with better years and 30 had comparable years. But in 37 cases, ... the next season was worse [16].*

9.9 Sir Francis Galton planted seeds with seven different diameters (in hundredths of an inch) and computed the average diameter of 100 seeds of their offspring [17]:

Parent	15	16	17	18	19	20	21
Offspring	15.3	16.0	15.6	16.3	16.0	17.3	17.5

After deciding which is the dependent variable, use least squares to estimate the regression equation $Y = \alpha + \beta X + \varepsilon$. Is there a statistically significant relationship at the 1 percent level? Plot these data in a scatter diagram with each axis going from 0 to 25 and draw in the least squares line. What would the line look like if the diameters did not regress toward the mean?

9.10 Use the data in Exercise 9.9 to estimate the equation $\text{Parent} = \alpha_1 + \beta_1\text{Offspring} + \varepsilon_1$ and to estimate the equation $\text{Offspring} = \alpha_2 + \beta_2\text{Parent} + \varepsilon_2$ by ordinary least squares. Let b_1 be your estimate of β_1 and b_2 be your estimate of β_2.

a. Does $b_1 = 1/b_2$?

b. If both equations are estimated by least squares, why doesn't b_1 always equal $1/b_2$?

9.11 Using monthly 2008 data for the rate of inflation (Y) and the unemployment rate (X), the equation $Y = \alpha + \beta X + \varepsilon$ was estimated by ordinary least squares with the data entered in a computer program in this order: January, February, March, and so on. How would the estimates be affected if the data were entered backwards: December, November, October, and so on?

9.12 Data were collected for 25 new cars on the cars' weights (X) and estimated highway miles per gallon (Y). The model $Y = \alpha + \beta X + \varepsilon$ was estimated by least squares in three ways: the data were arranged alphabetically by car name, the data were arranged numerically from the lightest car to the heaviest, and the data were arranged numerically from the lowest miles per gallon to the highest. Which procedure yielded the smallest estimate of the intercept? Which yielded the highest estimate?

9.13 A California college student wanted to see if there is a relationship between the weather and the color of clothes students wear. He collected data for 13 days in April on these two variables:

X: Weather, ranging from 1 (precipitation) to 6 (bright sunny day).
Y: Clothing brightness, ranging from 1 (black) to 6 (white).

Each value of X was recorded right before the start of a philosophy class; Y was the average brightness value for the approximately 20 students who attended the class. He then used least squares to estimate the following equation: $Y = 2.270 + 0.423X$. The standard error of the estimated slope was 0.092 and the value of R^2 was 0.657.

a. Explain why you either agree or disagree with this student's decision to let X be weather and Y be clothing, rather than the other way around.
b. Do his regression results indicate that students tend to wear brighter or darker clothing on sunny days?
c. Is the relationship statistically significant at the 5 percent level? How can you tell?
d. What problcms do you scc with this study?

9.14 Suppose that you meet a 25-year-old man who is 6 feet, 2 inches tall and he tells you that he has two brothers, one 23 years old and the other 27 years old. If you had to guess whether he is the shortest, tallest, or in-between of these three brothers, which would you select? Explain your reasoning.

9.15 A study used the data in Table 9.8 to investigate the effect of driving speed on traffic fatalities:

$$Y = \alpha + \beta X + \varepsilon$$

where Y = deaths per million miles driven and X = average speed, miles per hour.

a. The researcher argued that "supporters of lower speed limits would expect β to be close to 1, reflecting a strong, direct effect of X on Y." What is wrong with this argument?
b. Do you think that the least squares estimate of β using these data is positive or negative?
c. Why are you suspicious of these data?

9.16 The model $Y = \alpha + \beta X + \varepsilon$ was estimated using annual data for 1991 through 2000. If $X = 100$ and $Y = 100$ in 1991 and $X = 200$ and $Y = 200$ in 2000, can the least squares estimate of β be negative? Explain your reasoning.

9.17 A statistician recommended a strategy for betting on professional football games based on the logic of regression to the mean [18]. Do you think that he recommended betting for or against teams that had been doing poorly? Explain clearly and concretely how the logic of regression to the mean applies to football games.

Table 9.8: Exercise 9.15

Year	*X*	*Y*
1980	57.0	1.6
1984	58.2	1.4
1985	58.5	1.3

9.18 Professional baseball players who are batting 0.400 at the halfway point in the season almost always find that their batting average declines during the second half of the season. This slump has been blamed on fatigue, pressure, and too much media attention. Provide a purely statistical explanation. If you were a major league manager, would you take a 0.400 hitter out of the lineup in order to play someone who was batting only 0.200?

9.19 Howard Wainer, a statistician with the Educational Testing Service, recounted the following [19]:

My phone rang just before Thanksgiving. On the other end was Leona Thurstone; she is involved in program evaluation and planning for the Akebono School (a private school) in Honolulu. Ms. Thurstone explained that the school was being criticized by one of the trustees because the school's first graders who finish at or beyond the 90th percentile nationally in reading slip to the 70th percentile by 4th grade. This was viewed as a failure in their education.…I suggested that it might be informative to examine the heights of the tallest first graders when they reached fourth grade. She politely responded that I wasn't being helpful.

Explain his suggestion.

9.20 Some child-development researchers tabulated the amount of crying by 38 babies when they were 4 to 7 days old, and then measured each baby's IQ at 3 years of age [20]. Plot the data in Table 9.9 on a scatter diagram and use least squares to estimate the regression line $Y = \alpha + \beta X + \varepsilon$, where Y is IQ and X is cry count. Is the relationship statistically significant at the 5 percent level? If there is a positive, statistically significant relationship, does this mean that you should make your baby cry?

9.21 Studies have found that people who have taken driver training courses tend to have more traffic accidents than people who have not taken such courses.

Table 9.9: Exercise 9.20

Cry Count	IQ	Cry Count	IQ	Cry Count	IQ	Cry Count	IQ
9	103	13	162	17	94	22	135
9	119	14	106	17	141	22	157
10	87	15	112	18	109	23	103
10	109	15	114	18	109	23	113
12	94	15	133	18	112	27	108
12	97	16	100	19	103	30	155
12	103	16	106	19	120	31	135
12	119	16	118	20	90	33	159
12	120	16	124	20	132		
13	104	16	136	21	114		

Table 9.10: Exercise 9.23

Age	Test	Age	Test	Age	Test	Age	Test	Age	Test	Age	Test	Age	Test
15	95	9	91	18	93	20	94	10	83	10	100	17	121
26	71	15	102	11	100	7	113	11	84	12	105	11	86
10	83	20	87	8	104	9	96	11	102	42	57	10	100

How might you dispute the implication that driver training courses make people worse drivers?

9.22 A petition urging students to take Latin classes noted that, "Latin students score 150 points higher on the verbal section of the SAT than their non-Latin peers." How might a skeptic challenge this argument?

9.23 Table 9.10 shows data on the age (in months) at which 21 children said their first word and their score on a later test of mental aptitude [21]. Use a regression equation to see if there is a statistical relationship.

a. Which variable did you use as the dependent variable? Explain your reasoning.
b. What is the two-sided P value for the slope?
c. Plot the residuals for your regression equation with X on the horizontal axis and draw a horizontal line at zero on the vertical axis. What pattern or other peculiarity do you see in the residuals?

9.24 A researcher wanted to investigate the theory that better baseball players tend to come from warmer climates, where they can play baseball 12 months of the year. He gathered data for 16 MLB teams on the number of players who were from warm climates (the Dominican Republic, Mexico, Puerto Rico, and 10 sunbelt states of the United States). A least squares regression of the number of games won on the number of warm-climate players gave these results:

$$Y = 54.462 + 2.120X, \quad R^2 = 0.41$$
$$[5.79] \quad [3.08]$$

[ ]: t values

a. The researcher observed that "Both t values are significant; so there is a real relationship between the number of southern players on a baseball team and the number of wins." Explain what he means by "significant." Explain why both t values either do or do not matter to his conclusion.
b. Use a two-standard-deviations rule of thumb to calculate a 95 percent confidence interval for the predicted effect of another warm-climate player on the number of wins by a team.
c. The researcher calculated the predicted number of wins for a team in which all 25 players on the roster are from warmer climates. What is the predicted value? Why should it be used with caution?

d. Carefully explain why, even if it is true that better baseball players come from warmer climates, there may be no relationship whatsoever between the number of warm-climate players on a team's roster and the number of games it wins.

9.25 A regression [22] of cricket chirps per minute (Y) for one cricket species on temperature in degrees Fahrenheit (X) gave this estimated equation: $Y = -0.309 + 0.212X$. Which is the dependent and which the independent variable? What is the predicted value of Y when $X = 0$? When $X = 300$? Do these predictions seem plausible? Can we judge the reasonableness of this model by the plausibility of these predictions?

9.26 A researcher obtained the combined math and reading SAT scores (X) (scale 400–1600) and GPAs (Y) of 396 college students and reported these regression results:

$$\begin{aligned} Y = \underset{[5.86]}{1{,}146.0} + \underset{[0.02]}{1.4312}X, \quad R^2 = 0.01 \end{aligned}$$

[ ]: t values

a. There is an error in these results. What is it?
b. The researcher said "Therefore I reject the null hypothesis and conclude that SAT scores and grades are not related." What is the statistical basis for this conclusion and why is it wrong?
c. What other important explanatory variables do you think affect GPAs?

9.27 A 1957 *Sports Illustrated* cover had a picture of the Oklahoma football team, which had not lost in 47 games, with the caption "Why Oklahoma Is Unbeatable." The Saturday after this issue appeared, Oklahoma lost to Notre Dame 7–0, starting the legend of the *Sports Illustrated* cover jinx—that the performance of an individual or team usually declines after they are pictured on the cover of *Sports Illustrated*. Explain how regression toward the mean might be used to explain the *Sports Illustrated* cover jinx.

9.28 A random sample of 30 second-semester college sophomores was asked to report their GPA (on a four-point scale) and the average number of hours per week spent on organized extracurricular activities, such as work, sports teams, clubs, committee meetings, and volunteer activities. These data were used to estimate the following equation by least squares:

$$Y = \underset{(0.139)}{3.270} + \underset{(0.00823)}{0.00817}X, \quad R^2 = 0.07$$

( ): standard error

a. Is the relationship statistically significant at the 5 percent level?
b. Is the relationship substantial?
c. What is the null hypothesis?
d. What does R^2 measure and how would you interpret the R^2 value here?

e. The average value of the number of hours a week spent on extracurricular activities was 15.15; what was the average GPA?
f. If the statistical relationship were positive, substantial, and high statistically significant, why would we still be cautious about concluding that the way to raise your grades is to spend more time on extracurricular activities?

9.29 Table 9.11 shows annual U.S. beer production (X), in millions of barrels, and the number of married people (Y), in millions. Draw a scatter diagram and estimate the equation $Y = \alpha + \beta X + \varepsilon$ by least squares to see if there is a statistically significant relationship at the 5 percent level. If so, do you think that beer drinking leads to marriage, marriage leads to drinking, or what?

9.30 Estimate the parameters in the linear equation $Y = \alpha + \beta X + \varepsilon$ using the data in Table 9.12. Now plot these data and draw in your fitted line. Explain why you either do or do not think that the fitted line gives an accurate description of the relationship between X and Y.

9.31 Table 9.13 shows [23] the percent neonatal mortality rate in the United States in 1980, for babies grouped by birth weight (grams).
a. Estimate a linear regression model.
b. Which variable is your dependent variable?
c. Is the relationship statistically significant at the 5 percent level?
d. Plot your residuals as a function of birth weight. What least squares assumption seems to be violated?
e. How would you explain the problem identified in question d? How would you correct it?

Table 9.11: Exercise 9.29

	1960	1965	1970	1975	1980	1985
X	95	108	135	158	193	200
Y	84.4	89.2	95.0	99.7	104.6	107.5

Table 9.12: Exercise 9.30

X	Y
1.0	0.8
2.0	1.9
3.0	3.0
4.0	4.1
5.0	5.2
30.0	3.0

Table 9.13: Exercise 9.31

Birth weight	250	750	1,250	1,750	2,250	2,750	3,250	3,750	4,250
Mortality	100	64.76	18.65	5.39	1.6	0.4	0.19	0.14	0.15

9.32 What pattern do you see in Figure 9.14, a plot of the residuals from a least squares regression?

9.33 What pattern do you see in Figure 9.15, a plot of the residuals from a least squares regression? What do you think might have caused this pattern?

9.34 Based on the Figure 9.16 plot of least squares residuals, what assumption about the regression model's error term seems to be violated?

9.35 Based on the Figure 9.17 plot of least squares residuals, what assumption about the regression model's error term seems to be violated?

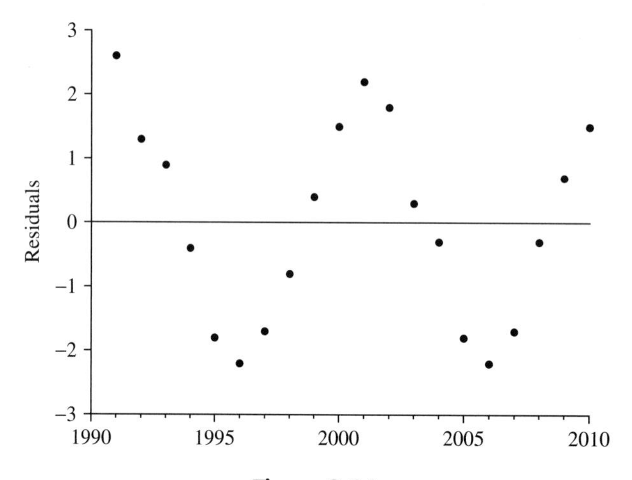

Figure 9.14
Exercise 9.32.

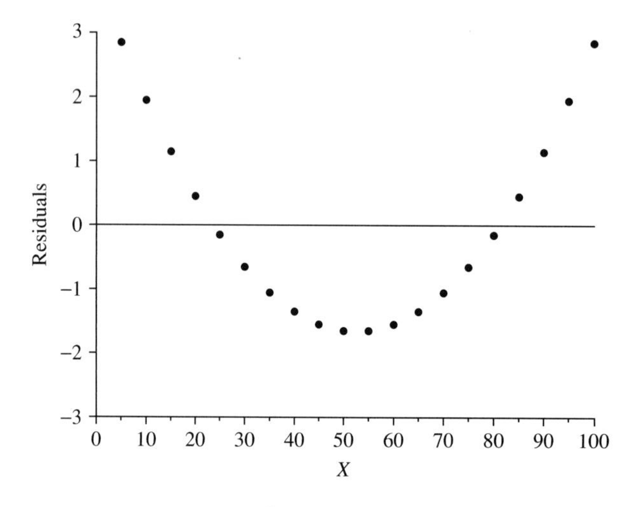

Figure 9.15
Exercise 9.33.

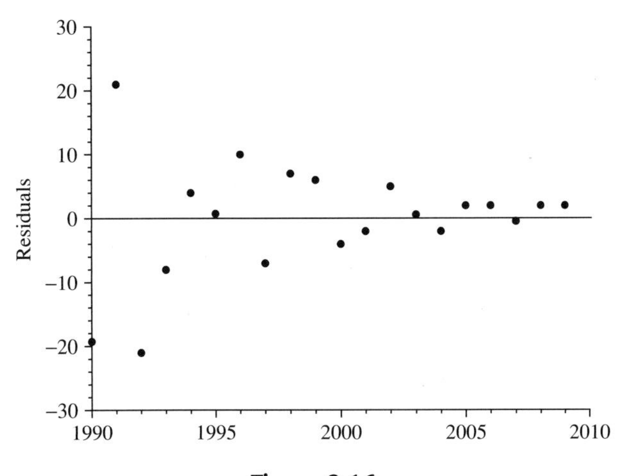

Figure 9.16
Exercise 9.34.

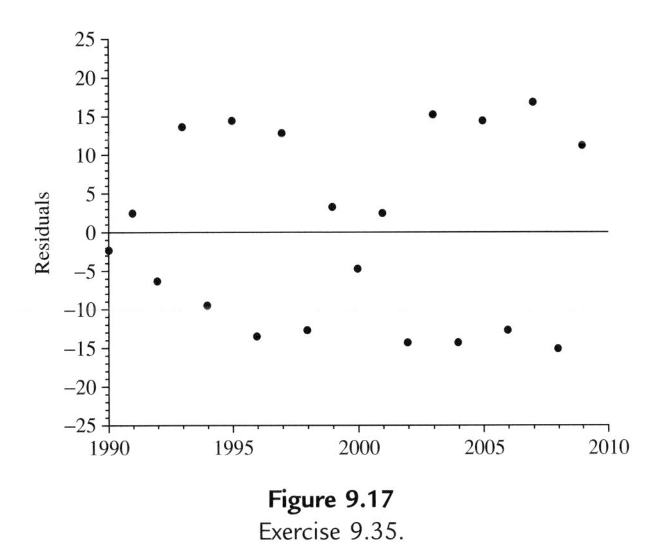

Figure 9.17
Exercise 9.35.

9.36 What is puzzling about Figure 9.18, a plot of the residuals from a least squares regression?

9.37 Consider the hypothetical income and spending data in Table 9.14.

a. Identify any important differences in the patterns in these data compared to the data in Table 9.3.
b. Estimate the simple regression model with spending as the dependent variable and income as the explanatory variable.
c. Calculate the annual percentage changes in income and spending and estimate the simple regression model with the percentage change in spending as the dependent variable and the percentage change in income as the explanatory variable.
d. Identify any important differences between your answers to questions b and c.

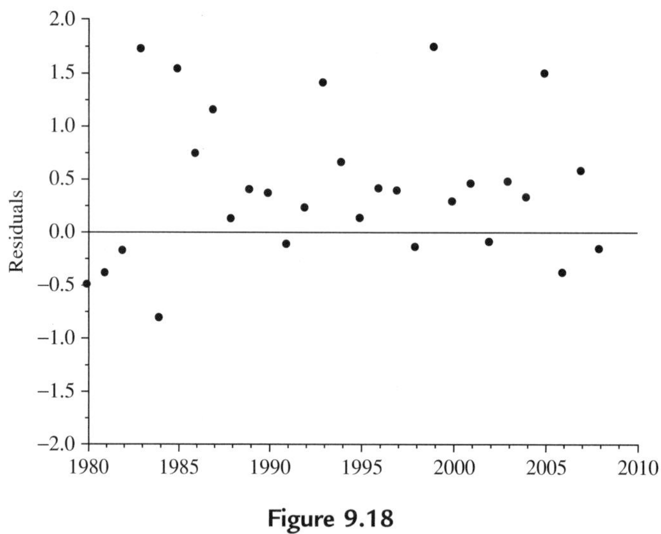

Figure 9.18
Exercise 9.36.

Table 9.14: Exercise 9.37

	Hypothetical Income	Hypothetical Spending
1929	100	90
1930	110	99
1931	121	107
1932	129	118
1933	132	115
1934	140	126
1935	148	135
1936	162	142
1937	173	151
1938	179	163
1939	194	169
1940	200	182

9.38 Use the data in Exercise 9.37 to answer these questions:

a. Estimate the simple regression model with spending as the dependent variable and income as the explanatory variable.

b. Estimate the equation $Y = \alpha + \beta X + \varepsilon$, where Y is income and X is the year. Use your estimates to calculate the predicted value of income each year, $\hat{Y} = a + bX$. The residuals $Y - \hat{Y}$ are the detrended values of income.

c. Repeat the procedure in question b using spending as the dependent variable.

d. Estimate a consumption function where the dependent variable is detrended spending and the explanatory variable is detrended income.

e. Identify any important differences between your answers to questions a and d.

Table 9.15: Exercise 9.39

	Y	C		Y	C		Y	C
1960	10.86	9.87	1977	18.02	16.05	1994	24.52	22.47
1961	11.05	9.91	1978	18.67	16.58	1995	24.95	22.80
1962	11.41	10.24	1979	18.90	16.79	1996	25.47	23.33
1963	11.67	10.51	1980	18.86	16.54	1997	26.06	23.90
1964	12.34	10.99	1981	19.17	16.62	1998	27.30	24.86
1965	12.94	11.54	1982	19.41	16.69	1999	27.80	25.92
1966	13.47	12.05	1983	19.87	17.49	2000	28.90	26.94
1967	13.90	12.28	1984	21.11	18.26	2001	29.30	27.39
1968	14.39	12.86	1985	21.57	19.04	2002	29.98	27.85
1969	14.71	13.21	1986	22.08	19.63	2003	30.45	28.37
1970	15.16	13.36	1987	22.25	20.06	2004	31.21	29.09
1971	15.64	13.70	1988	23.00	20.67	2005	31.34	29.79
1972	16.23	14.38	1989	23.38	21.06	2006	32.30	30.36
1973	17.17	14.95	1990	23.57	21.25	2007	32.68	30.87
1974	16.88	14.69	1991	23.45	21.00	2008	32.55	30.51
1975	17.09	14.88	1992	23.96	21.43			
1976	17.60	15.56	1993	24.04	21.90			

9.39 Table 9.15 shows U.S. data for C = real per capita consumption and Y = real per capita disposable personal income.

a. Draw a scatter diagram.

b. Estimate the consumption function $C = \alpha + \beta Y + \varepsilon$.

c. Calculate the residuals $e = C - (a + bY)$ for each year and make a scatter diagram with e on the vertical axis and the year on the horizontal axis.

d. Do you see anything in this residuals plot that suggests that the least squares assumptions might be violated?

9.40 Use the data in the Exercise 9.39 to calculate the annual change in income and consumption, for example, in 1961 $Y - Y_{-1} = 11.05 - 10.86 = 0.19$.

a. Draw a scatter diagram using these annual changes.

b. Estimate the consumption function $(C - C_{-1}) = \alpha + \beta(Y - Y_{-1}) + \varepsilon$.

c. Calculate the residuals $e = (C - C_{-1}) - [a + b(Y - Y_{-1})]$ for each year and make a scatter diagram with e on the vertical axis and the year on the horizontal axis.

d. Do you see anything in this residuals plot that suggests that the least squares assumptions might be violated?

9.41 Use the data in Exercise 9.39 to calculate the annual percentage change in income and consumption, for example, in 1961

$$\%Y = 100[(Y - Y_{-1})/Y_{-1}] = 100[(11.05 - 10.86)/10.86] = 1.75$$

a. Draw a scatter diagram using these annual percentage changes.

b. Estimate the consumption function $(\%C) = \alpha + \beta(\%Y) + \varepsilon$.

c. Calculate the residuals $e = (\%C) - [a + b(\%Y)]$ for each year and make a scatter diagram with e on the vertical axis and the year on the horizontal axis.
d. Do you see anything in this residuals plot that suggests that the least squares assumptions might be violated?

9.42 Explain why you either agree or disagree with this reasoning: "You can always identify an outlier in a regression model by looking at the residuals because an outlier has a large residual." Draw a scatterplot to illustrate your argument.

9.43 A stock portfolio consisting of the 10 companies identified by *Fortune* magazine each year as America's most admired companies was compared to the overall U.S. stock market [24]. A least squares regression was used to analyze the relationship between the risk-adjusted daily percentage return on the *Fortune* portfolio R_F with the risk-adjusted daily return on the market as a whole R_M:

$$\underset{(0.0075)}{R_F = 0.0270} + \underset{(0.0110)}{0.9421R_M}, \quad R^2 = 0.809$$

There were 5,547 observations. The standard errors are in parentheses.
a. The average value of R_M is 0.0522 percent (this is not 5 percent, but about 1/20 of 1 percent). What is the average value of R_F?
b. Is the difference between the average value of R_F and R_M substantial? Explain your reasoning.
b. Test the null hypothesis that the slope is equal to 1.
c. Test the null hypothesis that the intercept is equal to 0.

9.44 Two portfolios were followed: one portfolio consisting of stocks that were deleted from the Dow Jones Industrial Average, the other portfolio consisting of stocks that replaced them [25]. Whenever a Dow substitution was made, the deleted stock was added to the deletion portfolio and the replacement stock was added to the addition portfolio. Using 20,367 daily observations, the difference between the daily percentage return on the addition portfolio R_A and the daily percentage return on the deletion portfolio R_D was regressed on the daily percentage return for the stock market as a whole:

$$R_A - R_D = \underset{[2.78]}{-0.0163} + \underset{[3.29]}{0.0185R_M}, \quad R^2 = 0.0005$$

There were 20,367 observations. The t values are in brackets.
a. Explain why regression to the mean suggests that stocks removed from the Dow may outperform the stocks that replace them.
b. The average value of R_M is 0.0396 percent (this is not 4 percent, but 4/100 of 1 percent). What is the average value of $R_A - R_D$?
c. Is the average value of $R_A - R_D$ large or small? Explain your reasoning.

d. Test the null hypothesis that the slope is equal to 0.
e. Test the null hypothesis that the intercept is equal to 0.

9.45 Answer this question that was given to graduate students in psychology, which described the professor's experience when he advised the Israeli air force:

The instructors in a flight school adopted a policy of consistent positive reinforcement recommended by psychologists. They verbally reinforced each successful execution of a flight maneuver. After some experience with this training approach, the instructors claimed that contrary to psychological doctrine, high praise for good execution of complex maneuvers typically results in a decrement of performance on the next try. What should the psychologist say in response? [26]

9.46 Use these data on the average weight in pounds of U.S. men who are 68 inches tall to estimate the equation $Y = \alpha + \beta X + \varepsilon$. Plot the residuals and identify a pattern.

Age (X)	20	30	40	50	60	70
Weight (Y)	157	168	174	176	174	169

9.47 Eighteen plots of land in Indiana were given different applications of nitrogen. In each case, the previous crop had been soybeans and all other factors were roughly comparable. Use the 18 observations in Table 9.16 to estimate the equation $Y = \alpha + \beta X + \varepsilon$, where Y is the corn yield (bushels per acre) and X is nitrogen (pounds per acre). Is the relationship substantial and statistically significant at the 1 percent level? Now make a scatter diagram with the residuals on the vertical axis and X on the horizontal axis and a horizontal line drawn at 0. What pattern do you see and how do you explain it? Does this pattern make logical sense?

9.48 In educational testing, a student's ability is defined as this student's average score on a large number of tests that are similar with respect to subject matter and difficulty. A student's score on a particular test is equally likely to be above or below the student's ability. Suppose a group of 100 students takes two similar tests and the scores on both tests have a mean of 65 with a standard deviation of 14. If a student's score on the second test is 52, would you predict that this student's score on the first test was (a) below 52, (b) 52, (c) between 52 and 65, (d) 65, or (e) above 65? Explain your reasoning.

Table 9.16: Exercise 9.47

Nitrogen	Yield	Nitrogen	Yield	Nitrogen	Yield
50	91	90	129	170	160
130	155	250	175	250	183
90	138	210	185	170	177
170	172	130	170	90	124
130	155	250	166	50	95
210	183	50	84	210	171

9.49 To examine the relationship between the death penalty and murder rates, the following equation was estimated by least squares using U.S. cross-sectional data for 50 states:

$$Y = 4.929 + 1.755X, \quad R^2 = 0.063$$
$$(0.826) \quad (0.973)$$

where Y = murder rate (murders per 100,000 residents), X = 1 if the state has a death penalty or 0 if it does not, and the standard errors are in parentheses.

a. What does R^2 mean and how is it measured?
b. Is the relationship statistically significant at the 5 percent level?
c. The researcher added the District of Columbia to the data and obtained these results:

$$Y = 7.013 - 0.330X, \quad R^2 = 0.001$$
$$(1.366) \quad (1.626)$$

The murder rate in D.C. is 36.2; the state with the next highest murder rate is Michigan, with a murder rate of 12.2. Explain how a comparison of the two estimated equations persuades you that D.C. either does or does not have a death penalty.

d. Explain the error in this interpretation of the results including D.C.: "Those of us who oppose the death penalty can breathe a sigh of relief, now armed with statistically significant evidence that the death penalty is only cruel and not effective."

9.50 A psychologist once suggested this possible relationship between the IQs of children and their parents [27]:

$$(\text{child IQ} - 100) = 0.50(\text{adult IQ} - 100)$$

What statistical phenomenon would explain this psychologist's use of the coefficient 0.5, rather than 1.0? A few years later another researcher used the data in Table 9.17,

Table 9.17: Exercise 9.50

	Fathers	Children
Higher professional	139.7	120.8
Lower professional	130.6	114.7
Clerical	115.9	107.8
Skilled	108.2	104.6
Semi-skilled	97.8	98.9
Unskilled	84.9	92.6

relating the mean IQ scores of children and their fathers, grouped according to the father's occupation, to estimate the regression equation $Y = \alpha + \beta X + \varepsilon$. Draw a scatter diagram and estimate this equation by least squares.

a. Which variable did you use as the dependent variable? Explain your reasoning.
b. Is your estimated equation consistent with the theory that (child IQ – 100) = 0.50(adult IQ – 100)?
c. What is suspicious about your estimated equation?

CHAPTER 10

Multiple Regression

Chapter Outline

The idea is there, locked inside. All you have to do is remove the excess stone.
—Michelangelo

In the many fields, researchers can use laboratory experiments to isolate the effect of one variable on another. For instance, to study the effect of nitrogen on plant growth, an agricultural researcher can do a controlled experiment in which similar plants are given varying amounts of nitrogen, but the same amounts of light, water, and nutrients other than nitrogen. Because the other important factors are held constant, the researcher can attribute the observed differences in plant growth to the differing applications of nitrogen.

There will still be some variation among plants given the same amount of nitrogen because: not all plants are created equal; errors are made when measuring plant growth; light, water, and nonnitrogen nutrients are not exactly the same; and some factors may have been overlooked and consequently not held constant. However, if the most important influences have been identified and the controls imposed carefully, the experiment should yield useful data for estimating the simple regression model

$$Y = \alpha + \beta X + \varepsilon$$

where Y is a measure of plant growth, X is the amount of nitrogen, and the error term ε allows for experimental imperfections.

Outside the natural sciences, we usually cannot do laboratory experiments under controlled conditions. Instead we must make do with observational data—"nature's experiments"—and use statistical methods to unravel the effects of one variable on another when many variables are changing simultaneously. In this chapter, we learn how to do this statistical magic.

We begin with a description of a regression model with several explanatory variables. Then we see how the model's parameters can be estimated and used for statistical tests.

10.1 The Multiple Regression Model

In earlier chapters we looked at Keynes' simple consumption function:

$$C = \alpha + \beta Y + \varepsilon$$

where C is spending and Y is household income. It seems plausible that spending is also affected by wealth—households spend more when the stock market or the real estate market is booming and cut back on their spending when these markets crash.

If consumer spending depends on *both* income and wealth, can we determine the separate effects of each, holding the other factor constant? If we could do a controlled experiment, like the agricultural researcher, we would hold wealth constant while we manipulate income, and then hold income constant while we manipulate wealth. Since we obviously cannot do that, we have to make do with observational data and rely on statistical methods to distinguish the effects of income and wealth.

If we analyze a simple regression model relating consumption to income, our results might be very misleading. In the simple regression model

$$C = \alpha + \beta Y + \varepsilon$$

wealth is an omitted variable that is part of the error term ε.

Because we believe that an increase in income increases spending, we expect β to be positive. But, what if, during the historical period for which we have observational data, income happens to be rising (which increases consumption) but the stock market crashes (which reduces consumption) and the wealth effect is more powerful than the income effect, so that consumption declines? If we ignore wealth and look only at the increase in income and fall in spending, our least squares estimate of β will be negative—indicating that an increase in income reduces spending.

Another possibility is that, during this historical period, wealth and income both increase and boost consumer spending. Now our least squares estimate of β has the right sign (positive) but is too large, because we incorrectly attribute all of the increase in spending to the rise in income, when it is partly due to the rise in wealth.

A simple regression model that includes income but excludes wealth is misleading unless the observed changes in wealth either do not have a large effect on spending or happen to be uncorrelated with income. If, however, changes in wealth have a substantial effect on spending and are correlated (positively or negatively) with income, then the simple regression model yields biased estimates. To avoid these biases, we include both income and wealth in our model.

The General Model

The multiple regression model is

$$Y = \alpha + \beta_1 X_1 + \beta_2 X_2 + \cdots + \beta_k X_k + \varepsilon \qquad (10.1)$$

where the dependent variable Y depends on k explanatory variables and an error term ε that encompasses the effects of omitted variables on Y. The order in which the explanatory variables are listed does not matter. What does matter is which variables are included in the model and which are left out. A large part of the art of regression analysis is choosing explanatory variables that are important and ignoring those that are unimportant.

Each parameter β is the effect of one explanatory variable on Y, *holding the other explanatory variables constant*. Each β measures the same thing that controlled laboratory experiments seek—the effect on Y of an isolated change in one explanatory variable. But, while a controlled experiment estimates each effect, one at a time, by holding the other variables constant, multiple regression models estimate all the βs simultaneously, using data in which none of the explanatory variables are held constant. Maybe it should be called *miracle regression analysis*?

A multiple regression model with both income and wealth as explanatory variables could be written as

$$C = \alpha + \beta_1 Y + \beta_2 W + \varepsilon$$

where C is consumption, Y is income, W is wealth, and all variables are measured in billions of dollars. Econometricians often use longer variable names to help them remember each variable, for example:

$$\text{CONS} = \alpha + \beta_1 \text{INC} + \beta_2 \text{WLTH} + \varepsilon$$

The coefficient β_1 is the effect on consumption of a \$1 billion increase in income, holding wealth constant. The coefficient β_2 is the effect on consumption of a \$1 billion increase in wealth, holding income constant. We believe that both parameters are positive.

Before seeing how to estimate these coefficients, we look at how the multiple regression model handles categorical variables.

Dummy Variables

Many explanatory variables have numerical values—inches, pounds, and dollars. However other explanatory variables are categorical—male or female, white or nonwhite, Republican or Democrat. We can handle categorical variables by using 0–1 dummy variables, variables whose values are 0 or 1, depending on whether or not a certain characteristic is true.

Multiple regression models have often been used in court cases involving alleged discrimination against a group of people. Consider, for instance, the claim that a company's wage policy discriminates against women. To support this claim, the plaintiff might produce data showing that, on average, female employees earn less than male employees. In response, the company might argue that its salaries are based on experience, and female employees happen, on average, to be less experienced than male employees. It might even argue that, ironically, this situation is due to its recent successful efforts to hire more women. How might a jury evaluate such claims? With a multiple regression model, of course.

A multiple regression model using salary as the dependent variable and experience and gender as explanatory variables can be used to evaluate the firm's claim that its wages depend on experience but not gender. The model might be

$$Y = \alpha + \beta_1 D + \beta_2 X + \varepsilon$$

where Y is the annual salary, X is the number of years of experience, and D is a dummy variable that equals 1 if the employee is male and 0 if the employee is female. The model's parameters can be interpreted by considering the two possible values of the dummy variable:

$$\text{Men } (D = 1)\text{: } Y = (\alpha + \beta_1) + \beta_2 X + \varepsilon$$

$$\text{Women } (D = 0)\text{: } Y = \alpha + \beta_2 X + \varepsilon$$

The parameter β_1 measures the effect on salary, at any experience level, of being male. The employer's claim is that $\beta_1 = 0$, which can be tested statistically using the methods explained in the next section. Here, we focus on interpreting the coefficients.

Suppose that our estimated equation turns out to be

$$Y = 25{,}000 + 5{,}000D + 1{,}000X$$

These estimates indicate that, neglecting the error term, the starting salary for a female employee ($D = 0$) with no experience ($X = 0$) is \$25,000; salaries increase by \$1,000 with each additional year of experience; and for any given level of experience, men earn \$5,000 more than women. If statistically significant, the \$5,000 coefficient of D is evidence against the employer's claim. Figure 10.1 shows the implied structure of male and female salaries. By using a dummy variable, we have in effect estimated two separate equations with different intercepts and the same slope:

$$\text{Men } (D = 1)\text{: } Y = 30{,}000 + 1{,}000X$$
$$\text{Women } (D = 0)\text{: } Y = 25{,}000 + 1{,}000X$$

If we want to consider the possibility that gender affects both the intercept and the slope of the salary equation, we can use this multiple regression model:

$$Y = \alpha + \beta_1 D + \beta_2 X + \beta_3 DX + \varepsilon$$

Again, we can interpret the parameters by considering the possible values of the dummy variable:

$$\text{Men } (D = 1)\text{: } Y = (\alpha + \beta_1) + (\beta_2 + \beta_3)X + \varepsilon$$
$$\text{Women } (D = 0)\text{: } Y = \alpha + \beta_2 X + \varepsilon$$

A positive value for β_1 indicates that, independent of experience, men earn an additional amount β_1. A positive value for β_3 indicates that, for every additional year of experience, men earn an amount β_3 more than women. For instance, the estimated equation

$$Y = 25{,}000 + 5{,}000D + 1{,}000X + 500DX$$

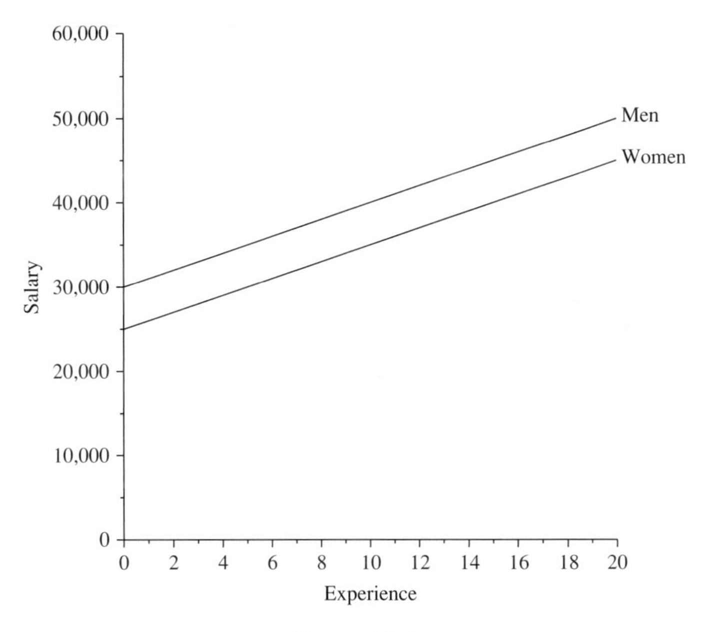

Figure 10.1
Men's and women's salaries with a dummy variable for the intercept.

implies these female and male equations:

$$\text{Men } (D = 1)\text{: } Y = 30{,}000 + 1{,}500X$$
$$\text{Women } (D = 0)\text{: } Y = 25{,}000 + 1{,}000X$$

The starting salary for someone with no experience is $25,000 for a female and $30,000 for a male, and each additional year of experience increases salaries by $1,000 for women and by $1,500 for men. Figure 10.2 shows that this model allows the implied male and female equations to have different intercepts and slopes.

In these examples, the dummy variable equals 1 if the employee is male and 0 if female. Would it have made any difference if we had done it the other way around? No. If the dummy variable D equals 1 if the employee is female and 0 if male, the estimated equation is

$$Y = 30{,}000 - 5{,}000D + 1{,}500X - 500DX$$

These coefficients indicate that the starting salary for a male employee with no experience is $30,000; men's salaries increase by $1,500 with each additional year of experience; women's starting salaries are $5,000 lower and increase by $500 less for each year of experience:

$$\text{Men } (D = 0)\text{: } Y = 30{,}000 + 1{,}500X$$
$$\text{Women } (D = 1)\text{: } Y = 25{,}000 + 1{,}000X$$

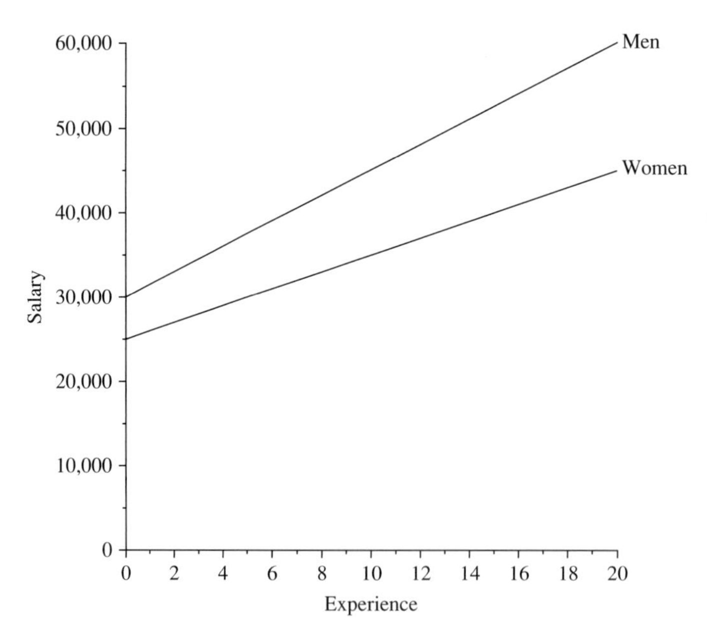

Figure 10.2
Men's and women's salaries with a dummy variable for the intercept and slope.

This is the same conclusion we reached before. If we are careful in our interpretation, it does not matter whether the dummy variable is 1 for male and 0 for female, or the other way around.

10.2 Least Squares Estimation

Multiple regression statistical analysis is a straightforward extension of least squares estimation for the simple regression model. Therefore, much of our discussion is familiar. The numerical calculations are more complex, but these are done by statistical software and we can focus on the statistical logic without getting bogged down in computational details.

We want to use n observations to estimate the parameters in the Equation 10.1 in order to predict the dependent variable:

$$\widehat{Y} = a + b_1X_1 + b_2X_2 + \cdots + b_kX_k \tag{10.2}$$

To estimate these $k + 1$ parameters, there must be at least $k + 1$ observations. The least squares procedure identifies the parameter estimates that minimize the sum of squared prediction errors $\Sigma(Y - \widehat{Y})^2$.

To illustrate, we use the U.S. consumption (C), income (Y), and wealth (W) data in Table 10.1. Our model is

$$C = \alpha + \beta_1 Y + \beta_2 W + \varepsilon$$

Because spending, income, and wealth all tend to increase over time, it might be better to use changes in the variables or detrended data, as explained in Chapter 9. We use unadjusted data here to focus on how multiple regression works, but two chapter exercises use changes and detrended data.

Statistical software is used to calculate the parameter estimates that minimize the sum of squared prediction errors:

$$C = -110.126 + 0.798Y + 0.026W$$

Holding wealth constant, a \$1 billion increase in income is predicted to increase consumption by \$0.798 billion. Holding income constant, a \$1 billion increase in wealth is predicted to increase consumption by \$0.026 billion.

The coefficient of wealth may seem small (a \$1 billion change in wealth is predicted to change consumption by \$26 million), but observed changes in wealth are often quite large. In 2008, U.S. household wealth dropped by more than \$1 trillion, which our model predicts would reduce consumer spending by about \$26 billion. For comparison, disposable income increased by about \$50 billion in 2008, which our model predicts would increase consumption by about \$40 billion.

Table 10.1: United States Consumption, Income, and Wealth

	C	Y	W		C	Y	W
1960	1,784.4	1,963.9	9,835.3	1985	4,540.4	5,144.8	22,175.0
1961	1,821.2	2,030.8	10,011.5	1986	4,724.5	5,315.0	23,930.9
1962	1,911.2	2,129.6	10,749.0	1987	4,870.3	5,402.4	25,742.4
1963	1,989.9	2,209.5	10,905.2	1988	5,066.6	5,635.6	26,443.0
1964	2,108.4	2,368.7	11,325.6	1989	5,209.9	5,785.1	27,714.8
1965	2,241.8	2,514.7	12,033.1	1990	5,316.2	5,896.3	29,028.3
1966	2,369.0	2,647.3	12,789.7	1991	5,324.2	5,945.9	28,212.3
1967	2,440.0	2,763.5	12,838.0	1992	5,505.7	6,155.3	29,348.1
1968	2,580.7	2,889.2	13,941.0	1993	5,701.2	6,258.2	29,810.6
1969	2,677.4	2,981.4	15,055.6	1994	5,918.9	6,459.0	30,928.2
1970	2,740.2	3,108.8	14,513.2	1995	6,079.0	6,651.6	31,279.4
1971	2,844.6	3,249.1	14,535.9	1996	6,291.2	6,870.9	33,974.1
1972	3,019.5	3,406.6	15,430.6	1997	6,523.4	7,113.5	35,725.2
1973	3,169.1	3,638.2	17,046.7	1998	6,865.5	7,538.8	39,271.9
1974	3,142.8	3,610.2	16,806.5	1999	7,240.9	7,766.7	43,615.4
1975	3,214.1	3,691.3	15,353.7	2000	7,608.1	8,161.5	48,781.8
1976	3,393.1	3,838.3	16,119.8	2001	7,813.9	8,360.1	47,286.3
1977	3,535.9	3,970.7	17,201.8	2002	8,021.9	8,637.1	46,023.0
1978	3,691.8	4,156.5	17,674.5	2003	8,247.6	8,853.9	44,130.5
1979	3,779.5	4,253.8	18,659.4	2004	8,532.7	9,155.1	49,567.5
1980	3,766.2	4,295.6	19,763.8	2005	8,819.0	9,277.3	54,063.0
1981	3,823.3	4,410.0	20,495.4	2006	9,073.5	9,650.7	58,823.3
1982	3,876.7	4,506.5	20,302.6	2007	9,313.9	9,860.6	61,642.3
1983	4,098.3	4,655.7	20,626.9	2008	9,290.9	9,911.3	60,577.8
1984	4,315.6	4,989.1	21,328.5				

Note: Consumption is personal consumption expenditures; income is disposable personal income (after taxes); wealth is household net worth, measured at the beginning of each year. All data are in billions of 2005 dollars.

It is interesting to compare this multiple regression model to estimates of simple regression models that omit one of the explanatory variables:

$$C = -258.213 + 0.957Y$$
$$C = 753.767 + 0.152W$$

The effects of income and wealth are positive in each model, but the magnitudes are smaller in the multiple regression model than in the simple regression models. How is this possible?

Suppose that our reasoning is correct, that changes in income and wealth both have positive effects on consumption. Now, what if income and wealth happen to be positively correlated because a strong economy increases both income and stock prices? If so, the omission of either variable from our model tends to bias upward the estimated coefficient of the variable that is included. If we look at the simple relationship between consumption and income, income appears to have an exaggerated effect on consumption, because we do not take wealth into account. If we look at the simple relationship between consumption and wealth, wealth appears to have an exaggerated effect because we do not take income into account. And that is, in fact, what happens here. The correlation between income and wealth is 0.98, and the estimated coefficients in simple regression models are biased upward.

Multiple regression models are very appealing because the inclusion of additional explanatory variables may improve the model's predictive accuracy and give us unbiased estimates of the effects of a change in an explanatory variable, holding other factors constant.

Confidence Intervals for the Coefficients

If the error term is normally distributed and satisfies the four assumptions detailed in the simple regression chapter, the estimators are normally distributed with expected values equal to the parameters they estimate:

$$a \sim N[\alpha, \text{standard deviation of } a]$$
$$b_i \sim N[\beta_i, \text{standard deviation of } b_i]$$

To compute the standard errors (the estimated standard deviations) of these estimators, we need to use the standard error of estimate (SEE) to estimate the standard deviation of the error term:

$$\text{SEE} = \sqrt{\frac{\sum(Y-\hat{Y})^2}{n-(k+1)}} \tag{10.3}$$

Because n observations are used to estimate $k + 1$ parameters, we have $n - (k + 1)$ degrees of freedom. After choosing a confidence level, such as 95 percent, we use the t distribution with $n - (k + 1)$ degrees of freedom to determine the value t^* that corresponds to this probability. The confidence interval for each coefficient is equal to the estimate plus or minus the requisite number of standard errors:

$$\begin{aligned} &a \pm t^*(\text{standard error of } a) \\ &b_i \pm t^*(\text{standard error of } b_i) \end{aligned} \tag{10.4}$$

For our consumption function, statistical software calculates SEE = 59.193 and these standard errors:

$$\text{standard error of } a = 27.327$$
$$\text{standard error of } b_1 = 0.019$$
$$\text{standard error of } b_2 = 0.003$$

With 49 observations and 2 explanatory variables, we have 49 − (2 + 1) = 46 degrees of freedom. Table A.2 gives t^* = 2.013 for a 95 percent confidence interval, so that 95 percent confidence intervals are

$$\alpha:\ a \pm t^*(\text{standard error of } a) = -110.126 \pm 2.013(27.327) = -110.126 \pm 55.010$$
$$\beta_1:\ b_1 \pm t^*(\text{standard error of } b_1) = 0.798 \pm 2.013(0.019) = 0.798 \pm 0.039$$
$$\beta_2:\ b_2 \pm t^*(\text{standard error of } b_2) = 0.026 \pm 2.013(0.003) = 0.026 \pm 0.006$$

Hypothesis Tests

The test statistic for a specified null hypothesis about β_i is

$$t = \frac{\text{estimated value of } \beta_i - \text{null hypothesis value of } \beta_i}{\text{standard error of the estimate of } \beta_i} \tag{10.5}$$

The most interesting null hypotheses are typically that the coefficients equal 0. For the null hypothesis $\beta_i = 0$, the t statistic simplifies to the ratio of the coefficient estimate to its standard error:

$$t = \frac{b_i}{\text{standard error of } b_i} \tag{10.6}$$

The null hypothesis is rejected if the t value is larger than the appropriate cutoff, using a t distribution with $n - (k + 1)$ degrees of freedom; with a large number of observations, the cutoff for a test at the 5 percent level is approximately 2. Instead of using a fixed cutoff, we can report the P values—the probability that a parameter estimate would be so far from the null hypothesis, if the null hypothesis were true.

In our consumption example, we have 46 degrees of freedom, and Table A.2 shows the cutoff for a test at the 5 percent level to be a t value larger than 2.013. Using the estimates and standard errors given previously, we have these t values and two-sided P values:

$$\beta_1\text{: } t = \frac{0.798}{0.019} = 41.48;\ \text{two-sided } P\text{-value} = 6.6 \times 10^{-11}$$

$$\beta_2\text{: } t = \frac{0.026}{0.003} = 8.45;\ \text{two-sided } P\text{-value} = 0.0011$$

Both explanatory variables are highly statistically significant.

The Coefficient of Determination, R^2

As with the simple regression model, the model's predictive accuracy can be gauged by the *multiple correlation coefficient* R or the coefficient of determination, R^2, which compares the sum of squared prediction errors to the sum of squared deviations of Y about its mean:

$$R^2 = 1 - \frac{\sum(Y - \hat{Y})^2}{\sum(Y - \bar{Y})^2} \tag{10.7}$$

In our consumption model, $R^2 = 0.999$, indicating that our multiple regression model explains an impressive 99.9 percent of the variation in the annual consumption.

Prediction Intervals for Y *(Optional)*

This model can be used to predict the value of consumption spending. Suppose, for instance, that we anticipate disposable income being 10,000 and wealth 60,000. By substituting these values into our model, we obtain a predicted value for consumption:

$$\begin{aligned} C &= -110.126 + 0.798(10{,}000) + 0.026(60{,}000) \\ &= 9{,}427.81 \end{aligned}$$

To gauge the precision of this prediction, we can use statistical software to determine a 95 percent prediction interval for the dependent variable and the expected value of the dependent variable:

$$\begin{aligned} 95\text{ percent prediction interval for } C &= 9{,}427.81 \pm 128.15 \\ 95\text{ percent confidence interval for } E[C] &= 9{,}427.81 \pm 47.19 \end{aligned}$$

Because of the error term, a prediction interval for consumption is wider than a confidence interval for the expected value of consumption.

The Fortune *Portfolio*

In Chapter 7, we saw that a stock portfolio of *Fortune*'s 10 most admired American companies beat the S&P 500 by a substantial and statistically persuasive margin. Maybe the *Fortune* portfolio's superior performance was due to factors other than inclusion on *Fortune*'s Top 10 list? To test this research hypothesis, a multiple regression model was estimated using daily data. The dependent variable is

$R - R_0$ = return on *Fortune* portfolio minus return on Treasury bills. This is the *Fortune* portfolio's return relative to a risk-free investment.

The explanatory variables are three factors that Fama and French [1] found explain differences among stock returns:

$R_M - R_0$ = stock market return minus return on Treasury bills. A stock price may go up simply because macroeconomic factors, such as a decline in unemployment, cause the stock market to go up.
$R_S - R_L$ = size factor, return on small stocks minus return on large stocks. Small stocks tend to outperform large stocks.
$R_V - R_G$ = value factor, return on value stocks minus return on growth stocks. Value stocks tend to outperform growth stocks.

The regression model is

$$R - R_0 = \alpha + \beta_1(R_M - R_0) + \beta_2(R_S - R_L) + \beta_3(R_V - R_G) + \varepsilon$$

The intercept α measures the extent to which the *Fortune* portfolio has an extra (or excess) return over Treasury bills that cannot be explained by the three Fama-French factors.

The estimated equation is

$$R - R_0 = \underset{[3.02]}{0.023} + \underset{[85.22]}{0.94}(R_M - R_0) - \underset{[24.94]}{0.94}(R_S - R_L) - \underset{[21.06]}{0.42}(R_V - R_G), \quad R^2 = 0.81$$

where the t values are in brackets. The substantial and statistically significant α value shows that these three factors do not fully explain the strong performance of the *Fortune* portfolio. (The annualized value of a 0.023 percent daily excess return is 6 percent.)

The Real Dogs of the Dow Revisited

In Chapter 9, we saw that stocks deleted from the Dow Jones Industrial Average have outperformed the stocks that replaced them. We can use the Fama-French three-factor model described in the preceding section to see whether the deletion portfolio's superior performance can be fully explained by these factors. Because the daily factor data go back to only 1963, we use monthly returns.

The estimated multiple regression equation is

$$R - R_0 = \underset{[2.11]}{0.195} + \underset{[58.23]}{1.04}(R_M - R_0) + \underset{[6.81]}{0.194}(R_S - R_L) + \underset{[17.53]}{0.452}(R_V - R_G), \quad R^2 = 0.85$$

where R is the monthly return on the deletion portfolio and the other variables are defined earlier.

The substantial and statistically significant α value shows that these three factors do not fully explain the strong performance of the deletion portfolio. (The annualized value of a 0.195 percent monthly excess return is 2.4 percent.)

Would a Stock by Any Other Ticker Smell as Sweet?

Stocks are identified by ticker symbols (so-called because trading used to be reported on ticker tape machines). Stocks traded on the New York Stock Exchange have one to three letters plus additional characters that can be used to identify the type of security; for example, Citigroup (C), General Electric (GE), and Berkshire class A (BRK.A). As in these examples, ticker symbols are usually abbreviations of a company's name, and companies sometimes become known by these abbreviations: GE, IBM, 3M.

In recent years, several companies have shunned the traditional name-abbreviation convention and chosen ticker symbols that are related to what the company does. Some are memorable for their cheeky cleverness; for example, Southwest Airlines' choice of LUV as a ticker symbol is related to its efforts to brand itself as an airline "built on love." Southwest is based at Love Field in Dallas and has an open-seating policy that reportedly can lead to romance between strangers who sit next to each other. Its on-board snacks were originally called "love bites" and its drinks "love potions." Other notable examples are MOO (United Stockyards), and GEEK (Internet America).

The efficient market hypothesis assumes that a stock's market price incorporates all publicly available information and implies that investors who beat the market are lucky, not skillful. A stock's ticker symbol is no secret and it would be surprising if a stock's performance were related to its ticker symbol. Surely, savvy investors focus on a company's profits, not its ticker symbol.

A study compared the daily returns from a portfolio of clever-ticker stocks to the overall stock market [2]. Once again, we estimate a multiple regression equation using the three Fama-French factors identified in the preceding examples (with the t values in brackets):

$$R - R_0 = \underset{[3.17]}{0.045} + \underset{[39.30]}{0.80}(R_M - R_0) + \underset{[22.62]}{0.63}(R_S - R_L) + \underset{[7.43]}{0.28}(R_V - R_G), \quad R^2 = 0.29$$

The estimated value of α implies an annual excess return of about 12 percent.

These clever tickers might be a useful signal of the company's creativity, a memorable marker that appeals to investors, or a warning that the company feels it must resort to gimmicks to attract investors. Surprisingly, a portfolio of clever-ticker stocks beat the market by a substantial and statistically significant margin, contradicting the efficient market hypothesis.

The Effect of Air Pollution on Life Expectancy

Many scientists believe that air pollution is hazardous to human health, but they cannot run laboratory experiments on humans to confirm this belief. Instead, they use observational data—noting, for example, that life expectancies are lower and the incidence of respiratory disease is higher for people living in cities with lots of air pollution. However, these grim statistics might be explained by other factors, for example, that city people tend to be older and live more stressful lives. Multiple regression can be used in place of laboratory experiments to control for other factors. One study used biweekly data from 117 metropolitan areas to estimate the following equation [3]:

$$Y = 19.61 + \underset{[2.5]}{0.041}X_1 + \underset{[3.2]}{0.071}X_2 + \underset{[1.7]}{0.001}X_3 + \underset{[5.8]}{0.41}X_4 + \underset{[18.9]}{6.87}X_5$$

where

Y = annual mortality rate, per 10,000 population (average = 91.4).
X_1 = average suspended particulate reading (average = 118.1).
X_2 = minimum sulfate reading (average = 4.7).

X_3 = population density per square mile (average = 756.2).
X_4 = percentage of population that is nonwhite (average = 12.5).
X_5 = percentage of population that is 65 or older (average = 8.4).
[ ] = t value.

The first two explanatory variables are measures of air pollution. The last three explanatory variables are intended to accomplish the same objectives as a laboratory experiment in which these factors are held constant to isolate the effects of air pollution on mortality.

With thousands of observations, the cutoff for statistical significance at the 5 percent level is a t value of approximately 1.96. The two pollution measures have substantial and statistically significant effects on the mortality rate. For a hypothetical city with the average values of the five explanatory variables, a 10 percent increase in the average suspended particulate reading, from 118.1 to 129.9, increases the mortality rate by 0.041(11.8) = 0.48, representing an additional 48 deaths annually in a city of 1,000,000. A 10 percent increase in the minimum sulfate reading, from 4.7 to 5.17, increases the mortality rate by 0.71(0.47) = 0.34 (representing an additional 34 deaths annually in a city of 1,000,000).

10.3 Multicollinearity

Each coefficient in a multiple regression model tells us the effect on the dependent variable of a change in that explanatory variable, holding the other explanatory variables constant. To obtain reliable estimates, we need a reasonable number of observations and substantial variation in each explanatory variable. We cannot estimate the effect of a change in income on consumption if income never changes. We will also have trouble if some explanatory variables are perfectly correlated with each other. If income and wealth always move in unison, we cannot tell which is responsible for the observed changes in consumption.

The good news is that few variables move in perfect unison. The bad news is that many variables are highly correlated and move in close unison. The consequence is that, while our model may have a gratifyingly high R^2, some parameter estimates may have disappointingly large standard errors and small t values. This malaise is known as the *multicollinearity* problem: High correlations among the explanatory variables prevent precise estimates of the individual coefficients.

If some explanatory variables are perfectly correlated, it is usually because the researcher has inadvertently included explanatory variables that are related by definition. For example, a model in which consumption depends on current income, last year's income, and the change in income,

$$C = \alpha + \beta_1 Y + \beta_2 Y(-1) + \beta_3 [Y - Y(-1)] + \varepsilon$$

sounds plausible until we realize that the third explanatory variable is exactly equal to the first explanatory variable minus the second explanatory variable. Any attempt to estimate this equation results in a software error message. This makes sense, because we cannot change any one of these explanatory variables while holding the other two constant. In this situation, where explanatory variables are related by their very definition, the solution is simple: Omit one of the redundant variables. In our example, we can simplify the model to

$$C = \alpha + \beta_1 Y + \beta_2 Y(-1) + \varepsilon$$

The situation is more difficult when the explanatory variables are not related by definition but happen to be correlated in our data. This kind of multicollinearity problem can be diagnosed by regressing each explanatory variable on the other explanatory variables to see if the R^2 values reveal high intercorrelations among the explanatory variables. While this diagnosis can explain the model's high standard errors and low t values, there is no statistical cure. A builder cannot make solid bricks without good clay, and a statistician cannot make precise estimates without informative data. The solution to the multicollinearity problem is additional data that are not so highly intercorrelated.

There is no firm rule for deciding when the correlations among the explanatory variables are large enough to constitute a "problem." Multicollinearity does not bias any of the estimates. It is more of an explanation for why the estimates are not more precise. In our consumption function example, the estimated equation seems fine (the t values are in brackets):

$$\begin{aligned} C = &-110.126 + 0.798Y + 0.026W, \quad R^2 = 0.993 \\ &\;\;[4.03] \qquad [41.48] \qquad [8.45] \end{aligned}$$

The coefficients have the expected positive signs, plausible magnitudes, and solid t values. The R^2 is impressive. Yet, the correlation between the two explanatory variables is 0.98. Income and wealth are strongly positively correlated. As explained earlier, this is a persuasive reason for *not* omitting either of these important variables. If the explanatory variables were less correlated, the estimates would be more precise. But they are what they are, and that is fine.

The Coleman Report

In response to the 1964 Civil Rights Act, the U.S. government sponsored the 1966 Coleman Report [4], which was an ambitious effort to estimate the degree to which academic achievement is affected by student backgrounds and school resources. One of the more provocative findings involved a multiple regression equation that was used to explain student performance. Some explanatory variables measured student background (including family income, occupation, and education) and others measured school resources (including annual budgets, number of books in the school library, and science facilities).

The researchers first estimated their equation using only the student background variables. They then reestimated the equation using the student background variables and the school resources variables. Because the inclusion of the resources variables had little effect on R^2, they concluded that school resources "show very little relation to achievement." The problem with this two-step procedure is that school resources in the 1950s and early 1960s depended heavily on the local tax base and parental interest in a high-quality education for their children. The student background variables were positively correlated with the school resources variables, and this multicollinearity problem undermines the Coleman Report's conclusions.

Multicollinearity explains why the addition of the school resources variables had little effect on R^2. It also suggests that, if they had estimated the equation with the school resources variables first and then added the student background variables, they again would have found little change in R^2, suggesting that it is student backgrounds, not school resources, that "show very little relation to achievement." More generally, this example shows why we should not add variables in stages but, instead, estimate a single equation using all the explanatory variables.

Exercises

10.1 The manager of a fleet of identical trucks used monthly data to estimate this model:

$$Y = 0.0 + 0.20X_1 + 1{,}000.0X_2$$

where Y is gallons of fuel consumed (average value 71,000); X_1 is miles traveled (average value 80,000); and X_2 is average speed (average value 55 miles per hour).

a. In plain English, interpret the 0.20 coefficient of X_1.
b. In plain English, interpret the 1,000.0 coefficient of X_2.
c. Can we conclude from the fact that the coefficient of X_2 is much larger than the coefficient of X_1 that average speed is more important than the number of miles traveled in determining fuel consumption?

10.2 Data collected by a county agricultural agency were used to estimate this model:

$$Y = -80 + 0.10X_1 + 5.33X_2$$

where Y is corn yield (bushels per acre, average value 90); X_1 is fertilizer (pounds per acre, average value 100); and X_2 is rainfall (inches, average value 30).

a. In plain English, interpret the value 0.10.
b. In plain English, interpret the value 5.33.
c. What does −80 measure and why does it not equal 0?

10.3 The following equation was used to predict the first-year grade point average (GPA) for students enrolling at a highly selective liberal arts college:

$$Y = 0.437 + 0.361X_1 + 0.0014X_2 + 0.0007X_3$$

where Y is the student's first-year college GPA (average 3.0); X_1 is the student's high school GPA (average 3.7); X_2 is the reading SAT score (average 591); and X_3 is the math SAT score (average 629).

a. What is the predicted first-year college GPA for a student with a 3.7 high school GPA, 591 reading SAT score, and 629 math SAT score?
b. What is the predicted first-year college GPA for a student with a 4.0 high school GPA, 800 reading SAT score, and 800 math SAT score?
c. Why should we be cautious in using this equation to predict the first-year college GPA for a student with a 1.0 high school GPA, 200 reading SAT, and 200 math SAT?
d. Using ordinary English, compare the coefficients of X_2 and X_3.

10.4 A study found that with simple regression equations, the increase in a human baby's length during the first 3 months of life is negatively correlated with the baby's length at birth and negatively correlated with the baby's weight at birth; however, a multiple regression equation showed the length gain to be related negatively to birth length and positively to birth weight. Explain these seemingly contradictory findings.

10.5 An economist believes that housing construction depends (positively) on household income and (negatively) on interest rates. Having read only the chapter on simple regression, he regresses housing construction on income, ignoring interest rates, and obtains a negative estimate of β, indicating that an increase in household income decreases housing construction. How might his omission of interest rates explain this incorrect sign? Be specific.

His sister, a chemist, proposes a controlled experiment, in which interest rates are held constant, while incomes are varied: "In this way, you will be able to see the effects of income changes on construction." Why does the economist resist this sisterly advice?

10.6 An economist believes that the percentage of the popular vote received by the incumbent party's candidate for president depends (negatively) on the unemployment rate and (negatively) on the rate of inflation. Having read only the chapter on simple regression, he estimates an equation with voting percent as the dependent variable and the unemployment rate as the explanatory variable and ignores the rate of inflation. He obtains a positive estimate of β, indicating that an increase in the unemployment rate increases the vote received by the incumbent party. How might the omission of the rate of inflation explain this wrong sign? Be specific.

10.7 An auto dealer claims that the resale value P of a 2005 Toyota Camry LE sedan depends on the number of miles M that it has been driven and whether it has an automatic or five-speed manual transmission. Market prices are used to estimate this equation, using the variable $D = 0$ if manual, $D = 1$ if automatic:

$$P = \alpha + \beta_1 D + \beta_2 M + \beta_3 DM + \varepsilon$$

Interpret each of the parameters α, β_1, β_2, and β_3. What advantages does a regression model have over a simple comparison of the average resale values of manual and automatic Camrys?

10.8 Use Table 10.2's 2009 prices of 2005 Camry LE sedans in excellent condition to estimate the equation $P = \alpha + \beta_1 D + \beta_2 M + \beta_3 DM + \varepsilon$, where P = price, M = mileage, and $D = 0$ if the car has a five-speed manual transmission, $D = 1$ if it has an automatic transmission.

a. Which of the estimated coefficients are statistically significant at the 5 percent level?

b. Are the signs and magnitudes of all the estimated coefficients plausible and substantial?

10.9 A study [5] of the housing market in Fishers, Indiana, looked at 15 matched pairs of houses that were sold or rented in 2005. Table 10.3 shows the sale price for each house, the size of the house (square feet), and the monthly rent for the house that is a matched pair.

Table 10.2: Exercise 10.8

Price ($)	Miles	Transmission	Price ($)	Miles	Transmission
11,500	10,000	Automatic	10,200	47,000	Manual
9,000	68,000	Automatic	9,000	100,000	Automatic
11,000	18,000	Manual	12,200	14,000	Manual
7,500	110,000	Automatic	9,990	72,000	Manual
7,000	125,000	Manual	13,500	12,000	Automatic
7,200	120,000	Manual	11,400	56,000	Automatic

Table 10.3: Exercise 10.9

Price ($)	Size	Rent ($)	Price ($)	Size	Rent ($)
137,000	1,300	1,100	143,000	2,213	1,250
208,000	2,146	1,200	127,500	1,122	1,000
123,000	1,390	875	134,000	1,917	1,250
167,500	2,008	1,250	315,000	3,014	2,150
181,400	2,160	1,500	164,900	2,008	1,550
344,950	2,931	2,100	136,000	1,696	1,150
132,500	1,316	1,150	200,000	2,301	2,800
130,000	1,352	1,100			

a. Estimate a simple regression equation using price as the dependent variable and size as the explanatory variable.
b. Estimate a simple regression equation using price as the dependent variable and rent as the explanatory variable.
c. Estimate a multiple regression equation using price as the dependent variable and size and rent as explanatory variables.
d. Explain any differences between your multiple regression equation (c) and the two simple regression equations (a and b).

10.10 A study [6] of the housing market in Mission Viejo, California, looked at 15 matched pairs of houses that were sold or rented in 2005. Table 10.4 shows the sale price for each house, the size of the house (square feet), and the monthly rent for the house that is a matched pair.
a. Estimate a simple regression equation using price as the dependent variable and size as the explanatory variable.
b. Estimate a simple regression equation using price as the dependent variable and rent as the explanatory variable.
c. Estimate a multiple regression equation using price as the dependent variable and size and rent as explanatory variables.
d. Explain any differences between your multiple regression equation (c) and the two simple regression equations (a and b).

10.11 Here are two regressions, one using annual data for 1980–1989, the other for 1970–1989. Make an educated guess about which is which and explain your reasoning.

$$C = -134.292 + 0.671Y + 0.045P, \quad \text{SEE} = 24.2$$
$$\quad [2.1] \quad [21.1] \quad [1.0]$$
$$C = -100.073 + 0.649Y + 0.076P, \quad \text{SEE} = 19.7$$
$$\quad [9.1] \quad [101.0] \quad [4.7]$$

where C = consumption spending, Y = gross domestic product, P = Dow Jones Industrial Average, and the t values are in brackets.

Table 10.4: Exercise 10.10

Price ($)	Size	Rent ($)	Price ($)	Size	Rent ($)
608,000	1,325	2,000	590,000	1,200	2,100
667,500	2,145	2,350	688,000	1,549	2,000
650,000	1,561	2,300	599,900	1,109	1,900
660,000	1,200	1,900	715,000	2,000	2,800
865,000	2,300	2,750	630,000	1,560	2,325
652,900	1,472	2,195	915,000	3,418	3,000
725,000	2,026	2,500	600,000	1,471	2,200

10.12 Fill in the missing numbers in this computer output:

$$Y = 134.292 + 0.671X_1 + 0.045X_2, \quad \text{SEE} = 24.2$$

Explan. Var.	Coefficient	Std. Error	*t* Value
Constant		65.641	
X_1		0.032	
X_2			1.04

a. Is the estimated coefficient of X_1 statistically significant at the 5 percent level?
b. What is the predicted value of Y when $X_1 = 4{,}000$ and $X_2 = 2{,}000$?

10.13 Use these numbers to fill in the blanks in this computer output: 0.01, 0.03, 0.14, 0.26, 1.95, and 2.69.

Explan. Var.	Coefficient	Std. Error	*t* Value	*P* Value
Constant	5.80	20.48	0.28	0.39
X_1	0.51			
X_2	0.38			
Degrees of freedom = 26		R2 = 0.42		

10.14 A linear regression looked at the relationship between the dollar prices P of 45 used 2001 Audi A4 2.8L sedans with manual transmission and the number of miles M the cars have been driven. A least squares regression yields

$$P = \underset{(2{,}553)}{16{,}958} - \underset{(0.0233)}{0.0677}M, \quad R^2 = 0.70$$

Suppose that the true relationship is $P = \alpha - \beta_1 M + \beta_2 C + \varepsilon$, where C is the condition of the car and α, β_1, and β_2 are all positive. If M and C are negatively correlated, does the omission of C from the estimated equation bias the estimate of β_1 upward or downward?

10.15 In American college football, Rivals.com constructs a widely followed index for rating colleges based on the high school players they recruit and Jeff Sagarin constructs a highly regarded system for rating college football teams based on their performance. Table 10.5 shows the 2008 performance ratings and the 2004–2008 recruiting ratings for the 20 colleges with the highest recruiting averages over this 5-year period. (Five years of recruiting data are used because players often "red shirt" their first year by practicing with the team but not playing, thereby maintaining their 4 years of college eligibility.) Use these data to estimate a multiple regression model with performance the dependent variable and the five recruiting scores the explanatory variables.

a. Do all of the recruiting coefficients have plausible signs?
b. Which, if any, of the recruiting coefficients are statistically significant at the 5 percent level?

Table 10.5: Exercise 10.15

	Performance	Recruiting 2008	2007	2006	2005	2004
Alabama	89.48	2836	1789	2127	1424	1366
Auburn	71.69	1335	2013	2148	1596	954
California	83.58	1070	1378	1512	1816	935
Clemson	78.87	2113	1555	1741	1431	364
Florida	98.74	2600	2959	2901	1467	2065
Florida State	83.18	2251	1394	2703	2582	2377
Georgia	84.81	2321	1895	2436	1814	2074
LSU	81.96	2134	2695	2238	1335	2492
Miami-FL	77.42	2467	1452	1785	1976	2329
Michigan	64.28	2220	1750	1974	1995	2116
Nebraska	80.70	1159	1734	1432	2178	866
Notre Dame	73.75	2744	1932	2189	661	820
Ohio State	84.83	2481	1577	2000	1654	1913
Oklahoma	94.15	2422	1704	2180	2441	2055
Penn State	88.26	700	1305	2241	1152	1430
South Carolina	77.06	1322	2190	1236	1310	799
Southern Cal	94.85	2312	2761	3018	2631	2908
Tennessee	71.95	1027	2726	1245	2403	1748
Texas	93.50	1919	2562	2283	1409	1864
Texas A&M	65.14	1482	795	1165	1839	1501

c. Which team overperformed the most, in that its performance was much higher than predicted by your estimated equation?
d. Which team underperformed the most, in that its performance was much lower than predicted by your estimated equation?

10.16 U.S. stock prices nearly doubled between 1980 and 1986. Was this a speculative bubble, based on nothing more than greed and wishful dreams, or can these higher prices be explained by changes in the U.S. economy during this period? While the fortunes of individual companies are buffeted by factors specific to their industry and the firm, the overall level of stock prices depends primarily on national economic activity and interest rates. When the economy is booming, corporations make large profits and stock prices should increase. For any given level of economic activity, an increase in interest rates makes bonds more attractive, putting downward pressure on stock prices. Table 10.6 shows the value of the S&P 500 index of stock prices, the percent unemployment rate, and the percent interest rate on long-term Treasury bonds.

a. Estimate a simple regression model with the S&P 500 the dependent variable and the unemployment rate the explanatory variable.
b. Estimate a simple regression model with the S&P 500 the dependent variable and the interest rate the explanatory variable.
c. Estimate a multiple regression model with the S&P 500 the dependent variable and the unemployment rate and interest rate the explanatory variables.

Table 10.6: Exercise 10.16

	S&P 500	Unemployment	Interest Rate
1980	133.48	7.2	11.89
1981	123.79	8.6	12.88
1982	139.37	10.7	10.33
1983	164.36	8.2	11.44
1984	164.48	7.2	11.21
1985	207.26	6.9	9.60
1986	249.78	6.9	7.65

d. Are the unemployment rate and interest rate positively or negatively correlated during this period? How does this correlation explain the differences in the estimated coefficients in questions a, b, and c?

10.17 (continuation) We will use the model you estimated in question c of the preceding exercise to explain the sharp increase in stock prices in 1985 and 1986.

a. Stock prices increased by 42.78 in 1985. What is the predicted effect on stock prices of the 0.3 drop in the unemployment rate that occurred? Of the 1.61 drop in the interest rate that occurred? According to your estimated model, was the increase in stock prices in 1985 mostly due to falling unemployment or falling interest rates?

b. In 1986 the interest rate fell by 1.95 and the unemployment rate was constant. Using the approach in question a, was the 42.52 increase in stock prices in 1986 mostly due to changes in unemployment or interest rates?

10.18 When OPEC's manipulation of oil supplies in the 1970s created an energy crisis, U.S. speed limits were reduced in an effort to reduce fuel consumption. Use the data in Table 10.7 to estimate this model of U.S. motor vehicle fuel usage, $G = \alpha + \beta_1 M + \beta_2 H + \beta_3 S + \varepsilon$, where G is gallons of fuel used per vehicle, M is miles traveled per vehicle, H is average horsepower, and S is average speed traveled (miles per hour). Which of the estimated coefficients have plausible signs and are statistically significant at the 5 percent level?

10.19 A random sample of 30 college students yielded data on Y = grade in an introductory college chemistry course (four-point scale), X_1 = high school GPA (four-point scale), and X_2 = SAT score (2400 maximum). The standard errors are in parentheses, the t values are in brackets, and $R^2 = 0.22$.

$$\begin{aligned} Y = -4.20 &+ 2.79X_1 + 0.0027X_2 \\ (5.88) &\quad (1.05) \quad (0.0033) \\ [0.72] &\quad [2.64] \quad [0.83] \end{aligned}$$

a. Explain why you either agree or disagree with this interpretation: "The primary implication of these results is that high school GPA and SAT score are poor predictors of a student's success in this chemistry course."

Table 10.7: Exercise 10.18

	G	M	H	S
1970	851.5	10,341	178.3	63.8
1971	863.4	10,496	183.5	64.7
1972	884.7	10,673	183.0	64.9
1973	879.1	10,414	183.2	65.0
1974	818.3	9,900	178.8	57.6
1975	820.2	10,008	178.7	57.6
1976	835.4	10,195	175.7	58.2
1977	839.9	10,372	175.7	58.8
1978	843.0	10,431	174.5	58.8
1979	804.3	10,072	175.3	58.3
1980	738.1	9,763	175.6	57.5

b. Suppose that an important omitted variable is $X_3 = 1$ if the student has taken a high school chemistry course, 0 otherwise; the true coefficients of X_1, X_2, and X_3 are positive; and X_3 is negatively correlated with X_1 and uncorrelated with X_2. If so, is the estimated coefficient of X_1 in the preceding equation biased upward or downward? Be sure to explain your reasoning.

10.20 A study of a microfinance program in South India estimated the following least squares regression:

$$\begin{array}{llll} Y = & 1.711.40 + & 11.27X_1 + & 0.0316X_2 \\ & (265.86) & (43.01) & (0.0168) \\ & [6.44] & [0.26] & [1.88] \end{array}$$

where Y = monthly family income, X_1 = years in the microfinance loan program, and X_2 = total amount loaned. The standard errors are in parentheses, the t values are in brackets, and $R^2 = 0.05$.

a. Is the coefficient of X_2 substantial?
b. How was the t value for the coefficient of X_2 calculated?
c. How would you decide whether to calculate a one-sided or two-sided P value for the coefficient of X_2?
d. Explain why you either agree or disagree with this interpretation of the results: "The one-sided P value for the total amount loaned is 0.032, which is not quite high enough to show that the amount loaned is statistically significant at the 5 percent level."

10.21 A dean claims that the salaries of humanities professors Y are determined solely by the number of years of teaching experience X. To see if there is a statistically significant difference in the salaries of male and female professors, the equation $Y = \alpha + \beta_1 D + \beta_2 X + \beta_3 DX + \varepsilon$ was estimated, using the variable $D = 0$ if male, 1 if female.

a. Interpret each of the parameters α, β_1, β_2, and β_3.
b. What advantages does a regression model have over a simple comparison of the average men's and women's salaries?
c. Describe a specific situation in which a comparison of men's and women's average salaries shows discrimination against women while a regression equation does not.
d. Describe a specific situation in which a comparison of men's and women's average salaries shows no discrimination while a regression equation indicates discrimination against women.

10.22 In March 1973, the world's industrialized nations switched from fixed to floating exchange rates. To investigate the effects on international trade, a multiple regression equation was estimated using annual data on world trade and gross domestic product (GDP) for the ten years preceding this change and for the ten years following the change:

$$\begin{array}{llllll} T = & 28.85 & -\ 314.91D & +\ 0.57Y & +\ 5.04DY, & R^2 = 0.977 \\ & (27.02) & (51.38) & (0.48) & (0.71) & \end{array}$$

where T = world trade index (1985 = 200); Y = world GDP index (1985 = 100); $D = 0$ in 1963–1972 and $D = 1$ in 1974–1983; and the standard errors are in parentheses.
a. Which of these four coefficients are statistically significant at the 5 percent level?
b. Interpret each of these estimated coefficients: 28.85, −314.91, 0.57, 5.04.
c. What happened to the relationship between world GDP and world trade after 1973?

10.23 On average, men are taller than women and therefore take longer strides. One way to adjust for this height difference in comparing men's and women's track records is to divide the distance run by the runner's height, giving a proxy for the number of strides it would take this runner to cover the distance. The division of this figure by the runner's time gives an estimate of the runner's strides per second. For instance, Leroy Burrell is a former world record holder for the 100 meters. Dividing 100 meters by his height (1.828 meters) gives 54.70, indicating that 100 meters takes 54.70 strides for him. For his world record time of 9.85 seconds, this converts to 54.70/9.85 = 5.55 strides per second. A study of the top three runners in several events over a 7-year period yielded the data in Table 10.8 on strides per second. Use these data to estimate a multiple regression model $S = \alpha + \beta_1 D + \beta_2 X + \beta_3 DX + \varepsilon$, where S = strides per second, X = natural logarithm of distance, and $D = 1$ if a woman, 0 if a man. Are the estimated values of the following parameters statistically significant at the 5 percent level? Are the signs plausible?
a. β_1.
b. β_2.
c. β_3.

Table 10.8: Exercise 10.23

Distance (meters)	Men	Women
100	5.52	5.38
200	5.49	5.30
400	4.90	4.68
800	4.24	4.11
1,500	4.04	3.77
5,000	3.71	3.33
10,000	3.57	3.26

10.24 Positively correlated residuals often occur when models are estimated using economic time series data. The residuals for the consumption function estimated in this chapter turn out to be positively correlated, in that positive prediction errors tend to be followed by positive prediction errors and negative prediction errors tend to be followed by negative prediction errors. This problem can often be solved by working with changes in the values of the variables rather than the levels:

$$[C - C(-1)] = \alpha + \beta_1[Y - Y(-1)] + \beta_2[W - W(-1)] + \varepsilon$$

Use the data in Table 10.1 to calculate the changes in C, Y, and W; for example, the value of $C - C(-1)$ in 1961 is 1,821.2 − 1,784.4 = 36.8.

a. Estimate this simple regression model: $[C - C(-1)] = \alpha + \beta[Y - Y(-)] + \varepsilon$.
b. Estimate this simple regression model: $[C - C(-1)] = \alpha + \beta[W - W(-1)] + \varepsilon$.
c. Estimate this multiple regression model: $[C - C(-1)] = \alpha + \beta_1[Y - Y(-1)] + \beta_2[W - W(-1)] + \varepsilon$.
d. How do the estimated coefficients of the changes in income and wealth in the multiple regression model c compare with the estimated coefficients in the simple regression models a and b? How would you explain the fact that they are larger or smaller?
e. How do the estimated coefficients of the changes in income and wealth in the multiple regression model c compare with the equation estimated in the chapter: $C = -110.126 + 0.798Y + 0.026W$?

10.25 Use the data in Table 10.1 to answer these questions:

a. Estimate the equation $C = \alpha + \beta X + \varepsilon$, where C is spending and X is the year. Use your estimates to calculate predicted spending each year, $\hat{C} = a + bX$. The residuals $C - \hat{C}$ are the detrended values of spending.
b. Repeat the procedure in question a using income as the dependent variable.
c. Repeat the procedure in question a using wealth as the dependent variable.
d. Estimate a multiple regression model where the dependent variable is detrended spending and the explanatory variables are detrended income and detrended wealth.
e. How do the estimated coefficients in question d compare with the equation estimated in the chapter: $C = -110.126 + 0.798Y + 0.026W$?

10.26 Use the hard drive data for 2006 and 2007 in Table 1.6 to estimate the model $Y = \alpha + \beta_1 D + \beta_2 X + \beta_3 DX + \varepsilon$, where Y = price, X = size, $D = 0$ in 2006, and $D = 1$ in 2007.
a. Interpret your estimate of the coefficient β_1.
b. Interpret your estimate of the coefficient β_2.
c. Interpret your estimate of the coefficient β_3.
d. Calculate a 95 percent confidence interval for β_3.

10.27 A researcher used data on these three variables for 60 universities randomly selected from the 229 national universities in *U.S. News & World Report*'s rankings of U.S. colleges and universities: Y = graduation rate (mean 59.65), X_1 = student body's median math plus reading SAT score (mean 1,030.5), and X_2 = percent of student body with GPAs among the top 10 percent at their high school (mean = 46.6). He found a statistically significant positive relationship between SAT scores and graduation rate:

$$\begin{array}{rll} Y = & -33.39 & +0.090X_1, \quad R^2 = 0.71 \\ & (14.30) & (0.014) \\ & [2.34] & [6.58] \end{array}$$

The standard errors are in parentheses and the t values are in brackets. He also found a statistically significant positive relationship between GPA and graduation rate:

$$\begin{array}{rll} Y = & 42.20 & +0.375X_2, \quad R^2 = 0.52 \\ & (4.73) & (0.084) \\ & [8.93] & [4.46] \end{array}$$

But when he included both SAT scores and GPAs in a multiple regression model, the estimated effect of GPAs on graduation rates was very small and not statistically significant at the 5 percent level:

$$\begin{array}{rlll} Y = & -26.20 & +0.081X_1 & +0.056X_2, \quad R^2 = 0.71 \\ & (21.08) & (0.025) & (0.118) \\ & [1.24] & [3.30] & [0.47] \end{array}$$

a. He suspected that there was an error in his multiple regression results. Is it possible for a variable to be statistically significant in a simple regression but not significant in a multiple regression?
b. Do you think that X_1 and X_2 are positively correlated, negatively correlated, or uncorrelated?
c. If your reasoning in b is correct, how would this explain the fact that the coefficients of X_1 and X_2 are lower in the multiple regression than in the simple regression equations?

10.28 Sales data for single-family houses in a Dallas suburb were used to estimate the following regression equation:

$$P = 60{,}076 + 75.1S + 36.4G - 3{,}295A + 4{,}473B - 14{,}632T, \quad R^2 = 0.84$$
$$(14.4) \quad (19.7) \quad (12.1) \quad (1{,}078) \quad (1{,}708) \quad (3{,}531)$$

where P = sale price, S = square feet of living area, G = garage square feet, A = age of house in years, B = number of baths, and T = 1 if two-story, 0 if not. The standard errors are in parentheses.

a. What is the predicted sale price of a two-story, 10-year-old house with 2,000 square feet of living space, three baths, and a 300-square-foot garage? Does this predicted sale price seem reasonable to you?
b. Which of the estimated coefficients in the regression equation are statistically significant at the 5 percent level? How can you tell?
c. Does the sign and size of each coefficient seem reasonable? In each case, explain your reasoning.
d. Do you think that the parameters of housing regression equations change over time? Why or why not? Be specific.

10.29 A survey of 100 English professors obtained data on their salaries and four explanatory variables: S = salary (thousands of dollars); B = number of published books; A = number of articles published in the last 2 years; E = number of years since degree; and J = 1 if they have received a recent job offer, 0 otherwise. The estimated equation (with t values in brackets) is

$$S = 87.65 + 0.42B + 2.10A + 1.21E + 1.36J, \quad R^2 = 0.68$$
$$[12.58] \quad [1.77] \quad [4.60] \quad [2.97] \quad [3.45]$$

a. Which of the explanatory variables are dummy variables?
b. Which, if any, of the estimated signs do you find surprising, perhaps implausible?
c. Which of the estimated coefficients are statistically significant at the 5 percent level?
d. Give a 95 percent confidence interval for the effect on a professor's annual salary of having an article published. Does your answer seem like a lot of money or a little?
e. According to the estimated equation, will a professor's salary be increased more by writing an article or a book?
f. What is the predicted annual salary for a professor who has 20 years of experience, one published book, two articles published in the last 2 years, and no recent job offers? Does this prediction seem plausible?

10.30 Table 10.9 shows data on state educational spending per pupil (E), percentage of high school seniors taking the SAT (S), and mean reading SAT score (R).

Table 10.9: Exercise 10.30

State	*E*	*S*	*R*	State	*E*	*S*	*R*	State	*E*	*S*	*R*
AK	7,877	42	443	KY	4,390	10	477	NY	8,500	69	419
AL	3,648	8	482	LA	4,012	9	473	OH	5,639	23	451
AR	3,334	7	471	MA	6,351	72	432	OK	3,742	8	479
AZ	4,231	23	452	MD	6,184	60	434	OR	5,291	63	443
CA	4,826	44	422	ME	5,894	59	431	PA	6,534	63	423
CO	4,809	29	458	MI	5,257	12	458	RI	6,989	63	429
CT	7,914	75	435	MN	5,260	15	477	SC	4,327	55	399
DC	8,210	67	407	MO	4,415	13	471	SD	3,730	6	498
DE	6,016	60	435	MS	3,322	4	472	TN	3,707	12	486
FL	5,154	47	419	MT	5,184	20	469	TX	4,238	43	415
GA	4,860	59	402	NC	4,802	57	397	UT	2,993	5	499
HI	5,008	52	406	ND	3,685	5	500	VA	5,360	59	430
IA	4,839	5	512	NE	4,381	10	487	VT	5,740	63	435
ID	3,200	17	465	NH	5,504	66	477	WA	5,045	39	448
IL	5,062	17	462	NJ	9,159	67	423	WI	5,946	12	477
IN	5,051	55	412	NM	4,446	11	483	WV	5,046	15	448
KS	5,009	10	495	NV	4,564	23	439	WY	5,255	14	462

a. Estimate the equation $R = \alpha + \beta E + \varepsilon$.
b. Is the estimate of β positive or negative?
c. Is the estimate of β statistically significant at the 1 percent level?
d. Now estimate the equation $R = \alpha + \beta_1 E + \beta_2 S + \varepsilon$.
e. Is the estimate of β_1 positive or negative?
f. Is the estimate of β_1 statistically significant at the 1 percent level?
g. Explain any noteworthy differences between your estimates of β in a and β_1 in d.

10.31 A researcher was interested in how a movie's box office revenue is affected by the audience suitability rating assigned by the Motion Picture Association of America (MPAA) and the quality of the film as judged by *Entertainment Weekly*'s survey of ten prominent film critics. Data for 105 films were used to estimate the following equation by least squares:

$$R = 6{,}800{,}024 - 25{,}232{,}790S + 6{,}453{,}431Q, \quad R^2 = 0.214$$
$$[0.61] \qquad [4.36] \qquad [3.74]$$

where R = domestic revenue, in dollars; S = MPAA suitability rating (0 if rated G, PG, or PG-13, 1 if rated R); Q = *Entertainment Weekly* quality rating (12 = A, 11 = A−, 10 = B+, …); and the t values are in brackets. Explain the errors in each of these critiques of the results:

a. "The R^2 value of 0.214 is close enough to 0 that it can be assumed that the null hypothesis can be rejected."
b. "The coefficient of S has the wrong sign and is also much too large."

c. "The estimated coefficient of Q is biased downward because S and Q are negatively correlated."

10.32 Use the data in Exercise 8.28 to estimate a multiple regression model with systolic blood pressure as the dependent variable and diastolic blood pressure the explanatory variable, allowing the slope and intercept to depend on whether the person is male or female. Does gender have a statistically significant effect on the slope or intercept at the 5 percent level?

10.33 Data from a random sample of 100 college students were used to estimate the following multiple regression equation:

$$\underset{}{Y = 4.589} - \underset{(0.402)}{4.608X_1} + \underset{(0.080)}{0.405X_2} + \underset{(0.079)}{0.568X_3}, \quad R^2 = 0.74$$

where Y is the student's height (inches), X_1 is a dummy variable equal to 1 if the student is female and 0 if male, X_2 is the biological mother's height (inches), X_3 is the biological father's height (inches), and the standard errors are in parentheses.

a. Which (if any) of the coefficients other than the intercept are statistically significant at the 5 percent level?

b. Interpret each of the coefficients other than the intercept. Does the magnitude of each seem plausible or implausible?

c. Interpret the value of R^2.

d. Explain why this conclusion is misleading: "The student's gender is important in determining height. In these data, females are taller than males, which is shown in the t value for the coefficient of X_1."

10.34 Least squares were used to estimate the relationship between miles per gallon M, engine size S (liters), and the weights W (pounds) of 436 automobiles:

$$M = 38.6765 - \underset{[13.2987]}{2.1414S} - \underset{[8.6200]}{0.0022W}, \quad R^2 = 0.65$$

The t values are in brackets.

a. Does the value −2.1414 seem reasonable?

b. Does the value −0.0022 seem reasonable?

c. Does the value 0.65 seem reasonable?

d. Is the coefficient of S statistically significant at the 5 percent level?

e. Is the coefficient of W statistically significant at the 5 percent level?

10.35 Data for 100 college students were used to estimate this multiple regression equation: $Y = -0.752 + 0.470F + 0.507M + 0.047A + 4.330D$, where Y = height of student (inches), F = height of student's biological father (inches), M = height of student's biological mother (inches), A = student's age (years), and D = 1 if the student is male and 0 if female.

a. What is the predicted effect on a student's height if the heights of the father and mother are both increased by 1 inch?
b. Explain why you either agree or disagree with this interpretation of the results: "Men were, on average, 4.33 inches taller than women."
c. Are the results invalidated by the fact that this equation predicts a negative height if all of the explanatory variables equal 0?

10.36 A survey of 47 sophomores investigated the effect of studying and extracurricular activities on grades:

$$\begin{aligned} Y = \; & 2.936 \; + 0.023X - 0.028D \\ & (0.113) \quad (0.006) \quad (0.096) \end{aligned}$$

where Y = grade point average on a four point scale; X = average hours per week spent studying; D = 1 if the person spends at least 10 hours a week on an extracurricular activity, such as work or sports, and 0 otherwise; and the standard errors are in parentheses. The researcher concluded "The effect of extracurricular activity is not significant and does not lend support to the commonly held notion that extracurricular activity negatively affects grades because it reduces available study time." Carefully explain why the coefficient of D does not measure the extent to which extracurricular activity affects grades by reducing available study time.

10.37 A study collected data on the 15 best-selling fiction and 15 best-selling nonfiction hardcover books (according to the *New York Times* bestseller list): Y = price, X_1 = number of pages, and X_2 = category (0 if fiction, 1 if nonfiction). The standard errors are in parentheses and the t values are in brackets.

$$\begin{aligned} Y = \; & 15.89 \; + 0.022X_1 \; + 0.290X_2, \quad R^2 = 0.59 \\ & (1.27) \quad (0.0035) \quad (0.071) \\ & [12.73] \quad [6.20] \quad [0.41] \end{aligned}$$

This equation was then estimated, where $X_3 = X_1X_2$:

$$\begin{aligned} Y = \; & 18.49 + 0.0144X_1 - 3.90X_2 + 0.0119X_3, \quad R^2 = 0.63 \\ & (2.00) \quad (0.0053) \quad (2.52) \quad (0.0069) \\ & [9.25] \quad [2.71] \quad [1.55] \quad [1.73] \end{aligned}$$

a. Explain the conceptual difference between these two models, using the parameter estimates to illustrate your reasoning.
b. Using the second equation, what is the predicted price of a 300-page nonfiction book?
c. The t values are for testing the null hypothesis that the coefficients equal 0. Use the second equation to test the null hypothesis that the coefficient of X_1 is 0.05. What is the alternative hypothesis?

10.38 A random sample of 30 new hardcover textbooks at a college bookstore was used to estimate the following multiple regression equation:

$$\begin{aligned} Y = {} & 16.65 + 0.04X + 22.12D_1 + 0.43D_2, \quad R^2 = 0.70 \\ & (5.11) \quad (0.01) \quad (5.60) \quad (5.31) \end{aligned}$$

where Y is the book's price (in dollars), X is the number of pages in the book, D_1 is a dummy variable that equals 1 if it is a natural sciences book and 0 if it is a social sciences or humanities book, and D_2 is a dummy variable that equals 1 if it is a social sciences book and 0 if it is a natural sciences or humanities book; the standard errors are in parentheses.

a. Which (if any) of the coefficients other than the intercept are statistically significant at the 5 percent level? What does *statistically significant at the 5 percent level* mean?
b. Interpret each of the coefficients other than the intercept. Does the magnitude of each seem plausible or implausible?
c. Interpret the value of R^2.

10.39 A researcher commented on this estimated equation: "Multicollinearity could be a factor as indicated by the very high standard error and the small t value."

$$\begin{aligned} Y = {} & 1{,}146.0 + 1.4312X, \quad R^2 = 0 \\ & (195.6) \quad (71.6) \\ & [5.86] \quad [0.02] \end{aligned}$$

( ): standard errors
[ ]: t values

a. In general, how does multicollinearity affect the standard errors and t values?
b. Why is it clear that multicollinearity is not a problem here?

10.40 Financial theory says that stock prices are negatively related to the unemployment rate and interest rates. The following equation was estimated using U.S. data on the S&P index of stock prices (P), the unemployment rate (U), and the Treasury bond rate (R):

$$\begin{aligned} P = {} & 502.50 - 12.12U - 25.85R, \quad R^2 = 0.904 \\ & (42.64) \quad (4.17) \quad (5.03) \end{aligned}$$

The standard errors are in parentheses. During the period studied, the Federal Reserve used high interest rates to weaken the economy (and reduce inflation), and used low interest rates to stimulate the economy. As a consequence, the unemployment rate and interest rate were highly positively correlated. To cure this multicollinearity problem, the researcher omitted the interest rate and obtained this result:

$$\begin{aligned} P = {} & 482.77 - 65.09U, \quad R^2 = 0.702 \\ & (39.85) \quad (3.60) \end{aligned}$$

a. Explain why this cure for multicollinearity is worse than the problem.
b. Explain why the coefficient of the unemployment rate is lower in the multiple regression equation than in the simple regression.

10.41 Exercises 8.41 and 8.42 give height and weight data for 18- to 24-year-old U.S. men and women. Use these data to estimate the equation

$$Y = \alpha + \beta_1 D + \beta_2 X + \beta_3 DX + \varepsilon$$

where Y = weight, X = height, and D = 0 if male, 1 if female.
a. Interpret the coefficient α.
b. Interpret the coefficient β_1.
c. Interpret the coefficient β_2.
d. Interpret the coefficient β_3.
e. Is there a statistically significant difference at the 5 percent level between the effect of height on weight for men and women?
f. Explain why you either agree or disagree with this reasoning: "A person with zero height has zero weight. Therefore, a regression of weight on height must predict zero weight for zero height."

10.42 Exercise 8.44 shows the results of separate regressions for foreign-born and California-born mother-daughter pairs. Predict the estimated coefficients if this regression is run, using all 10,619 observations:

$$Y = \alpha + \beta_1 D + \beta_2 X + \beta_3 DX + \varepsilon$$

where D = 0 if the mother is California born and D = 1 if the mother is foreign born. What advantage is there to running a single regression instead of two separate regressions?

10.43 What difference would it make in Exercise 10.42 if we set D = 0 if the mother is foreign born and D = 1 if the mother is California born? How would you choose between these two specifications?

10.44 Data for the top 100 money winners on the Professional Golf Association (PGA) tour were used to estimate the following equation:

$$\begin{array}{rcccccc} Y = & -8{,}023{,}111 & +\,15{,}992D & +\,12{,}003A & +\,27{,}084G & +\,91{,}306P & +\,11{,}357S \\ & (1{,}103{,}835) & (3{,}547) & (6{,}571) & (10{,}078) & (19{,}649) & (3{,}705) \\ & [7.26] & [4.51] & [1.83] & [2.69] & [4.65] & [3.07] \end{array}$$

where Y = dollars won on the PGA tour; D = average length of drives from the tee, yards; A = driving accuracy (percent of drives landing in fairway); G = percent of greens that were reached in par minus two strokes; P = percent of those greens reached in par minus two that required only one put; S = percent of times player made par after hitting into sand; the standard errors are in parentheses; the t values are in brackets; and $R^2 = 0.468$.

a. Which, if any, of the estimated coefficients of the five explanatory variables are statistically significant at the 5 percent level?
b. Which, if any, of the estimated coefficients of the five explanatory variables have plausible signs?
c. Does the fact that the estimated coefficient of P is larger than the estimated coefficient of D imply that, other things being equal, a player who is a good putter wins more money than a person who hits long drives?
d. Does the fact that the t value for S is larger than the t value for G imply that, other things being equal, a golfer who plays well out of the sand wins more money than a player who does well reaching the green in par minus 2?

10.45 Is the multicollinearity problem a characteristic of the population or the sample? For example, if you are trying to estimate the effect of the unemployment rate and interest rates on stock prices, do you have a multicollinearity problem if the unemployment rate and interest rates tend to be correlated in general or if they happen to be correlated in the particular data you have?

10.46 In an IBM antitrust case, an economist, Franklin Fisher, estimated multiple regression equations predicting computer prices based on memory, speed, and other characteristics:

> *Despite the fact that t-statistics on the order of 20 were obtained for all of the regression coefficients, Alan K. McAdams, appearing as an expert for the government, testified that collinearity made it impossible reliably to separate the effects of the different independent variables and hence that little reliance could be placed on the results [7].*

Explain why you either agree or disagree with McAdams's logic.

10.47 College dining halls use attendance data to predict the number of diners at each meal. Daily weekday lunch data for 11 weeks were used to estimate the following regression equation:

$$\begin{array}{cccccc} \hat{Y} = 220.58 & -\,29.13X_1 & +\,6.71X_2 & -\,37.49X_3 & +\,7.69X_4 & -\,2.25X_5 \\ (10.6) & (10.3) & (9.6) & (21.2) & (6.4) & (1.2) \end{array}$$

where Y = number of diners; X_1 = 1 if Tuesday, 0 otherwise; X_2 = 1 if Wednesday, 0 otherwise; X_3 = 1 if Thursday, 0 otherwise; X_4 = 1 if Friday, 0 otherwise; X_5 = week of semester (X_5 = 1 during first week, X_5 = 2 during second week, etc.); and the standard errors are in parentheses.

a. What is predicted attendance on Wednesday during the tenth week of the semester?
b. Interpret the estimated coefficient of X_3.
c. Which of the explanatory variables have statistically significant effects at the 5 percent level?

10.48 Table 10.10 shows hypothetical abilities and test scores. These 20 students have abilities ranging from 75 to 95, which we assume to be constant between third and fourth grade. For the four students at each ability level, two score five points above their ability and two score five points below their ability. Whether they score above or below their ability on the fourth grade test is independent of whether they scored above or below their ability on the third grade test. The uniform distribution of abilities and error scores is unrealistic, as are the severely limited values. This stark simplicity is intended to clarify the argument.

a. Estimate the simple linear regression model where third grade score is the dependent variable and ability is the explanatory variable.
b. Estimate the simple linear regression model where ability is the dependent variable and third grade score is the explanatory variable.
c. Explain how the difference between your answers to questions a and b reflect regression to the mean.

10.49 Use the data in Exercise 10.48:

a. Estimate the simple linear regression model where ability is the dependent variable and third grade score is the explanatory variable.
b. Estimate the simple linear regression model where fourth grade score is the dependent variable and third grade score is the explanatory variable.
c. Explain how your answers to questions a and b reflect regression to the mean.

Table 10.10: Exercise 10.48

Ability	Third Grade Scores	Fourth Grade Scores
75	70	70
75	70	80
75	80	70
75	80	80
80	75	75
80	75	85
80	85	75
80	85	85
85	80	80
85	80	90
85	90	80
85	90	90
90	85	85
90	85	95
90	95	85
90	95	95
95	90	90
95	90	100
95	100	90
95	100	100

10.50 Use the data in Exercise 10.48:

a. Estimate the simple linear regression model where ability is the dependent variable and third grade score is the explanatory variable.

b. Use the equation you estimated in question a to calculate the predicted ability $\hat{Y}$ of each student.

c. Estimate the simple linear regression model where fourth grade score is the dependent variable and predicted ability $\hat{Y}$ is the explanatory variable.

d. Explain how your answers to questions a, b, and c reflect regression to the mean.

CHAPTER 11

Modeling (Optional)

Chapter Outline

Econometricians, like artists, tend to fall in love with their models.

—Edward Leamer

An explicit model has many advantages. First, equations clarify our beliefs in ways that mere words do not. Second, data can be used to estimate and test an explicit model. Finally, an explicit model can be used to make specific predictions.

11.1 Causality

A model should have a theoretical underpinning. Before we write down an equation that says spending depends on income, we should have a persuasive reason for believing this. The same with a model that says elections are affected by the economy or that stock prices depend on corporate earnings. When we say that one variable *depends* on another, we are saying that there is a causal relationship—not a coincidental correlation. We believe that spending depends on income because it makes sense.

When we have a causal relationship, we typically have an expectation about the sign of the relationship. We expect that an increase in household income will cause spending to increase. We then use data to confirm our theory and to estimate the size of the effect.

Sometimes, there are competing theoretical arguments and we are agnostic about which effect is stronger. When hourly wages increase, people might choose to work more because they will be paid more, or they might work less because they need not work as many hours to pay their bills. In cases like this, where there are competing arguments, the data may tell us which effect is stronger. The important point is that we have good reasons for believing there is a relationship, even if we are unsure of the sign. We should not use a model that relates stock prices to football scores, because we have no good reason for believing that stock prices are influenced by football games.

11.2 Linear Models

The most convenient model is linear. For example, we might assume that aggregate consumer spending C is a linear function of income Y, each in trillions of dollars:

$$C = \alpha + \beta Y \tag{11.1}$$

In mathematics, the symbols Y and X are used repeatedly to show relationships among variables; for example, $Y = 2 + 3X$. In applied fields like economics, chemistry, and physics, variables represent real things (like income, hydrogen, or velocity) and a variety of symbols are used to help us remember what the variables represent. In economics, C is typically consumer spending and Y is usually income, and this is why these symbols are used in Equation 11.1.

A graph of Equation 11.1, as in Figure 11.1, is a straight line with an intercept α and a slope β that describes the constant effect on spending of an increase in income. If $\beta = 0.50$, then every \$1 trillion increase in income causes a \$0.50 trillion increase in spending.

It is unlikely that the real world is exactly linear, but we can often use a linear model as a useful approximation. For example, Figure 11.2 shows a hypothetical true relationship and a linear approximation. The slope of the true relationship falls as income increases because an

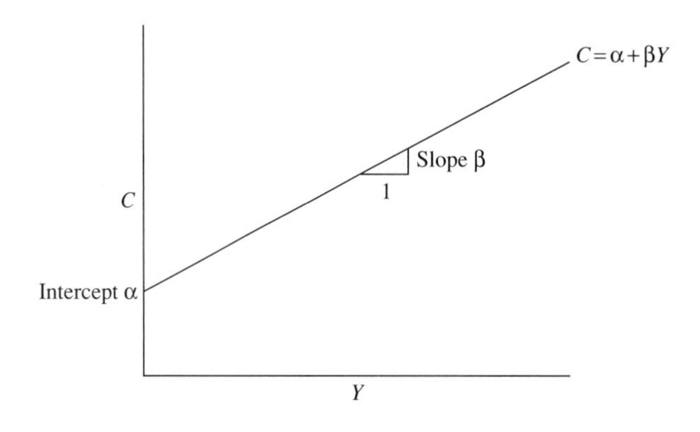

Figure 11.1
A linear relationship between consumer spending C and income Y.

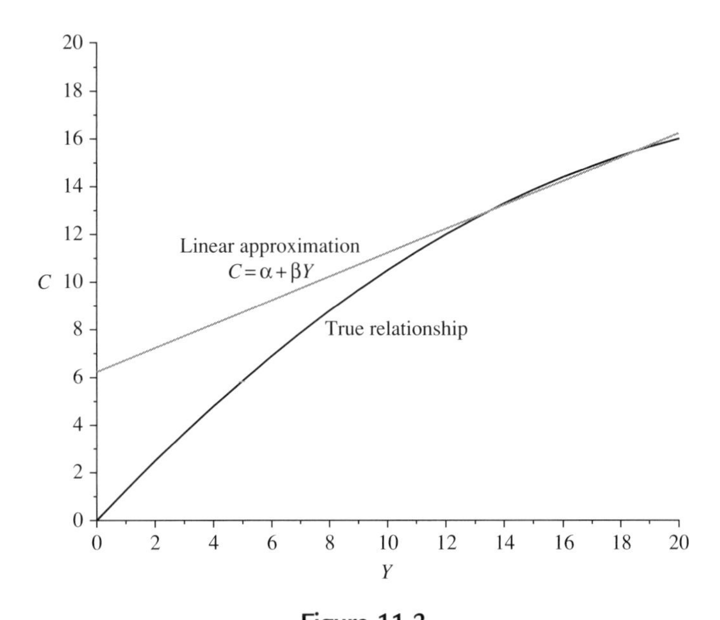

Figure 11.2
A linear approximation to a nonlinear relationship.

increase in income has a smaller effect on spending when income is high than when income is low. The linear approximation assumes that the slope is constant.

The linear approximation in Figure 11.2 is acceptable as long as we confine our attention to realistic and relatively modest variations in income, say, between 12 and 20. It is incautious extrapolation to use our linear approximation for extreme values of income (such as 0 or 100).

Because our model is a linear approximation, the Y intercept should not be used to predict spending when income is 0. No one knows what spending would be in such a catastrophic situation. Our model is not intended for and should not be applied to catastrophes. The intercept α is simply used to place the line so that our linear approximation works for realistic values of Y, say, between 12 and 20.

Similarly, we should be wary of using a linear model of the relationship between hours spent studying and test scores to predict the test score of someone who studies 168 hours a week or using a model of the relationship between rainfall and corn yield to predict corn yield when fields are flooded 2-feet deep. Linear approximations are often useful as long as we do not try to extrapolate them to drastically different situations.

11.3 Polynomial Models

One alternative to a linear model is a *polynomial function*:

$$Y = \alpha + \beta_1 X + \beta_2 X^2 + \cdots + \beta_n X^n \tag{11.2}$$

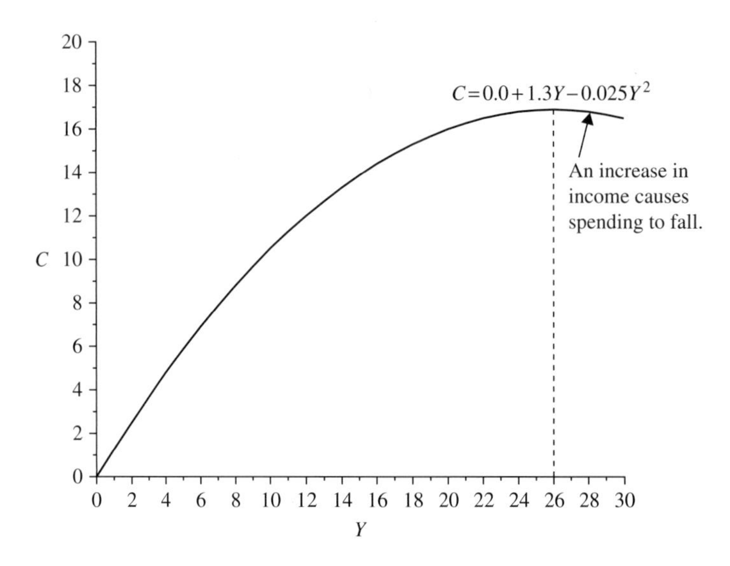

Figure 11.3
Quadratic models can have incautious extrapolation, too.

Some specific examples are

$$\begin{aligned} \text{linear: } Y &= \alpha + \beta_1 X \\ \text{quadratic: } Y &= \alpha + \beta_1 X + \beta_2 X^2 \\ \text{cubic: } Y &= \alpha + \beta_1 X + \beta_2 X^2 + \beta_3 X^3 \end{aligned}$$

The true relationship between spending and income graphed in Figure 11.2 is, in fact, this quadratic equation:

$$C = 0.0 + 1.3Y - 0.025Y^2 \qquad (11.3)$$

Polynomial equations are a reasonably simple way to model nonlinear relationships. However, they, too, can be susceptible to incautious extrapolation. Figure 11.3 shows that, as income increases past 26, Equation 11.3 predicts an implausible decline in spending.

Interpreting Quadratic Models

The general formula for a quadratic consumption function is

$$C = \alpha + \beta_1 Y + \beta_2 Y^2 \qquad (11.4)$$

Figure 11.4 shows a specific example to help us interpret the coefficients. The parameter α is the value of spending when income equals 0. This is where the consumption function intercepts the vertical axis; if $\alpha = 0$, the consumption function intersects the vertical axis at $C = 0$. The parameter β_1 is the slope of the equation when income equals 0; if $\beta_1 = 1.3$, the slope at the origin is 1.3. The parameter β_2 determines whether the slope of the equation increases or

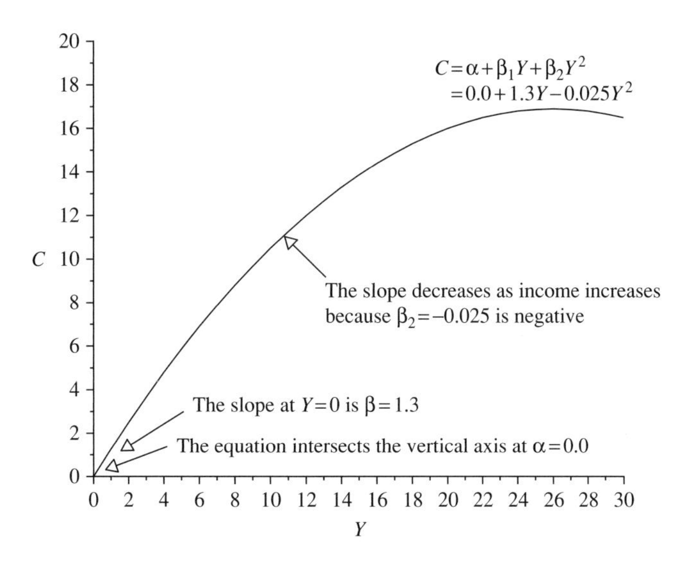

Figure 11.4
The parameters of a quadratic model.

decreases as income increases. Here, β_2 is negative (−0.025) and the slope decreases as income increases. If β_2 were positive, the slope would increase as income increases.

11.4 Power Functions

The equation

$$Y = AX^{\beta} \tag{11.5}$$

is called a *power function*, because X is raised to the power β. The parameter A is a scaling factor. The parameter β is called the *elasticity*, because it is (approximately) equal to the percentage change in Y resulting from a 1 percent increase in X.

If $\beta = 1$, then Y is simply a linear function of X with a 0 intercept:

$$Y = 0 + AX$$

The parameter β can be a fraction; for example, if $\beta = ½$, then Y is related to the square root of X:

$$\begin{aligned} Y &= AX^{1/2} \\ &= A\sqrt{X} \end{aligned}$$

Figure 11.5 shows three power functions, for values of β equal to 1.2, 1.0, and 0.8. When β equals 1, the equation is a straight line with a slope equal to 1. When β is larger than 1, the slope continuously increases as X increases. When β is less than 1, the reverse is true: the slope continuously declines as X increases.

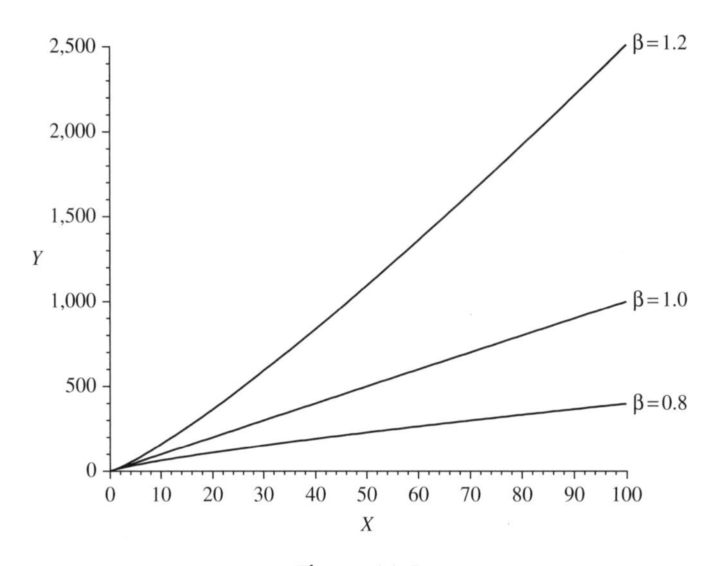

Figure 11.5
Three power functions.

Power functions with positive parameters A and β have several features that often make them appealing:

1. The Y intercept is at the origin; when X equals 0, Y equals 0.
2. Unlike a linear model, where changes in the level of X have a constant effect on the *level* of Y, in a power function, *percentage* changes in X have a constant *percentage* effect on Y.
3. Unlike quadratic functions, the slope is always positive, so that an increase in X always increases the value of Y. In Figure 11.5, the curve never bends backward for $\beta = 1.2$ and the slope never turns negative for $\beta = 0.8$.

We looked at a linear consumption function $C = \alpha + \beta Y$ and a quadratic function $C = \alpha + \beta_1 Y + \beta_2 Y^2$. We could instead use a power function $C = AY^{\beta}$. Figure 11.6 compares these three consumption equations, all scaled to be reasonably similar for values of income between 12 and 20. Over this limited range, it matters little which equation we use, and the linear approximation is fine. However, over wider ranges of income, the three equations diverge and the power function has some clear advantages over the other formulations.

In the linear and quadratic models, we cannot set spending equal to 0 when income is 0 unless we set the parameter α equal to 0, and this may well hinder our attempts to obtain a good approximation. (Imagine what would happen in Figure 11.6 if we set the linear intercept equal to 0.) Also, in the linear model, a \$1 increase in income increases spending by β, no matter what the level of income. In the quadratic model, we can force the slope to decline as income increases by making the parameter β_2 negative, but the slope eventually turns negative, which implies than an increase in income leads to a *decline* in spending.

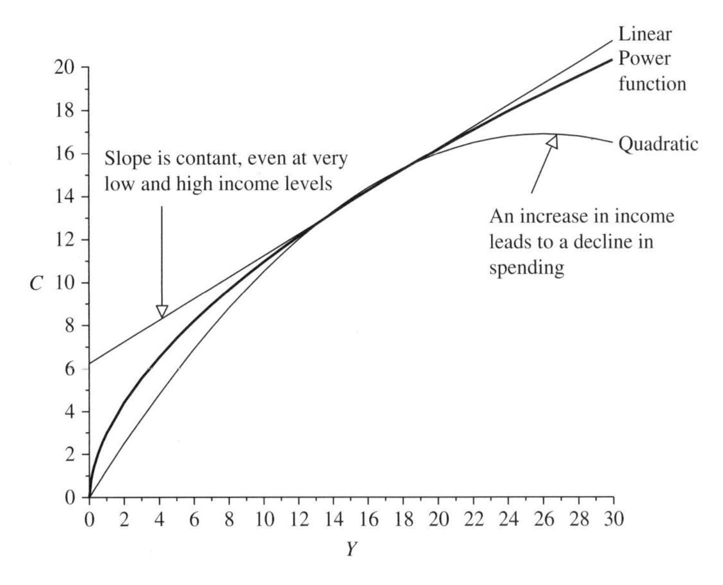

Figure 11.6
Three consumption functions.

In the power function model, a 1 percent increase in income increases spending by β percent, no matter what the level of income, and the slope never turns negative.

Negative Elasticities

The parameter β in a power function can be negative; for example, if $\beta = -1$, then

$$Y = AX^{-1}$$
$$= A\frac{1}{X}$$

Figure 11.7 shows the decline in the prices of 200 gigabyte (GB) hard drives between 2003 and 2007. A simple linear model does not fit these data very well and, in addition, has the implausible prediction that the price will be negative after 2007. Will hard drive manufacturers pay us to take their hard drives?

Figure 11.8 shows that a power function with a negative parameter β does a much better job modeling the path of hard drive prices and, in addition, never predicts a negative price.

Similarly, Figure 11.9 shows a linear demand function. The slope is negative because the quantity demanded falls as the price increases. (Economists traditionally put demand on the horizontal axis and price on the vertical axis; but we do the opposite, because demand is the dependent variable and, in statistics, the dependent variable is put on the vertical axis.)

A linear demand function may be a reasonable approximation for small price changes. However, it is implausible in other situations. Does a $1 price increase have the same effect on demand

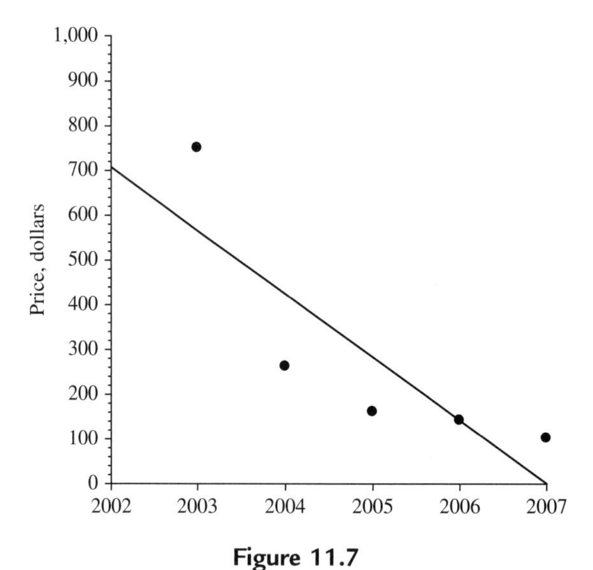

Figure 11.7
A linear model does not fit the data very well.

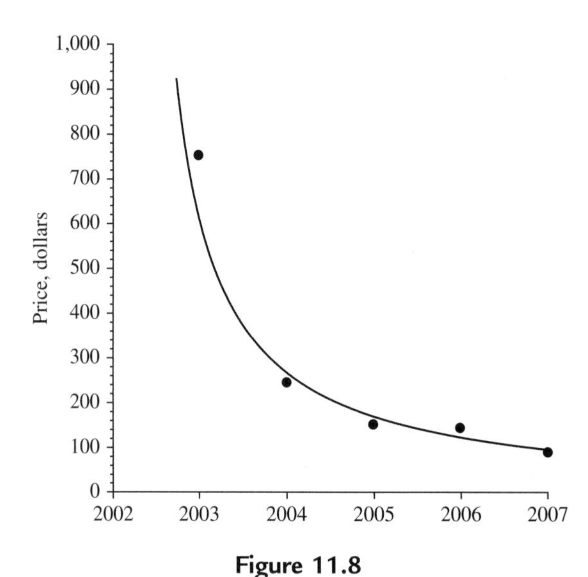

Figure 11.8
A power function does better.

when the price is \$1 and when the price is \$10? A linear demand function assumes the effect is the same. Not only that, a linear demand function implies that, when the price goes to 0, demand hits a maximum. It is more plausible that if something were free, demand would be virtually unlimited. A linear demand function also assumes that there is a maximum price where demand equals 0. Beyond that price, demand is either undefined or (nonsensically) negative.

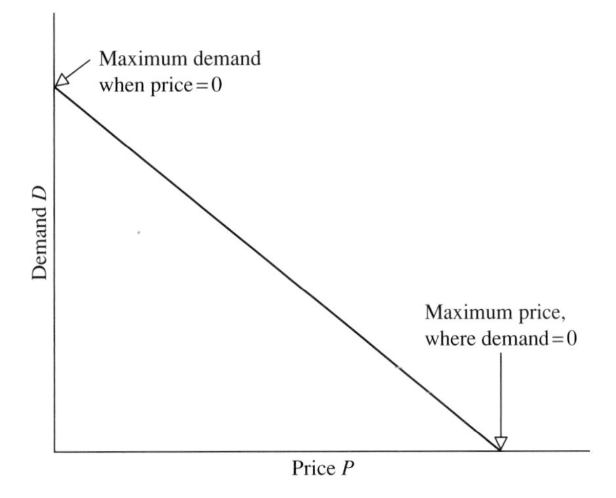

Figure 11.9
A linear demand function.

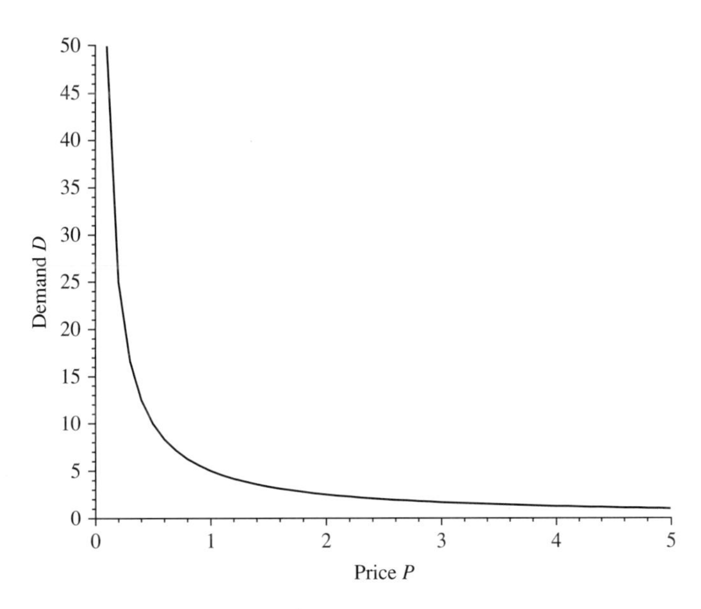

Figure 11.10
A power function demand equation.

In contrast, a power function assumes that the percentage increase in price is what matters. A $1 price increase is a 100 percent increase when the price is $1, but only a 10 percent increase when the price is $10. A 100 percent price increase should affect demand more than does a 10 percent price increase. Also, Figure 11.10 shows that, with a power function, demand increases indefinitely as the price approaches 0 and demand approaches 0 as the price becomes very large.

The Cobb-Douglas Production Function

Equation 11.2 shows that a polynomial equation is the *sum* of power functions. We can also use a model based on the *product* of power functions.

A production function shows how the output Q produced by a firm, an industry, or the entire economy is related to the amount of capital K, labor L, and other inputs used in production. For simplicity, we assume here that there are only two inputs: capital (including buildings, machines, and equipment) and labor (measured in the number of hours worked). More complex models distinguish different types of capital and labor and add other inputs, such as raw materials.

The marginal product of labor is equal to the change in output resulting from a 1-hour increase in labor, holding capital constant. The elasticity of output with respect to labor is equal to the percentage change in output resulting from a 1 percent increase in labor, holding capital constant. The marginal product of capital and the elasticity of output with respect to capital are defined analogously.

A linear production function is

$$Q = \alpha + \beta_1 K + \beta_2 L$$

The marginal product of labor is β_2 and is constant no matter how much labor is already employed. This is generally thought to be an implausible assumption; the *law of diminishing marginal returns* states that the marginal product declines as the input increases. Imagine a factory with a fixed work area and a fixed number of machines (remember, we are holding capital constant). As the amount of labor increases, workers increasingly get in each other's way and must share machines.

One way to allow for diminishing marginal returns is to use a quadratic production function:

$$Q = \alpha + \beta_1 K + \beta_2 K^2 + \beta_3 L + \beta_4 L^2$$

For any value of K, a positive value for β_3 and a negative value for β_4 imply a positive marginal product for small values of L and a diminishing marginal product for all values of L. Figure 11.11 shows an example using this particular function, holding capital constant at $K = 50$:

$$Q = 65K - 0.3K^2 + 78L + 0.2L^2$$

This specification might be a useful approximation for many values of L but has the unfortunate property that, for values of L larger than 195, output declines as labor increases. It would seem that the worst possible scenario is additional workers are told to stay home—which does not increase output, but does not decrease it either.

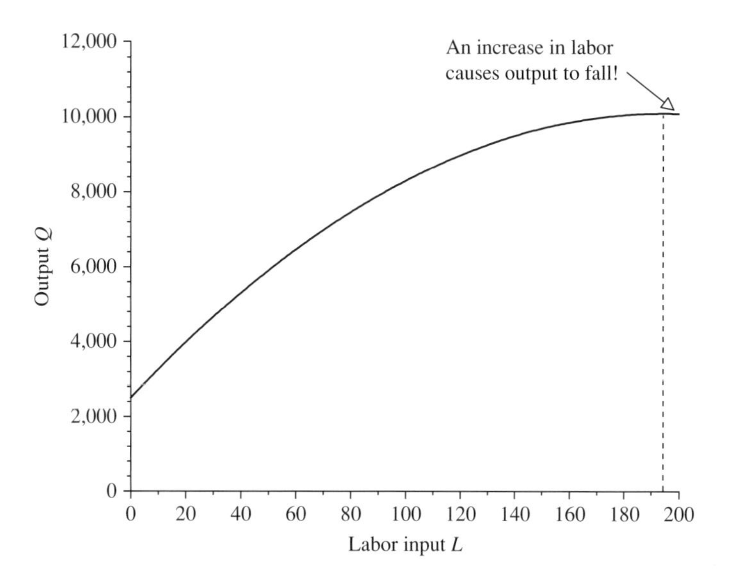

Figure 11.11
A quadratic production function, holding capital constant.

The *Cobb-Douglas production function* uses the following equation to describe how output Q depends on the amount of capital K and labor L used in production:

$$Q = AK^{\beta_1}L^{\beta_2} \tag{11.6}$$

where A, β_1, and β_2 are positive parameters. Figure 11.12 shows how labor affects output using the parameter values $A = 100$, $\beta_1 = 0.3$, and $\beta_2 = 0.7$, and holding capital K constant at 50. Similarly, Figure 11.13 shows how capital affects output, holding labor L constant at 100.

The Cobb-Douglas function has several appealing properties:

1. Output equals 0 when either input equals 0.
2. The slope is always positive, so that output always increases when an input increases.
3. If β_1 and β_2 are less than 1, then an increase in either input has less effect on output when the input is large than when the input is small (the law of diminishing returns).
4. The elasticities are constant. The elasticity of output with respect to capital is β_1 and the elasticity of output with respect to labor is β_2. For example, if $\beta_1 = 0.3$, then a 1 percent increase in capital increases output by 0.3 percent.

The Cobb-Douglas model is not restricted to production functions. A Cobb-Douglas model can be used, for example, to describe money demand, labor supply, or spending. In general, it can be used for any situation where its properties are appropriate.

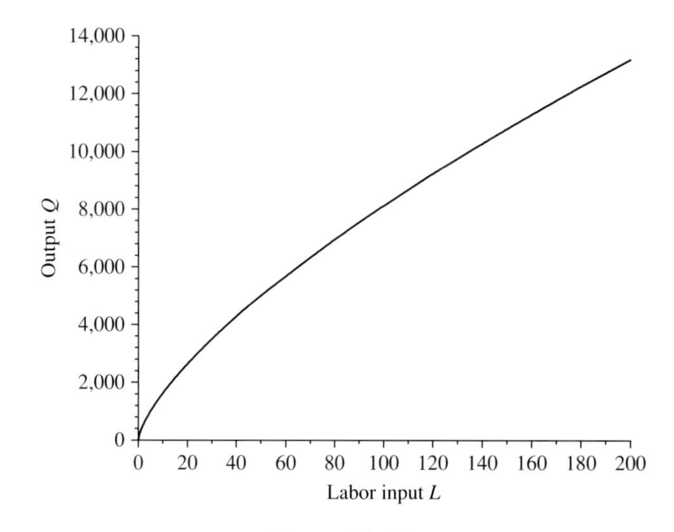

Figure 11.12
A Cobb-Douglas production function, holding capital constant.

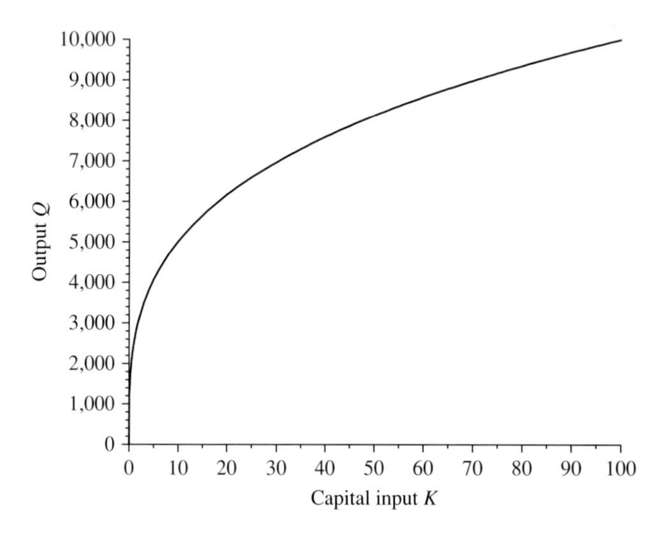

Figure 11.13
A Cobb-Douglas production function, holding labor constant.

11.5 Logarithmic Models

The general form for a power function is given by Equation 11.5: $Y = AX^{\beta}$. If we take logarithms of both sides of this equation, we obtain

$$\ln[Y] = \ln[A] + \beta \ln[X]$$

or

$$U = \alpha + \beta V$$

where $U = \ln[Y]$, $\alpha = \ln[A]$, and $V = \ln[X]$. So, if Y and X are related to each other by a power function, then the logarithm of Y is a linear function of the logarithm of X, and there is said to be a *log-linear relationship* between Y and X.

For the Cobb-Douglas function, there is a log-linear relation between output, labor, and capital:

$$Q = AK^{\beta_1}L^{\beta_2}$$
$$\ln[Q] = \ln[A] + \beta_1 \ln[K] + \beta_2 \ln[L]$$

This equation can be estimated by multiple regression procedures with the logarithm of output the dependent variable and the logarithms of capital and labor the explanatory variables.

11.6 Growth Models

Many variables (such as income, prices, or population) change over time, and a useful model should take time into account. We can do so by using a variable t, which marks the passage of time. If time is measured in years, we might set $t = 0$ in 2000, $t = 1$ in 2001, and so on. If we want to measure time in months, then we might set $t = 0$ in January 1980 and $t = 1$ in February 1980.

The variable t can be used as a subscript to show the value of a variable at different points in time. For example, if we denote income by Y and time is measured in years, then Y_t is equal to the value of income in year t. If $t = 0$ in 2000, then Y_0 is the value of income in 2000 and Y_1 is the value of income in 2001.

So far, so good, but we do not have a model yet, just labels. A very simple model is that Y grows at a constant rate g. Starting with $t = 0$, the relationship between Y_1 and Y_0 is

$$Y_1 = Y_0(1 + g)$$

For example, $g = 0.03$ means that income increases by 3 percent in a year's time, perhaps from $Y_0 = 100$ to $Y_1 = 103$:

$$\begin{aligned} Y_1 &= Y_0(1 + g) \\ &= 100(1 + 0.03) \\ &= 103 \end{aligned}$$

If Y increases at the same rate g between $t = 1$ and $t = 2$, then

$$\begin{aligned} Y_2 &= Y_1(1 + g) \\ &= Y_0(1 + g)^2 \end{aligned}$$

Continuing in this fashion, we can write our growth model as

$$Y_t = Y_0(1 + g)^t \quad (11.7)$$

This model can be applied to many phenomena. If a country's population is growing at a rate g, then Equation 11.7 shows the population in year t. If prices are increasing at a rate g, Equation 11.7 shows the price level in year t. If the value of an investment is increasing at a rate g, Equation 11.7 shows the value in year t.

In the growth model in Equation 11.7, the logarithm of Y is linearly related to time:

$$\ln[Y_t] = \ln[Y_0] + t\ln[1+g] \tag{11.8}$$

or

$$U_t = \alpha + \beta t$$

where $U_t = \ln[Y_t]$, $\alpha = \ln[Y_0]$, and $\beta = \ln[1+g]$. This equation can be estimated by simple regression procedures with the logarithm of Y the dependent variable and time the explanatory variable.

Figure 11.14 shows Y and the logarithm of Y for $g = 0.03$ (a 3 percent growth rate). The variable Y is curved upward with a slope that increases as Y increases; the logarithm of Y is a straight line with slope $\beta = \ln[1+g]$. For small values of g, the slope β is approximately equal to g. Therefore, when the natural logarithm of a variable is plotted in a time series graph, the slope is (approximately) equal to the variable's growth rate.

The Miracle of Compounding

When Albert Einstein was asked the most important concept he had learned during his life, he immediately said "compounding." Not statistical mechanics. Not quantum theory. Not relativity. Compounding.

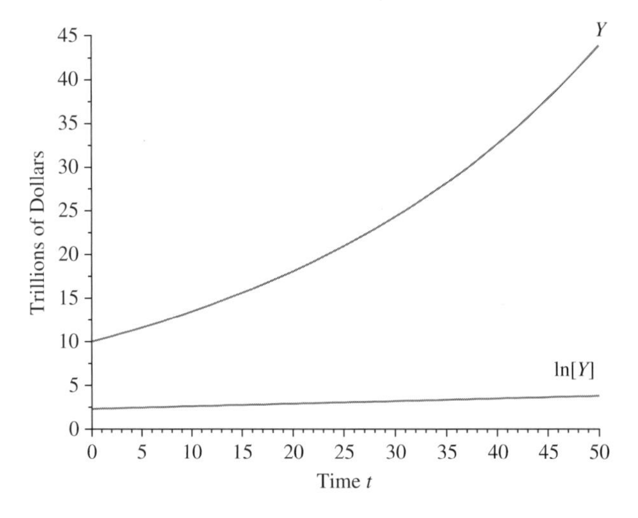

Figure 11.14
The growth model using logarithms.

Suppose that you invest \$1,000 at a 10 percent annual interest rate. The first year, you earn \$100 interest on your \$1,000 investment. Every year after that, you earn 10 percent interest on your \$1,000 and also earn *interest on your interest*, which is what makes compounding so powerful, bordering on the miraculous. Equation 11.7 shows that after 50 years of compound interest, your \$1,000 grows to \$117,391:

$$\begin{aligned} Y_t &= (1+g)^t Y_0 \\ &= (1+0.10)^{50}\$1{,}000 \\ &= \$117{,}391 \end{aligned}$$

A seemingly modest rate of return, compounded many times, turns a small investment into a fortune. The miracle of compounding applies to anything growing at a compound rate: population, income, prices.

Equation 11.7 can also be used to solve for the growth rate if we know the values of Y_0 and Y_t. For example, a very concerned citizen wrote a letter to his local newspaper complaining that houses that sold for tens of thousands of dollars decades ago were now selling for hundreds of thousands of dollars: "That house can't be worth \$400,000; my grandfather bought it for \$40,000!" Suppose that a house sold for \$40,000 in 1950 and for \$400,000 in 2000. Equation 11.7 shows that this is a 4.7 percent annual rate of increase:

$$\begin{aligned} Y_t &= (1+g)^t Y_0 \\ &= (1+0.047)^{50}\$40{,}000 \\ &= \$400{,}000 \end{aligned}$$

If the letter writer had been told that home prices had increased by 4.7 percent a year, would he have been so flabbergasted?

Frequent Compounding

In the 1960s, banks and other financial institutions began using compound interest to relieve some of the pressures they felt during credit crunches. At the time, the maximum interest rates that banks could pay their depositors were set by the Federal Reserve Board, under Regulation Q. In the 1980s, Regulation Q was phased out at the urging of consumer groups (who felt that rate wars are good for depositors) and banks (who were losing depositors to other investments that did not have interest rate ceilings).

In the 1960s, whenever market interest rates jumped the above deposit rates allowed by Regulation Q, banks lost deposits and could do little more than offer toasters and electric blankets to their customers and pressure the Fed to raise ceiling rates. Then someone somewhere noticed that Regulation Q was phrased in terms of annual rates, which could be turned into higher effective rates by compounding more often than once a year.

Suppose that the quoted annual rate is 12 percent (more than banks were paying, but a number easily divisible in the arithmetic to come). If there is no compounding during the year, then \$1 grows to \$1.12 by year's end. With semiannual compounding, the deposit is credited with 12/2 = 6 percent interest halfway through the year, and then, at the end of the year, with another 6 percent interest on the initial deposit and on the first 6 months' interest, giving an effective rate of return S for the year of 12.36 percent:

$$1 + S = (1.06)(1.06) = 1.1236$$

This is a 12.36 percent effective rate of return in the sense that 12 percent compounded semiannually pays as much as 12.36 percent compounded annually.

With quarterly compounding, 12/4 = 3 percent interest is credited every three months, raising the effective rate to 12.55 percent:

$$1 + S = (1.03)(1.03)(1.03)(1.03) = 1.1255$$

Monthly compounding, paying 12/12 = 1 percent each month, pushes the effective return up to 12.68 percent and daily compounding increases the effective annual return to 12.7468 percent. In general, \$1 invested at an annual rate of return R, compounded m times a year, grows to $(1 + R/m)^m$ after 1 year and to $(1 + R/m)^{mn}$ after n years:

$$1 + S = \left(1 + \frac{R}{m}\right)^{mn} \tag{11.9}$$

Continuous Compounding

As the examples have shown, more frequent compounding increases the effective return. Theoretically, we can even have continuous compounding, taking the limit of $(1 + R/m)^m$ as the frequency of compounding becomes infinitely large (and the time between compounding infinitesimally small):

$$1 + S = \lim_{m \to \infty} \left(1 + \frac{R}{m}\right)^{mn} = e^{Rn} \tag{11.10}$$

where $e = 2.718\ldots$ is the base of natural logarithms. In our example, with $R = 12$ percent, continuous compounding pushes the effective annual rate up to 12.7497 percent. Although advertisements trumpeting "continuous compounding" convey the feeling that the bank is doing something marvelous for you and your money, the improvement over daily compounding is slight.

On the other hand, continuous compounding is very useful for modeling growth because it simplifies many computations. If the continuously compounded rate of growth of X is 3 percent and the continuously compounded rate of growth of Y is 2 percent, then the continuously compounded rate of growth of the product XY is 3 + 2 = 5 percent, and the continuously compounded rate of growth of the ratio X/Y is 3 − 2 = 1 percent. It is as simple as that.

In Chapter 1, we looked at the distinction between nominal data (denominated in dollars, euros, or another currency) and real data (adjusted for changes in the cost of living). Using continuous compounding, if nominal income Y is increasing by 5 percent and prices P are increasing by 3 percent, how fast is real income Y/P increasing? By $5 - 3 = 2$ percent. If real income Z is increasing by 2 percent and prices P are increasing by 1 percent, how fast is nominal income ZP increasing? By $2 + 1 = 3$ percent.

Continuously compounded models are written like this:

$$Y_t = Y_0 e^{gt} \tag{11.11}$$

where Y_t is the value of Y at time t and g is the continuously compounded growth rate.

Suppose, for example, that Y is nominal income and that Y initially equals 3,000 and grows at a continuously compounded rate of 5 percent:

$$Y_t = 3{,}000e^{0.05t}$$

We let the price level P initially equal 1 and grow at a continuously compounded rate of 3 percent:

$$P_t = 1.0e^{0.03t}$$

Real income initially equals 3,000 and grows at 2 percent:

$$\frac{Y}{P_t} = \frac{3{,}000e^{0.05t}}{1.0e^{0.03t}}$$
$$= 3{,}000e^{0.02t}$$

Figure 11.15 shows a graph of the growth of nominal income and real income over time.

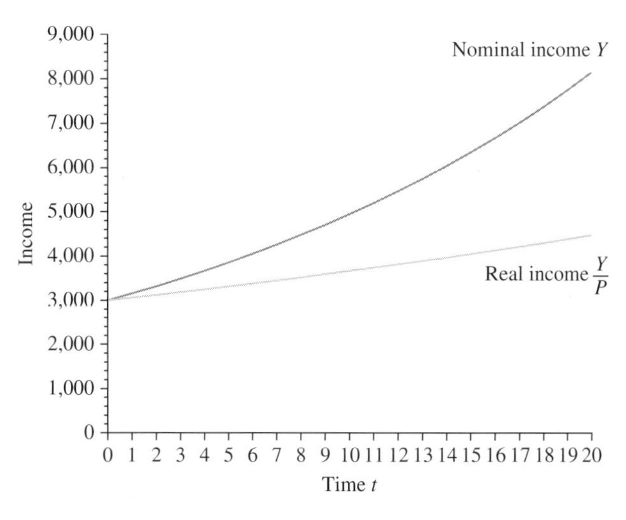

Figure 11.15
Continuously compounded growth.

The continuously compounded growth rate can be estimated by taking the logarithm of both sides of Equation 11.11,

$$\ln[Y_t] = \ln[Y_0] + gt$$

and using simple regression procedures with the logarithm of Y the dependent variable and time the explanatory variable.

11.7 Autoregressive Models

Another way to introduce time into a model is with a first-order autoregressive equation in which the value of a variable is influenced by its value in the immediately preceding period:

$$Y_t = \alpha + \beta Y_{t-1} \tag{11.12}$$

The subscripts t and $t-1$ indicate the values of Y in period t and the previous period, $t-1$. A period of time might be a day, week, month, or year.

In the special case where $\beta = 1$, the change in Y is simply equal to α:

$$\begin{aligned} Y_t &= \alpha + Y_{t-1} \\ Y_t - Y_{t-1} &= \alpha \end{aligned}$$

If β does not equal 1, there is a dynamic equilibrium value Y^e where Y is constant. The equilibrium value of Y can be determined by substituting $Y_t = Y_{t-1} = Y^e$ into Equation 11.12:

$$\begin{aligned} Y^e &= \alpha + \beta Y^e \\ Y^e - \beta Y^e &= \alpha \end{aligned}$$

Rearranging,

$$Y^e = \frac{\alpha}{1-\beta} \tag{11.13}$$

The actual value of Y may or may not converge to this equilibrium value, depending on whether the absolute value of β is less than 1 or larger than 1. For example, if $\alpha = 8$ and $\beta = 0.9$, the equilibrium is at $Y = 80$:

$$\begin{aligned} Y^e &= \frac{\alpha}{1-\beta} \\ &= \frac{8}{1-0.9} \\ &= 80 \end{aligned}$$

Figure 11.16 shows that Y gets closer and closer to 80 each period. In this case, the model is *monotonically stable* because Y converges directly to its equilibrium value.

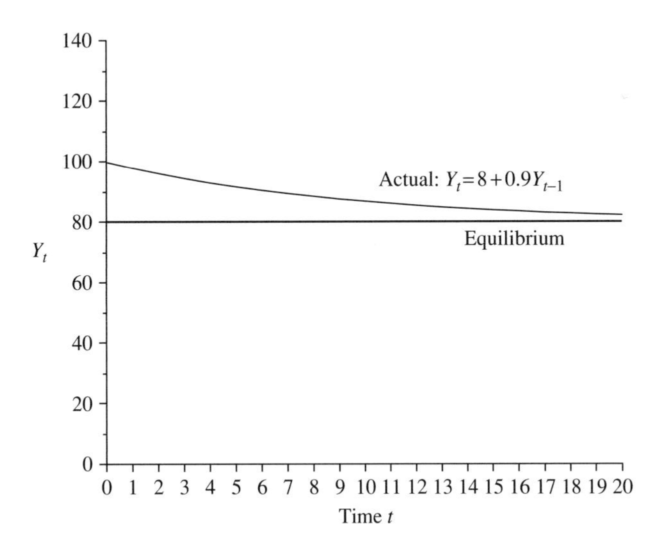

Figure 11.16
A monotonically stable autoregressive model.

In contrast, if $\alpha = -4$ and $\beta = 1.05$, the equilibrium is again at $Y = 80$:

$$Y^e = \frac{\alpha}{1-\beta}$$
$$= \frac{-4}{1-1.05}$$
$$= 80$$

but Figure 11.17 shows that Y gets further from equilibrium each period. This model is *monotonically unstable* because Y diverges directly from its equilibrium value.

If β is negative, the autoregressive model is cyclical, in that Y cycles alternately above and below its equilibrium value, in some cases getting closer to equilibrium each period and in other cases getting further from equilibrium. For example, if $\alpha = 152$ and $\beta = -0.9$, the equilibrium is at $Y = 80$ and Figure 11.18 shows that Y gets closer and closer to 80 each period. This model is *cyclically stable* because Y cycles but still converges to its equilibrium value.

On the other hand, if $\alpha = 164$ and $\beta = -1.05$, the equilibrium is at $Y = 80$ but Figure 11.19 shows that Y gets further from equilibrium each period. In this case, the model is said to be *cyclically unstable*.

In general, autoregressive models have two main characteristics: whether the model is stable and whether the model is cyclical. The crucial parameter is β. The model is stable if the absolute value of β is less than 1 and unstable if the absolute value of β is larger than 1.

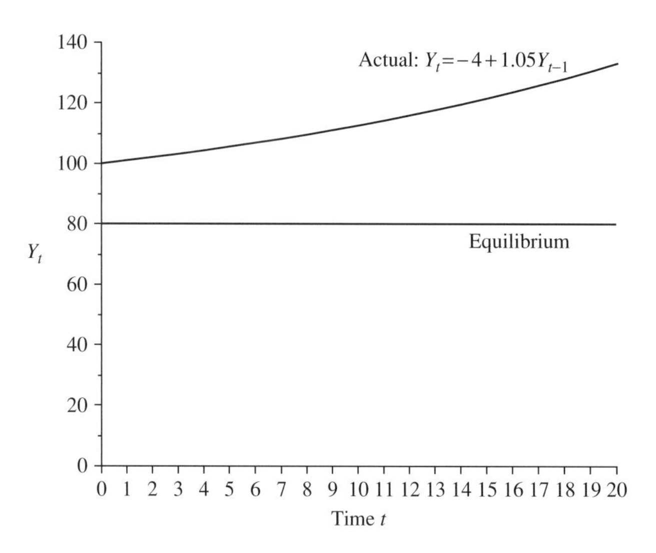

Figure 11.17
A monotonically unstable autoregressive model.

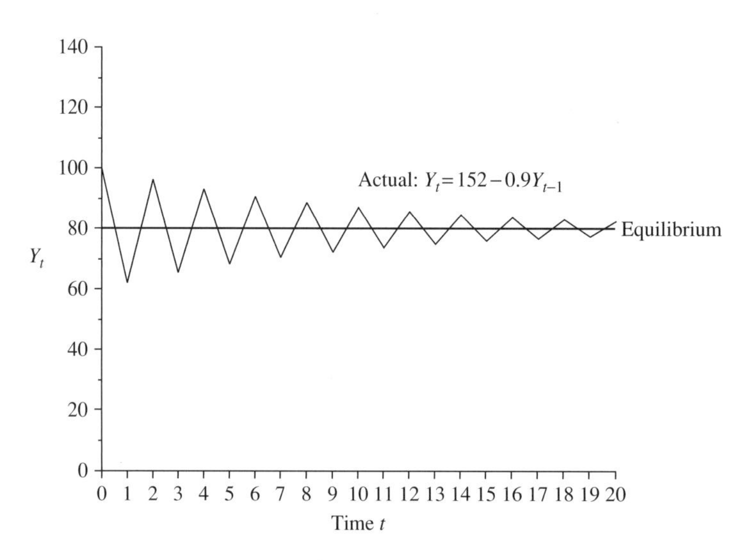

Figure 11.18
A cyclically stable autoregressive model.

The model is cyclical if β is negative and monotonic if β is positive. Table 11.1 summarizes the various cases (for simplicity, the special cases $\beta = -1$, $\beta = 0$, and $\beta = 1$ are not shown).

Autoregressive models may be an appropriate way to describe a variety of situations. This year's government spending depends on government spending last year. This month's unemployment rate depends on the unemployment rate last month. Today's value of the Dow Jones Industrial Average depends on its value yesterday.

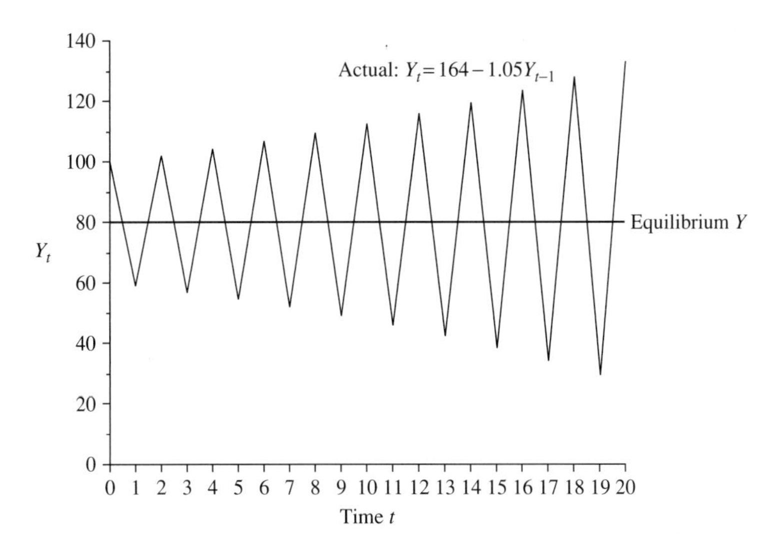

Figure 11.19
A cyclically unstable autoregressive model.

Table 11.1: The Crucial Parameter β

Value of β	Model
$\beta < -1$	*Y* is cyclically unstable
$-1 < \beta < 0$	*Y* is cyclically stable
$0 < \beta < 1$	*Y* is monotonically stable
$\beta > 1$	*Y* is monotonically unstable

Autoregressive models can include other explanatory variables. If we believe that *Y* depends on its value the previous period and also depends on two other variables, *X* and *Z*, we can expand Equation 11.8 to

$$Y_t = \alpha + \beta_1 Y_{t-1} + \beta_2 X_t + \beta_3 Z_t \tag{11.14}$$

For example, the unemployment rate might depend on its value the previous period and also the current level of GDP. Equation 11.14 can be directly estimated by multiple regression procedures.

Exercises

11.1 Okun's law says that changes in the unemployment rate are related to the percentage change in output. Suppose that this relationship is described by the equation $U = 1.3 - 0.4Y$, where *U* is the change in the unemployment rate and *Y* is the percentage change in real GDP.

a. What is the change in the unemployment rate when real GDP is constant?
b. How much does the unemployment rate increase when there is a 1 percent increase in real GDP?

11.2 The Fed's stock valuation model says that the earnings/price ratio E/P is related to the interest rate R on 10-year Treasury bonds. Suppose that this relationship is described by the equation $E/P = 1.2 + 0.8R$.
a. What is the earnings/price ratio when the interest rate is 0?
b. How much does the earnings/price ratio increase when the interest rate increases by 1?

11.3 Suppose that Okun's law in Exercise 11.1 is the quadratic equation $U = 1.4 - 0.6Y + 0.02Y^2$.
a. What is the change in the unemployment rate when real GDP is constant?
b. What is the slope of the line at $Y = 0$?
c. Does the slope of the line increase or decrease as Y increases?

11.4 Suppose that the stock valuation model in Exercise 11.2 is a quadratic equation $E/P = 1.5 + 0.7R + 0.005R^2$.
a. What is the earnings/price ratio when the interest rate is 0?
b. What is the slope of the line at $R = 0$?
c. Does the slope of the line increase or decrease as R increases?

11.5 You want to write down a simple linear model relating college grade point average (GPA) to SAT scores.
a. Which variable is the dependent variable?
b. Do you expect the slope to be positive, negative, or 0?
c. How do you determine the value of the Y intercept?

11.6 You want to write down a simple linear model relating miles per gallon (MPG) to car weight.
a. Which variable is the dependent variable?
b. Do you expect the slope to be positive, negative, or 0?
c. How do you determine the value of the Y intercept?

11.7 You want to write down a simple linear model relating used car prices to the number of miles the car has been driven.
a. Which variable is the dependent variable?
b. Do you expect the slope to be positive, negative, or 0?
c. How do you determine the value of the Y intercept?

11.8 Suppose that cigarette sales in the United States between 1910 and 2010 are described by the equation $S = 800 + 164t - 1.4t^2$, where S is per capita sales and $t = 0$ in 1910, $t = 1$ in 1911, and so on.
a. Are per capita sales increasing or decreasing in 1910?
b. Does the annual *increase* in per capita cigarette sales rise or fall as time passes?

11.9 Suppose that cigarette sales in the United States between 1910 and 2010 are described by the equation $S = 130 + 18t - 0.12t^2$, where S is total sales (in billions) and $t = 0$ in 1910, $t = 1$ in 1911, and so on.

a. Are sales increasing or decreasing in 1910?
b. Does the annual *increase* in sales rise or fall as time passes?

11.10 Suppose consumption C is related to income Y by the equation $C = 10 + 0.9Y - 0.02Y^2$.

a. Where does this equation intersect the vertical axis?
b. What is the slope of the line at $Y = 0$?
c. Does the slope of the line increase or decrease as income increases?

11.11 Suppose that labor supply L is related to wages W by the equation $L = 1.2 + 3.5W + 0.01W^2$.

a. Where does the equation intersect the vertical axis?
b. What is the slope of the line at $W = 0$?
c. Does the slope of the line increase or decrease as wages increase?

11.12 In the quadratic model $Y = \alpha + \beta_1 X + \beta_2 X^2$, which parameter determines

a. The slope of the equation at $X = 0$?
b. Whether the slope of the equation increases or decreases as X increases?

11.13 A firm has this cubic cost function:

$$C = 200Q - 20Q^2 + Q^3$$

where Q is output (the number of items produced) and C is the total cost of producing these items. The marginal cost MC is a quadratic function:

$$\text{MC} = 200 - 40Q + 3Q^2$$

a. What is the marginal cost at $Q = 0$?
b. At $Q = 0$, is the slope of the marginal cost function positive or negative?
c. Does the slope of the marginal cost function increase or decrease as Q increases?

11.14 A firm has this cost function:

$$C = 1{,}200Q - 45Q^2 + 0.6Q^3$$

where Q is output (the number of items produced) and C is the total cost of producing these items. The marginal cost MC is

$$\text{MC} = 1{,}200 - 90Q + 1.8Q^2$$

a. What is the marginal cost at $Q = 0$?
b. At $Q = 0$, is the slope of the marginal cost function positive or negative?
c. Does the slope of the marginal cost function increase or decrease as Q increases?

11.15 Use the equations in Exercise 11.13 to make rough sketches of cost and marginal cost as functions of output Q.

11.16 Use the equations in Exercise 11.14 to make rough sketches of cost and marginal cost as functions of output Q.

11.17 Suppose that the relationship between hard drive prices P and size G (in gigabytes) is $P = 490 + 0.46G$ in 2006 and $P = 505 + 0.24G$ in 2007.

a. Which equation intersects the vertical axis at a higher value?
b. Which equation has a higher slope?
c. Is the price of a 100 gigabyte drive higher in 2006 or 2007?

11.18 You are considering using a linear equation to model how egg demand D is related to the price P: $D = \alpha + \beta P$.

a. Do you expect the value of the parameter β to be positive, negative, or 0? Explain.
b. Do you expect the value of the parameter α to be positive, negative, or 0? Explain.

11.19 You are considering using a linear equation to model how home prices P are related to home construction costs C: $P = \alpha + \beta C$.

a. Do you expect the value of the parameter β to be positive, negative, or 0? Explain.
b. Do you expect the value of the parameter α to be positive, negative, or 0? Explain.

11.20 You are considering using a quadratic equation to model how the supply of wheat S is related to the price P: $S = \alpha + \beta_1 P + \beta_2 P^2$.

a. Do you expect the value of the parameter β_1 to be positive, negative, or 0? Explain.
b. Do you expect the value of the parameter β_2 to be positive, negative, or 0? Explain.

11.21 You are considering using a quadratic equation to model how the number of hours a week spent studying S is related to the dollar value of a prize P given to students with 4.0 grade point averages: $S = \alpha + \beta_1 P + \beta_2 P^2$.

a. Do you expect the value of the parameter β_1 to be positive, negative, or 0? Explain.
b. Do you expect the value of the parameter β_2 to be positive, negative, or 0? Explain.

11.22 You are considering using a quadratic equation to model how cost C of raising a family is related to the number of children in the family N: $C = \alpha + \beta_1 N + \beta_2 N^2$.

a. Do you expect the value of the parameter β_1 to be positive, negative, or 0? Explain.
b. Do you expect the value of the parameter β_2 to be positive, negative, or 0? Explain.

11.23 Consider this Cobb-Douglas production function: $Q = 100K^{0.3}Y^{0.7}$. What is the value of Q if $K = 50$ and $L = 100$? What is the (approximate) percentage increase in Q if both K and L increase by 10 percent, to $K = 55$ and $L = 110$?

11.24 Consider the Cobb-Douglas production function $Q = 100K^{0.4}L^{0.6}$. What is the value of Q if $K = 100$ and $L = 50$? What is the (approximate) percentage increase in Q if both K and L increase by 10 percent, to $K = 110$ and $L = 55$?

11.25 Consider this equation for the demand for money: $M = 250R^{-0.5}L^{0.8}$, where M is the amount of money demanded, R is an interest rate, and Y is income. What is the elasticity of money demand with respect to income? With respect to the interest rate?

11.26 Consider this equation for the demand for orange juice: $D = 100Y^{0.7}P^{-0.8}$, where D is demand, Y is income, and P is the price of orange juice. What is the elasticity of demand with respect to income? With respect to price?

11.27 Suppose that real income Y and the price level P are described by these exponential growth equations:

$$Y = 160e^{0.02t}$$
$$P = 100e^{0.03t}$$

a. What is the continuously compounded rate of growth of real income Y?
b. What is the continuously compounded rate of growth of the price level P?
c. What is the continuously compounded rate of growth of nominal income $Z = YP$?

11.28 Suppose that nominal income Y and the price level P are described by these exponential growth equations:

$$Y = 16{,}000e^{0.05t}$$
$$P = 100e^{0.03t}$$

a. What is the continuously compounded rate of growth of nominal income Y?
b. What is the continuously compounded rate of growth of the price level P?
c. What is the continuously compounded rate of growth of real income Y/P?

11.29 Suppose that income Y and population P are described by these exponential growth equations:

$$Y = 16{,}000e^{0.05t}$$
$$P = 400e^{0.02t}$$

a. What is the continuously compounded rate of growth of income Y?
b. What is the continuously compounded rate of growth of the population P?
c. What is the continuously compounded rate of growth of per capita income Y/P?

11.30 Suppose that productivity q and labor L are described by these exponential growth equations:

$$q = 1.5e^{0.03t}$$
$$L = 170e^{0.02t}$$

a. What is the continuously compounded rate of growth of productivity q?
b. What is the continuously compounded rate of growth of labor L?
c. What is the continuously compounded rate of growth of output qL?

11.31 In 1870, real per capita GDP was $1,800 in England and $1,400 in the United States (30 percent higher in England). In 1970, real per capita GDP was $7,000 in England and $11,000 in the United States (60 percent higher in the United States). What were the respective annual growth rates over this 100-year period?

11.32 U.S. real GDP was $1,643.2 billion in 1948 and $11,652.8 billion in 2008. The U.S. population was 146.6 million in 1948 and 304.5 million in 2008.
a. What was the annual growth rate of real GDP over this period?
b. What was the annual growth rate of real per capita GDP over this period?

11.33 The cubic square feet of marketable timber from a certain species of pine tree is related to the age of the tree by this equation:

$$F = 200t - t^2$$

where F is the cubic footage and t is the tree's age, in years. What is the value of F when the tree is planted ($t = 0$)? When $t = 50$? When $t = 100$? When $t = 150$? Draw a rough sketch of the relationship between F and t. Are there any values of t for which F seems implausible? If there are values of t for which F is implausible, is this equation useless?

11.34 Using data for 1901 through 1999, annual batting averages (BAs) were tabulated for all Major League Baseball (MLB) players for each season in which the player had at least 50 times at bat. For each season, each player's BA was converted to a standardized Z value by subtracting the average BA for all players that year and dividing by the standard deviation of BAs that year [1]. The annual Z values were grouped according to whether it was the player's first year in the major leagues, second year, and so on. The average Z value was then calculated for each career year. For example, looking at each player's first year in the major leagues, the average Z value is -0.337.
a. Use the data in Table 11.2 to estimate the multiple regression equation $Z = \alpha + \beta_1 t + \beta_2 t^2$.
b. Are the coefficients of t and t^2 statistically significant at the 5 percent level?
c. What do the signs of t and t^2 tell you about the shape of the fitted line?

Table 11.2: Exercise 11.34

Year t	Mean Z	Year t	Mean Z	Year t	Mean Z	Year t	Mean Z
1	−0.337	6	0.151	11	0.112	16	0.125
2	−0.180	7	0.095	12	0.079	17	0.131
3	−0.019	8	0.125	13	0.122	18	0.009
4	0.067	9	0.146	14	0.127	19	0.280
5	0.098	10	0.193	15	0.154	20	−0.079

Table 11.3: Exercise 11.35

Year t	Mean Z	Year t	Mean Z	Year t	Mean Z	Year t	Mean Z
1	−0.176	6	0.084	11	0.138	16	0.048
2	−0.109	7	0.067	12	0.206	17	0.266
3	−0.050	8	0.129	13	0.128	18	0.162
4	0.008	9	0.096	14	0.087	19	0.205
5	0.044	10	0.138	15	0.253	20	−0.002

d. Explain why a comparison of the predicted Z value for $t = 1$ and $t = 2$ does not tell us the predicted change in the Z value if all first-year players were to play a second year in the majors.

11.35 Using data for 1901 through 1999, annual earned run averages (ERAs) were tabulated for all Major League Baseball (MLB) pitchers for each season in which the player pitched at least 25 innings. For each season, each player's ERA was converted to a standardized Z value by subtracting the average ERA for all players that year and dividing by the standard deviation of ERAs that year [2]. The annual Z values were grouped according to whether it was the player's first year in the major leagues, second year, and so on. The average Z value was then calculated for each career year. For example, looking at each player's first year in the major leagues, the average Z value is −0.176.

a. Use the data in Table 11.3 to estimate the multiple regression equation $Z = \alpha + \beta_1 t + \beta_2 t^2$.
b. Are the coefficients of t and t^2 statistically significant at the 5 percent level?
c. What do the signs of t and t^2 tell you about the shape of the fitted line?
d. What worrisome pattern do you see in the residuals about the fitted line?

11.36 A firm's production is characterized by this Cobb-Douglas production function: $Q = 12K^{0.4}L^{0.6}$.

a. If capital K increases by 5 percent, what is the approximate percentage change in output Q?
b. If capital K increases by 3 percent and labor L increases by 4 percent, what is the approximate percentage change in the capital/labor ratio K/L?

c. If price P increases by 5 percent and output Q falls by 4 percent, what is the approximate percentage change in revenue PQ?
d. If capital K and labor L both increase by 5 percent, what is the approximate percentage increase in output Q?

11.37 The following variables might appear in a macroeconomic growth model: Y = real GDP; M = money supply; P = price level; and L = labor supply. Suppose that the continuously compounded rates of growth are 2 percent for Y, 4 percent for M, 3 percent for P, and 1 percent for L. What are the continuously compounded growth rates of

a. Nominal GDP?
b. The real money supply?
c. Real per capita GDP?
d. Nominal per capita GDP?

11.38 Consider a firm whose assets and earnings at time t are described by these equations:

$$A_t = 32e^{0.05t}$$
$$E_t = 0.2A_t$$

a. What is the value of assets at time $t = 0$?
b. What is the percentage rate of growth of assets?
c. What is the percentage rate of growth of earnings?
d. What is the percentage rate of growth of the ratio of earnings to assets E_t/A_t?

11.39 When you graduate from college, your annual salary will be Y_0.

a. Write down an equation for determining your annual salary t years after graduation if your salary grows at a continuously compounded rate of 5 percent.
b. If you spend 90 percent of your salary every year, what is the continuously compounded rate of growth of your spending?
c. If you save 10 percent of your salary every year, what is the continuously compounded rate of growth of your saving?

11.40 In the cobweb model of farm output, the demand D for a crop depends on the current price P, but the supply S depends on the price the previous period because crops must be planted before they can be harvested:

$$D_t = \alpha_1 - \beta_1 P_t$$
$$S_t = \alpha_2 + \beta_2 P_{t-1}$$

where β_1 and β_2 are positive parameters. If demand is equal to supply, what must be true of β_1 and β_2 for prices to converge to an equilibrium value? If the model is stable, will it be monotonically stable or cyclically stable?

11.41 In an inflation-augmented Phillips curve, the rate of inflation P depends on the unemployment rate U and the anticipated rate of inflation A, which in turn depends on the rate of inflation the previous period:

$$P_t = \alpha_1 - \beta_1 U_t + \beta_2 A_t$$
$$A_t = \alpha_2 + \beta_3 P_{t-1}$$

where β_1, β_2, and β_3 are positive parameters. If the unemployment rate is constant, what must be true of β_1, β_2, and β_3 for inflation to converge to an equilibrium value? If the model is stable, is it monotonically stable or cyclically stable?

11.42 Consider the following model of wage adjustment in the labor market:

$$W_t - W_{t-1} = (1-\lambda)(W^e - W_{t-1})$$
$$W_t = (1-\lambda)W^e + \lambda W_{t-1}$$

where W^e is the equilibrium wage rate, and λ is a positive parameter. If $\lambda = 0.6$, is the model monotonically stable, cyclically stable, monotonically unstable, or cyclically unstable? Explain why your answer makes sense logically.

11.43 Redo the preceding exercise, this time assuming that $\lambda = -0.2$.

11.44 Consider this income-expenditure model:

$$Y_t = E_t$$
$$E_t = 100 + 0.5Y_t + 0.2Y_{t-1}$$

where Y is aggregate income and E is aggregate spending. Is this model monotonically stable, cyclically stable, monotonically unstable, or cyclically unstable?

11.45 Redo Exercise 11.44, this time assuming $E_t = 100 + 1.5Y_t + 0.2Y_{t-1}$.

11.46 Redo Exercise 11.44, this time assuming $E_t = 100 + 0.8Y_t + 0.5Y_{t-1}$.

11.47 Consider this dynamic income-expenditure model:

$$Y_t = E_t$$
$$E_t = 100 + 1.2(1-c)Y_{t-1}$$

where Y is national income, E is spending, and c is the tax rate on income ($0 < c < 1$). Is this model more likely to be dynamically stable if c is large or small? Derive your answer mathematically and also explain the economic logic without using mathematical symbols.

11.48 The data in Exercise 9.39 were used to calculate the annual percentage change in real per capita disposable personal income, $g_t = 100(Y_t - Y_{t-1})/Y_{t-1}$ (Table 11.4).

Table 11.4: Exercise 11.48

	g_t	g_{t-1}		g_t	g_{t-1}		g_t	g_{t-1}
1962	3.26	1.75	1978	3.61	2.39	1994	2.00	0.33
1963	2.28	3.26	1979	1.23	3.61	1995	1.75	2.00
1964	5.74	2.28	1980	−0.21	1.23	1996	2.08	1.75
1965	4.86	5.74	1981	1.64	−0.21	1997	2.32	2.08
1966	4.10	4.86	1982	1.25	1.64	1998	4.76	2.32
1967	3.19	4.10	1983	2.37	1.25	1999	1.83	4.76
1968	3.53	3.19	1984	6.24	2.37	2000	3.96	1.83
1969	2.22	3.53	1985	2.18	6.24	2001	1.38	3.96
1970	3.06	2.22	1986	2.36	2.18	2002	2.32	1.38
1971	3.17	3.06	1987	0.77	2.36	2003	1.57	2.32
1972	3.77	3.17	1988	3.37	0.77	2004	2.50	1.57
1973	5.79	3.77	1989	1.65	3.37	2005	0.42	2.50
1974	−1.69	5.79	1990	0.81	1.65	2006	3.06	0.42
1975	1.24	−1.69	1991	−0.51	0.81	2007	1.18	3.06
1976	2.98	1.24	1992	2.17	−0.51	2008	−0.40	1.18
1977	2.39	2.98	1993	0.33	2.17			

a. Use least squares to estimate the relationship between this year's growth rate and the previous year's growth rate: $g_t = \alpha + \beta g_{t-1} + \varepsilon$.

b. Based on your estimates of α and β, what is the dynamic equilibrium value of g?

c. Based on your estimates of α and β, is the growth rate monotonically stable, cyclically stable, monotonically unstable, or cyclically unstable?

d. Why do you think that the model $g_t = \alpha + \beta g_{t-1}$ is not an adequate guide to whether g converges to a dynamic equilibrium value?

11.49 Consider this model of government fiscal policy:

$$G = G_0 e^{\alpha t}$$
$$H = \theta Y$$
$$Z = G - H$$
$$Y = Y_0 e^{\beta t}$$

where G is government spending, H is taxes, Z is the government deficit, Y is national income, and t is time ($t = 0$ in the current year). All the variables G, H, Z, and Y are in real terms; G_0, Y_0, α, β, and θ are constant parameters.

a. Interpret the parameter G_0.

b. Interpret the parameter α.

c. Find the equation for determining Z/Y, the ratio of the deficit to national income.

d. If Z is currently positive, determine the conditions under which Z/Y falls over time.

e. State the conclusion in question d in plain English, using no mathematical symbols.

11.50 Consider this model in which national income Y depends on the previous period's income and the current and previous period's money supply M:

$$Y_t = \alpha + \beta_1 Y_{t-1} + \beta_2 M_t + \beta_3 M_{t-1}$$

where the parameters α, β_1, β_2, and β_3 are positive.

a. If the money supply is held constant, show the conditions under which the model is stable.

b. If the money supply is adjusted in each period to hold income constant, show the conditions under which the model is stable.

Appendix

Table A.1: Standardized Normal Distribution

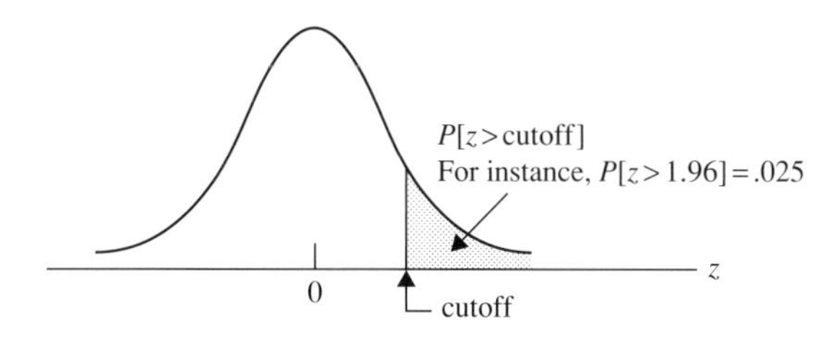

Cutoff	.00	.01	.02	.03	.04	.05	.06	.07	.08	.09
.0	.5000	.4960	.4920	.4880	.4840	.4801	.4761	.4721	.4681	.4641
.1	.4602	.4562	.4522	.4483	.4443	.4404	.4364	.4325	.4286	.4247
.2	.4207	.4168	.4129	.4090	.4052	.4013	.3974	.3936	.3897	.3859
.3	.3821	.3783	.3745	.3707	.3669	.3632	.3594	.3557	.3520	.3483
.4	.3446	.3409	.3372	.3336	.3300	.3264	.3228	.3192	.3156	.3121
.5	.3085	.3050	.3015	.2981	.2946	.2912	.2877	.2843	.2810	.2776
.6	.2743	.2709	.2676	.2643	.2611	.2578	.2546	.2514	.2483	.2451
.7	.2420	.2389	.2358	.2327	.2296	.2266	.2236	.2206	.2177	.2148
.8	.2119	.2090	.2061	.2033	.2005	.1977	.1949	.1922	.1894	.1867
.9	.1841	.1814	.1788	.1762	.1736	.1711	.1685	.1660	.1635	.1611
1.0	.1587	.1562	.1539	.1515	.1492	.1469	.1446	.1423	.1401	.1379
1.1	.1357	.1335	.1314	.1292	.1271	.1251	.1230	.1210	.1190	.1170
1.2	.1151	.1131	.1112	.1093	.1075	.1056	.1038	.1020	.1003	.0985
1.3	.0968	.0951	.0934	.0918	.0901	.0885	.0869	.0853	.0838	.0823
1.4	.0808	.0793	.0778	.0764	.0749	.0735	.0722	.0708	.0694	.0681
1.5	.0668	.0655	.0643	.0630	.0618	.0606	.0594	.0582	.0571	.0559
1.6	.0548	.0537	.0526	.0516	.0505	.0495	.0485	.0475	.0465	.0455
1.7	.0446	.0436	.0427	.0418	.0409	.0401	.0392	.0384	.0375	.0367
1.8	.0359	.0352	.0344	.0336	.0329	.0322	.0314	.0307	.0301	.0294
1.9	.0287	.0281	.0274	.0268	.0262	.0256	.0250	.0244	.0239	.0233
2.0	.0228	.0222	.0217	.0212	.0207	.0202	.0197	.0192	.0188	.0183
2.1	.0179	.0174	.0170	.0166	.0162	.0158	.0154	.0150	.0146	.0143
2.2	.0139	.0136	.0132	.0129	.0125	.0122	.0119	.0116	.0113	.0110
2.3	.0107	.0104	.0102	.0099	.0096	.0094	.0091	.0089	.0087	.0084
2.4	.0082	.0080	.0078	.0075	.0073	.0071	.0069	.0068	.0066	.0064
2.5	.0062	.0060	.0059	.0057	.0055	.0054	.0052	.0051	.0049	.0048
2.6	.0047	.0045	.0044	.0043	.0041	.0040	.0039	.0038	.0037	.0036
2.7	.0035	.0034	.0033	.0032	.0031	.0030	.0029	.0028	.0027	.0026
2.8	.0026	.0025	.0024	.0023	.0023	.0022	.0021	.0021	.0020	.0019
2.9	.0019	.0018	.0017	.0017	.0016	.0016	.0015	.0015	.0014	.0014
3.0	.0013	.0013	.0013	.0012	.0012	.0011	.0011	.0011	.0010	.0010
4.0	.000 0317									
5.0	.000 000 287									
6.0	.000 000 000 987									
7.0	.000 000 000 001 28									
8.0	.000 000 000 001									

Table A.2: Student's t Distribution

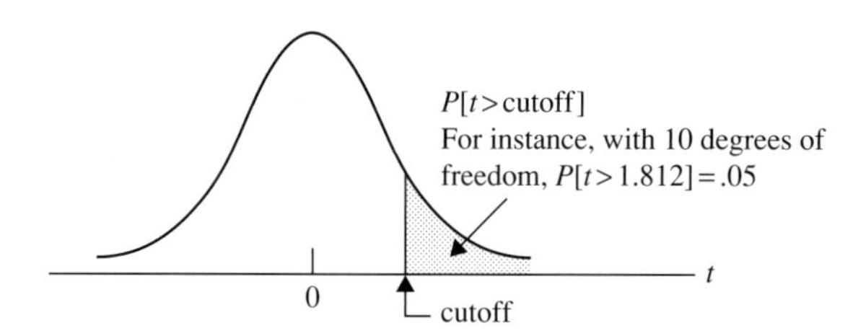

	Probability of a t Value Larger Than the Indicated Cutoff				
Degrees of Freedom	0.10	0.05	0.025	0.01	0.005
1	3.078	6.314	12.706	31.821	63.657
2	1.886	2.920	4.303	6.965	9.925
3	1.638	2.353	3.182	4.541	5.841
4	1.533	2.132	2.776	3.747	4.604
5	1.476	2.015	2.571	3.365	4.032
6	1.440	1.943	2.447	3.143	3.707
7	1.415	1.895	2.365	2.998	3.499
8	1.397	1.860	2.306	2.896	3.355
9	1.383	1.833	2.262	2.821	3.250
10	1.372	1.812	2.228	2.764	3.169
11	1.363	1.796	2.201	2.718	3.106
12	1.356	1.782	2.179	2.681	3.055
13	1.350	1.771	2.160	2.650	3.012
14	1.345	1.761	2.145	2.624	2.977
15	1.341	1.753	2.131	2.602	2.947
16	1.337	1.746	2.120	2.583	2.921
17	1.333	1.740	2.110	2.567	2.898
18	1.330	1.734	2.101	2.552	2.878
19	1.328	1.729	2.093	2.539	2.861
20	1.325	1.725	2.086	2.528	2.845
21	1.323	1.721	2.080	2.518	2.831
22	1.321	1.717	2.074	2.508	2.819
23	1.319	1.714	2.069	2.500	2.807
24	1.318	1.711	2.064	2.492	2.797
25	1.316	1.708	2.060	2.485	2.787
26	1.315	1.706	2.056	2.479	2.779
27	1.314	1.703	2.052	2.473	2.771
28	1.313	1.701	2.048	2.467	2.763
29	1.311	1.699	2.045	2.462	2.756
30	1.310	1.697	2.042	2.457	2.750
40	1.303	1.684	2.021	2.423	2.704
60	1.296	1.671	2.000	2.390	2.660
120	1.289	1.661	1.984	2.358	2.626
∞	1.282	1.645	1.960	2.326	2.576

References

Chapter 1 Data, Data, Data

[1] Herbert Hoover. State of the Union address, December 3, 1929.
[2] Stephen Feinstein. The 1930s: From the Great Depression to the Wizard of Oz. Revised edition, Berkeley Heights, NJ: Enslow Publishers, 2006.
[3] Herbert Hoover. Speech to the annual dinner of the Chamber of Commerce of the United States, May 1, 1930.
[4] John Steele Gordon. A fiasco that fed the Great Depression. Barron's, December 15, 2008.
[5] Edmund Wilson. New Republic, February 1933.
[6] Herbert Hoover. Memoirs. London: Hollis and Carter, 1952, p. 30.
[7] Franklin Roosevelt, quoted in Burton Folsom, Jr. New Deal or raw deal? How FDR's economic legacy has damaged America. New York: Threshold Editions, 2008, p. 40.
[8] Arthur M. Schlesinger, Jr. The crisis of the old order, 1919–1933. Boston: Houghton Mifflin, 1956, p. 457.
[9] Ben S. Bernanke. Federal Reserve Board speech: Remarks by Governor Ben S. Bernanke, at the Conference to Honor Milton Friedman, University of Chicago, Chicago, Illinois, November 8, 2002.
[10] John Maynard Keynes. Essays in biography. London: Macmillan, 1933, p. 170.
[11] Bureau of Economic Analysis, U.S. Department of Commerce.
[12] Dorothy Brady. Family saving, 1888 to 1950. In: R. W. Goldsmith, D. S. Brady, H. Menderhausen, editors, A study of saving in the United States, III. Princeton, NJ: Princeton University Press, 1956, p. 183.
[13] The Big Mac index. The Economist, July 5, 2007.
[14] Arthur Schlesinger, Jr. Inflation symbolism vs. reality. Wall Street Journal, April 9, 1980.
[15] Dan Dorfman. Fed boss banking on housing slump to nail down inflation. Chicago Tribune, April 20, 1980.
[16] Availble at: http://www.mattscomputertrends.com/.
[17] Dow Jones press release. Changes announced in components of two Dow Jones indexes, December 20, 2000.
[18] Jack W. Wilson, Charles P. Jones. Common stock prices and inflation: 1857–1985. Financial Analysts Journal, July–August 1987:67–71.
[19] Inflation to smile about. Los Angeles Times, January 11, 1988.
[20] Trenton Times, March 24, 1989.
[21] Lil Phillips. Phindex shows horrific inflation. Cape Cod Times, July 10, 1984.
[22] John A. Johnson. Sharing some ideas. Cape Cod Times, July 12, 1984.
[23] Ann Landers. Cape Cod Times, June 9, 1996.

Chapter 2 Displaying Data

[1] U.S. Census Bureau.
[2] Thomas Gilovich, Robert Vallone, Amos Tversky. The hot hand in basketball: On the misperception of random sequences. Cognitive Psychology, 1985;17:295–314.
[3] Reid Dorsey-Palmateer, Gary Smith. Bowlers' hot hands. American Statistician, 2004;58:38–45.
[4] James Gleick. Hole in ozone over South Pole worries scientists. New York Times, July 29, 1986:C1.

[5] Bureau of Labor Statistics; Richard K. Vedder, Lowell E. Gallaway. Out of work: Unemployment and government in twentieth-century America. Teaneck, NJ: Holmes & Meier, 1993.
[6] Alan Greenspan. The challenge of central banking in a democratic society, speech at the Annual Dinner and Francis Boyer Lecture of the American Enterprise Institute for Public Policy Research, Washington, DC, December 5, 1996.
[7] Tom Petruno. Getting a clue on when to buy and when to bail out. Los Angeles Times, May 16, 1990.
[8] Thomas L. Friedman. That numberless presidential chart. New York Times, August 2, 1981.
[9] Thomas L. Friedman. That numberless presidential chart. New York Times, August 2, 1981.
[10] Washington Post, 1976, from Howard Wainer. Visual revelations: Graphical tales of fate and deception from Napoleon Bonaparte to Ross Perot. New York: Copernicus, 1997, p. 49.
[11] National Science Foundation. Science indicators, 1974. Washington, DC: General Accounting Office, 1976, p. 15.
[12] Abraham Lincoln. Second state of the union address. December 1, 1862.
[13] Darrell Huff. How to lie with statistics. New York: Norton, 1954.
[14] Edward R. Tufte. The visual display of quantitative information. Cheshire, CT: Graphics Press, 1983, pp. 107–122.
[15] Adapted from Arthur H. Miller, Edie N. Goldenberg, Lutz Erbring. Type-set politics: impact of newspapers on public confidence. American Political Science Review, 1979;73:67–84.
[16] Adapted from Edward R. Tufte. The visual display of quantitative information. Cheshire, CT: Graphics Press, 1983, p. 121.
[17] Lynn Rapaport. The cultural and material reconstruction of the Jewish communities in the Federal Republic of Germany. Jewish Social Studies, Spring 1987:137–154.
[18] Robert Cunningham Fadeley. Oregon malignancy pattern physiographically related to Hanford, Washington, radioisotope storage. Journal of Environmental Health, 1965;28:883–897.
[19] J. M. Barnola, D. Raynaud, C. Lorius, Y. S. Korotkevich. Historical CO_2 record from the Vostok ice core. In: T. A. Boden, D. P. Kaiser, R. J. Sepanski, F. W. Stoss, editors, Trends '93: A compendium of global change. Oak Ridge, TN: Carbon Dioxide Information Analysis Center, Oak Ridge National Laboratory, 1993, pp. 7–10.
[20] R. J. Hoyle. Decline of language as a medium of communication. In: George H. Scherr, editor, The best of the journal of irreproducible results. New York: Workman, 1983, pp. 134–135.
[21] Brian J. Whipp, Susan A. Ward. Will women soon outrun men? Nature, January 2, 1992:25.
[22] Belinda Lees, Theya Molleson, Timothy R. Arnett, John C. Stevenson, Differences in proximal femur bone density over two centuries. The Lancet, March 13, 1993:673–675; I am grateful to the authors for sharing their raw data with me.
[23] The *New York Times* made a similar error in a graph that accompanied a story about how discount fares were reducing travel agent commissions: Air Travel boom makes agents fume. New York Times, August 8, 1978.
[24] The *Washington Post* distorted the data in a similar way when they published a figure with shrinking dollar bills showing the faces of five U.S. Presidents: Washington Post, October 25, 1978.
[25] These data were cited by David Frum. Welcome, nouveaux riches. New York Times, August 14, 1995: A-15; he argued that, "Nothing like this immense crowd of wealthy people has been seen in the history of the planet."

Chapter 3 Descriptive Statistics

[1] William A. Spurr, Charles P. Bonini. Statistical analysis for business decisions, revised edition. Homewood, IL: Irwin, 1973, p. 219.
[2] Joint Economic Committee of the United States Congress. The concentration of wealth in the United States. July 1986, pp. 7–43.
[3] F. Y. Edgeworth. The choice of means. London, Edinburgh, and Dublin Philosophical Magazine and Journal of Science, July–December 1887;24:269.
[4] Francis Galton. Natural inheritance. New York: Macmillan, 1889, p. 62.

[5] Michael J. Saks, Reid Hastie. Social psychology in court. New York: Van Nostrand Reinhold, 1978, p. 100.
[6] Newsweek, January 16, 1967, p. 6.
[7] New York Times, January 24, 1995.
[8] Wall Street Journal, January 24, 1995.
[9] W. Allen Wallis, Harry V. Roberts. Statistics: A new approach. Glencoe, IL: Free Press, 1956, p. 83.
[10] Susan Milton. Wellfleet the victim in statistical murder mystery. Cape Cod Times, December 12, 1994.
[11] S. M. Stigler. Do robust estimators work with real data? Annals of Statistics, 1977:1055–1078.
[12] Henry Cavendish. Experiments to determine the density of earth. In: A. Stanley MacKenzie, editor, Scientific Memoirs, vol. 9, The laws of gravitation. New York: American Book Company, 1900, pp. 59–105.
[13] Francesca Levy. America's Most expensive ZIP codes 2010: In these neighborhoods $4 million homes are the norm. Forbes.com, September 27, 2010.
[14] Ann Landers. *Boston Globe*, August 14, 1976.
[15] Gary Smith. Horseshoe pitchers' hot hands. Psychonomic Bulletin and Review, 2003;10:753–758.
[16] Edward Yardeni. Money and Business Alert, Prudential-Bache Securities, November 11, 1988.
[17] Running shoes. Consumer Reports, May 1995:313–317.
[18] Risteard Mulcahy, J. W. McGilvray, Noel Hickey. Cigarette smoking related to geographic variations in coronary heart disease mortality and to expectations of life in the two sexes. American Journal of Public Health, 1970:1516.
[19] New York Times, November 28, 1994.
[20] Wall Street Journal, November 28, 1994.

Chapter 4 Probability

[1] Pierre-Simon Laplace. Théorie analytique des probabilités. 1820, introduction.
[2] Geoffrey D. Bryant, Geoffrey R. Norman. Expressions of probability: Words and numbers. Letter to the *New England Journal of Medicine*, February 14, 1980, p. 411. Also see George A. Diamond, James S. Forrester. Metadiagnosis. American Journal of Medicine, July 1983:129–137.
[3] David Premack, Ann James Premack. The mind of an ape. New York: Norton, 1983.
[4] The great expectations of a royal reporter. Daily Mail, April 20, 1994:50.
[5] John Maynard Keynes. A treatise on probability. London: Macmillan, 1921, pp. 362–363.
[6] Leo Gould. You bet your life. Hollywood, CA: Marcel Rodd, 1946.
[7] Len Deighton. Bomber. New York: Harper, 1970.
[8] David Eddy. Probabilistic reasoning in clinical medicine: Problems and opportunities. In: Daniel Kahneman, Paul Slovak, and Amos Tversky. Judgment under uncertainty: Heuristics and biases. Cambridge, England: Cambridge University Press, 1982, pp. 249–267.
[9] David Eddy. Probabilistic reasoning in clinical medicine: Problems and opportunities. In: Daniel Kahneman, Paul Slovak, and Amos Tversky. Judgment under uncertainty: Heuristics and biases. Cambridge, England: Cambridge University Press, 1982:249–267.
[10] Clement McQuaid, editor. Gambler's digest. Northfield, IL: Digest Books, 1971, pp. 24–25.
[11] Quoted in Clement McQuaid, editor. Gambler's digest. Northfield, IL: Digest Books, 1971, p. 287.
[12] Supreme Court of California. *People v. Collins*. See also William B. Fairly, Frederick Mosteller. A conversation about Collins. In: Fairly and Mosteller, Statistics and public policy. Reading, MA: Addison-Wesley, 1977, pp. 355–379.
[13] George D. Leopold. Mandatory unindicated urine drug screening: Still chemical McCarthyism. Journal of the American Medical Association, 1986;256(21):3003–3005.
[14] W. J. Youdon (1956), from Stephen M. Stigler. Statistics on the table. Cambridge, MA: Harvard University Press, 1999, p. 415.
[15] James Roy Newman. The world of mathematics. New York: Simon and Schuster, 1956.
[16] Darrell Royal, quoted in Royal is chosen coach of year. New York Times, January 9, 1964.
[17] Marilyn vos Savant. Ask Marilyn. Parade, September 25, 1994.
[18] Darrell Huff. How to take a chance. New York: Norton, 1959, p. 110.

[19] Richard Scammon. Odds on virtually everything. New York: Putnam, 1980, p. 135.
[20] John Allen Paulos. Orders of magnitude. Newsweek, November 24, 1986.
[21] David Lykken. Polygraph interrogation. Nature, February 23, 1984:681–684. Another careful study of six polygraph experts found that 18 of 50 innocent people were classified as liars, while 12 of 50 confessed thieves were judged truthful: Benjamin Kleinmuntz, Julian J. Szucko. A field study of the fallibility of polygraph lie detection. Nature, March 29, 1984:449–450.
[22] James E. Haddow, Glenn E. Palomaki, George J. Knight, George C. Cunningham, Linda S. Lustig, Patricia A. Boyd. Reducing the need for amniocentesis in women 35 years of age or older with serum markers for screening. New England Journal of Medicine, April 21, 1994;330:1114–1118.
[23] Marilyn vos Savant. Ask Marilyn. Parade, July 1, 1990.
[24] Marilyn vos Savant. Ask Marilyn. Parade, July 12, 1992.
[25] Harold Jacobs. Mathematics: A human endeavor. San Francisco: W. H. Freeman, 1982, p. 570.
[26] Charlotte, West Virginia, Gazette, July 29, 1987.
[27] Tim Molloy. Gatemen, Athletics Cape League picks. Cape Cod Times, July 19, 1991.
[28] Marilyn vos Savant. Ask Marilyn. Parade, January 3, 1999.
[29] P. A. Mackowiak, S. S. Wasserman, M. M. Levine. A critical appraisal of 98.6° F, the upper limit of the normal body temperature, and other legacies of Carl Reinhold August Wunderlich. Journal of the American Medical Association, September 23–30, 1992:1578–1580.
[30] Galileo Galilei. Sopra le Scoperte Dei Dadi. Opere Firenze, Barbera, 8, 1898:591–594.
[31] Marilyn vos Savant. Ask Marilyn. Parade, October 29, 2000.
[32] Hans Zeisel. Dr. Spock and the case of the vanishing women jurors. University of Chicago Law Review, Autumn 1969;37:12.

Chapter 5 Sampling

[1] Gary Smith, Margaret Hwang Smith, Like mother, like daughter? A socioeconomic comparison of immigrant mothers and their daughters. Unpublished manuscript.
[2] R. Clay Sprowls. The admissibility of sample data into a court of law: A case history. UCLA Law Review, 1957;1:222–232.
[3] R. Clay Sprowls. The admissibility of sample data into a court of law: A case history. UCLA Law Review, 1957;1:222–232.
[4] Gary Smith, Michael Levere, Robert Kurtzman. Poker player behavior after big wins and big losses. Management Science, 2009;55:1547–1555.
[5] Arnold Barnett. How numbers can trick you. Technology Review, October 1994, p. 40.
[6] J. R. Berrueta-Clement, L. Schweinhart, W. W. Barnett, W. Epstien, D. Weikart. Changed lives: The effects of the Perry Preschool Program on youths through age 19. Monographs of the High/Scope Educational Research Foundation, Number 8. Ypsilanti, MI: High/Scope Educational Research Foundation, 1984.
[7] Associated Press. Anger doubles risk of attack for heart disease victims. New York Times, March 19, 1994.
[8] John Snow. On the mode of communication of cholera, 2nd edition. London: John Churchill, 1855.
[9] Philip Cole. Coffee-drinking and cancer of the lower urinary tract. The Lancet, June 26, 1971;7713:1335–1337.
[10] Catherine M. Viscoli, Mark S. Lachs, Ralph I. Horowitz. Bladder cancer and coffee drinking: A summary of case-control research. The Lancet, June 5, 1993;8858:1432–1437.
[11] W. Allen Wallis, Harry V. Roberts. Statistics: A new approach. New York: Free Press, 1956, pp. 479–480.
[12] Shere Hite. The Hite report: A national study of female sexuality. New York: Macmillan, 1976.
[13] Philip Zimbardo. Discovering psychology with Philip Zimbardo (Episode 2: Understanding Research), video program.
[14] Gary Smith, Margaret Hwang Smith. Like mother, like daughter? A socioeconomic comparison of immigrant mothers and their daughters. Unpublished manuscript.
[15] Big changes made in hunger report. San Francisco Chronicle, September 6, 1994.
[16] National vital statistics reports. U.S. National Center for Health Statistics.

[17] *Cape Cod Times*, August 28, 1984. The study was published in E. Scott Geller, Nason W. Russ, Mark G. Altomari. Naturalistic observations of beer drinking among college students. Journal of Applied Behavior Analysis, 1986;19:391–396.
[18] David Leonhardt. Colleges are failing in graduation rates. New York Times, September 9, 2009.
[19] The Sun, March 4, 1996.
[20] Doctors follow the nose (ring) to learn more about youth risks. University of Rochester Medical Center Newsroom, June 18, 2003.
[21] David Cole, John Lamberth, The fallacy of racial profiling. New York Times, May 13, 2001.
[22] Natalie Engler, Boys who use computers may be more active: Study. Reuters Health, October 22, 2001. This report was based on M. Y. Ho and T. M. C. Lee. Computer usage and its relations with adolescent lifestyle in Hong Kong. Journal of Adolescent Health, 2001;29:258–266.
[23] Chicago Daily News, April 8, 1955.
[24] *Boykin v. Georgia Power* 706 F.2d 1384, 32 FEP Cases 25 (5th Cir. 1983).
[25] L. L. Bairds. The graduates. Princeton, NJ: Educational Testing Service, 1973.
[26] Ann Landers. Ask Ann Landers. November 3, 1975. See also Ann Landers, If you had it to do over again, would you have children? Good Housekeeping, June 1976;182:100–101.
[27] U.S. Department of Commerce. Statistical abstract of the United States. Washington, DC. U.S. Government Printing Office, 1981, Table 202, p. 123.
[28] Available at: http://stufs.wlu.edu/~hodgsona/bingodeaths.html.
[29] SAT coaching disparaged. New York Times, February 10, 1988, Section II, p. 8.
[30] Cynthia Crossen. Studies galore support products and positions, but are they reliable? Wall Street Journal, November 14, 1991.
[31] Cape Cod Times, July 17, 1983.

Chapter 6 Estimation

[1] Lord Justice Matthews, quoted in Michael J. Saks, Reid Hastie. Social psychology in court. New York: Van Nostrand Reinhold, p. 100.
[2] Student, The probable error of a mean. Biometrika 1908;6:1–25.
[3] This example is from W. Allen Wallis, Harry V. Roberts. Statistics: A new approach. New York: Free Press, 1956, p. 471.
[4] *Reserve Mining Co. v. EPA* (1975), cited in David W. Barnes. Statistics as proof. Boston: Little, Brown, 1983, p. 244.
[5] Gary Smith, Margaret Hwang Smith. Like mother, like daughter? A socioeconomic comparison of immigrant mothers and their daughters. Unpublished manuscript.
[6] S. M. Stigler. Do robust estimators work with real data? Annals of Statistics, 1977, pp. 1055–1078.
[7] C. Wunderlich. Das Verhalten der Eiaenwarme in Krankenheiten. Leipzig, Germany: Otto Wigard, 1868.
[8] P. A. Mackowiak, S. S. Wasserman, and M. M. Levine. A critical appraisal of 98.6° F, the upper limit of the normal body temperature, and other legacies of Carl Reinhold August Wunderlich. Journal of the American Medical Association, September 23–30, 1992:1578–1580; I am grateful to Steven Wasserman for sharing the data with me.
[9] Wall Street Journal, July 6, 1987.
[10] Jerry E. Bishop. Statisticians occupy front lines in battle over passive smoking. Wall Street Journal, July 28, 1993.
[11] James T. McClave, P. George Benson, Statistics for business and economics, 2nd edition. San Francisco: Dellen, 1982, p. 279.
[12] Charles Seiter. Forecasting the future. MacWorld, September 1993:187.
[13] Margaret Harris, Caroline Yeeles, Joan Chasin, Yvonne Oakley. Symmetries and asymmetries in early lexical comprehension and production. Journal of Child Language, February 1995:1–18.
[14] Robert J. Samuelson, The strange case of the missing jobs. Los Angeles Times, October 27, 1983.
[15] The science of polling. Newsweek, September 28, 1992:38–39.

Chapter 7 Hypothesis Testing

[1] R. A. Fisher. The arrangement of field experiments. Journal of the Ministry of Agriculture of Great Britain, 1926;8:504.
[2] F. Arcelus, A. H. Meltzer. The effect of aggregate economic variables on congressional elections. American Political Science Review, 1975;69:1232–1239.
[3] Gary Smith, Margaret Hwang Smith. Like mother, like daughter? A socioeconomic comparison of immigrant mothers and their daughters. Unpublished manuscript.
[4] Jeff Anderson, Gary Smith. A great company can be a great investment. Financial Analysts Journal, 2006; 62:86–93.
[5] Stilian Morrison, Gary Smith. Monogrammic determinism? Psychosomatic Medicine, 2005;67:820–824.
[6] William Feller. Are life scientists overawed by statistics? Scientific Research, February 3, 1969:24–29.
[7] T. D. Sterling. Publication decisions and their possible effects on inferences drawn from tests of significance—or vice versa. Journal of the American Statistical Association, 1959;54:30–34.
[8] Sir Arthur Conan Doyle. A study in scarlet. London: Ward Lock & Co., 1887, Part 1, p. 27.
[9] Sue Avery. Market investors will be high on redskins today; and Morning briefing: Wall Street 'skinnish' on big game. Los Angeles Times, January 22, 1983.
[10] Jason Zweig, Super Bowl Indicator: The secret history. Wall Street Journal, January 28, 2011.
[11] Martin Gardner. Fads and fallacies in the name of science. New York: Dover, 1957, p. 303.
[12] Martin Gardner. Fads and fallacies in the name of science. New York: Dover, 1957, p. 305.
[13] Gail Howard. State lotteries: How to get in it ... and how to win it! 5th edition. Ben Buxton, Fort Lee, New Jersey, 1986.
[14] Sporting Edge, 1988.
[15] Gary Smith. Another look at baseball player initials and longevity. Perceptual and Motor Skills, 2011; 112:211–216.
[16] News briefs. Cape Cod Times, July 5, 1984.
[17] A. R. Feinstein, A. R. Horwitz, W. O. Spitzer, R. N. Batista. Coffee and pancreatic cancer. Journal of the America Medical Association, 1981;256:957–961.
[18] B. Macmahon, S. Yen, D. Trichopoulos, K. Warren, G. Nardi. Coffee and cancer of the pancreas. New England Journal of Medicine, 1981;304:630–633.
[19] D. P. Phillips, T. E. Ruth, L. M. Wagner. Psychology and survival. The Lancet, 1993;342:1142–1145.
[20] Gary Smith. The five elements and Chinese-American mortality. Health Psychology, 2006;25:124–129.
[21] Ernest L. Abel, Michael L. Kruger. Athletes, doctors, and lawyers with first names beginning with "D" die sooner. Death Studies, 2010;34:71–81.
[22] Gary Smith. Do people whose names begin with "D" really die young? Death Studies, 2011.
[23] David B. Allison, Stanley Heshka, Dennis Sepulveda, Steven B. Heymsfield. Counting calories—caveat emptor. Journal of the American Medical Association, 1993;270:1454–1456.
[24] Sanders Frank. Aural sign of coronary-artery disease. New England Journal of Medicine, August 1973;289: 327–328. However, see T. M. Davis, M. Balme, D. Jackson, G. Stuccio, and D. G. Bruce. The diagonal ear lobe crease (Frank's sign) is not associated with coronary artery disease or retinopathy in type 2 diabetes: The Fremantle Diabetes Study. Australian and New Zealand Journal of Medicine, October 2000; 30:573–577.
[25] Sheldon Blackman, Don Catalina. The moon and the emergency room. Perceptual and Motor Skills, 1973;37:624–626.
[26] Letter to the Editor, Sports Illustrated, January 30, 1984.
[27] Newsweek, February 4, 1974.
[28] Kidder, Peabody & Co. Portfolio Consulting Service. May 20, 1987.
[29] Texas Monthly, January 1982:83.
[30] Robert Sullivan. Scorecard. Sports Illustrated, February 24, 1986:7.
[31] Steven J. Milloy. The EPA's Houdini Act. Wall Street Journal, August 8, 1996.
[32] Francis Iven Nye. Family Relationships and Delinquent Behavior. New York: John Wiley & Sons, 1958, p. 29.

[33] Floyd Norris. Predicting victory in Super Bowl. New York Times, January 17, 1989.
[34] Jeffrey Laderman. Insider Trading. Business Week, April 29, 1985:78–92.
[35] Science at the EPA. Wall Street Journal, October 2, 1985.
[36] *Allen v. Prince George's County, MD* 538 F. Supp. 833 (1982), affirmed 737 F.2d 1299 (4th Cir. 1984).

Chapter 8 Simple Regression

[1] Johann Heinrich Lambert. Beyträge zum Gebrauche der Mathematik und deren Anwendung, Berlin, 1765, quoted in Laura Tilling. Early experimental graphs. British Journal for the History of Science, 1975;8:204–205.
[2] Arthur M. Okun. Potential GNP: Its measurement and significance. Proceedings of the Business and Economics Statistics Section of the American Statistical Association, 1962:98–104.
[3] Robert Cunningham Fadeley. Oregon malignancy pattern physiographically related to Hanford, Washington, radioisotope storage. Journal of Environmental Health, 1965;28:883–897.
[4] Peter Passell. Probability experts may decide vote in Philadelphia. New York Times, April 11, 1994:A-10.
[5] The data were provided by Shaun Johnson of Australia's National Climate Centre, Bureau of Meteorology.
[6] Philip N. Baker, Ian R. Johnson, Penny A. Gowland, Jonathan Hykin, Paul R. Harvey, Alan Freeman, Valerie Adams, Brian S. Worthington, Peter Mansfield. Fetal weight estimation by echo-planar magnetic resonance imaging. The Lancet, March 12, 1994;343:644–645.
[7] J. W. Kuzma, R. J. Sokel. Maternal drinking behavior and decreased intrauterine growth. Alcoholism: Clinical and Experimental Research, 1982;6:396–401.
[8] Janette B. Benson. Season of birth and onset of locomotion: Theoretical and methodological implications. Infant Behavior and Development, 1993;16:69–81.
[9] Sven Cnattingius, Michele R. Forman, Heinz W. Berendes, Leena Isotalo. Delayed childbearing and risk of adverse perinatal outcome. Journal of the American Medical Association, August 19, 1992;268:886–890.
[10] James F. Jekel, David L. Katz, Joann G. Elmore. Epidemiology, biostatistics, and preventive medicine, 2nd edition. Philadelphia: W. B. Saunders, 2001, p. 157.
[11] Frederick E. Croxton, Dudley J. Cowdon. Applied general statistics, 2nd edition. Englewood Cliffs, NJ: Prentice-Hall, 1955, pp. 451–454.
[12] Belinda Lees, Theya Molleson, Timothy R. Arnett, John C. Stevenson. Differences in proximal femur bone density over two centuries. The Lancet, March 13, 1993;8846:673–675. I am grateful to the authors for sharing their raw data with me.
[13] Risteard Mulcahy, J. W. McGilvray, Noel Hickey. Cigarette smoking related to geographic variations in coronary heart disease mortality and to expectations of life in the two sexes. American Journal of Public Health, 1970;60:1515–1521.
[14] Rick Hutchinson, Yellowstone National Park's research geologist, kindly provided these data.
[15] James Shields. Monozygotic twins. London: Oxford University Press, 1962. Three similar, separate studies by Cyril Burt all reported the same value of R^2 (0.594). A logical explanation is that the data were flawed; see Nicholas Wade. IQ and heredity: Suspicion of fraud beclouds classic experiment. Science, 1976;194: 916–919.

Chapter 9 The Art of Regression Analysis

[1] Lawrence S. Ritter, William F. Silber. Principles of money, banking, and financial markets. New York: Basic Books, 1986, p. 533.
[2] John Llewellyn, Roger Witcomb, letters to The Times, London, April 4–6, 1977; and David Hendry, quoted in The New Statesman, November 23, 1979:793–795.
[3] G. Rose, H. Blackburn, A. Keys, et al. Colon cancer and blood-cholesterol. The Lancet, 1974;7850:181–183.
[4] G. Rose, M. J. Shipley. Plasma lipids and mortality: a source of error. The Lancet, 1980;8167:523–526.
[5] Robert L. Thorndike. The concepts of over- and under-achievement. New York: Teacher's College, Columbia University, 1963, p. 14.

[6] Francis Galton. Regression towards mediocrity in hereditary stature. Journal of the Anthropological Institute, 1886;15:246–263.
[7] Teddy Schall, Gary Smith. Baseball players regress toward the mean. American Statistician, November 2000;54:231–235.
[8] Horace Secrist. The triumph of mediocrity in business. Evanston, IL: Northwestern University, 1933.
[9] William F. Sharpe. Investments, 3rd edition. Englewood Cliffs, NJ: Prentice-Hall, 1985, p. 430.
[10] Anita Aurora, Lauren Capp, Gary Smith. The real dogs of the Dow. Journal of Wealth Management, 2008;10:64–72.
[11] Richard W. Pollay, S. Siddarth, Michael Siegel, Anne Haddix, Robert K. Merritt, Gary A. Giovino, Michael P. Eriksen. The last straw? Cigarette advertising and realized market shares among youths and adults, 1979–1993. Journal of Marketing, April 1966;60:1–16.
[12] Gary Smith, Margaret Hwang Smith. Like mother, like daughter? A socioeconomic comparison of immigrant mothers and their daughters. Unpublished manuscript.
[13] David Upshaw of Drexel Burnham Lambert, quoted in John Andrew. Some of Wall Street's favorite stock theories failed to foresee market's slight rise in 1984. Wall Street Journal, January 2, 1985.
[14] Fred Schwed, Jr. Where are the customers' yachts? New York: Simon and Schuster, 1940, p. 47.
[15] P. C. Rosenblatt, M. R. Cunningham. Television watching and family tensions. Journal of Marriage and the Family, 1976;38:105–111.
[16] Hoyt says Cy Young Award a jinx. Cape Cod Times, August 4, 1984.
[17] Francis Galton. Regression towards mediocrity in hereditary stature. Journal of the Anthropological Institute, 1886;1:246–263.
[18] Marcus Lee, Gary Smith. Regression to the mean and football wagers. Journal of Behavioral Decision Making, 2002;15:329–342.
[19] Howard Wainer. Is the Akebono School failing its best students? A Hawaii adventure in regression. Educational Measurement: Issues and Practice, 1999;18:26–31, 35.
[20] S. Karelitz, V. R. Fisichelli, J. Costa, R. Kavelitz, L. Rosenfeld. Relation of crying in early infancy to speech and intellectual development at age three years. Child Development, 1964;35:769–777.
[21] Max R. Mickey, Olive Jean Dunn, Virginia Clark. Note on the use of stepwise regression in detecting outliers. Computers and Biomedical Research, July 1967;1:105–111.
[22] Frederick E. Croxton, Dudley J. Cowdon. Applied general statistics, 2nd edition. Englewood Cliffs, NJ: Prentice-Hall, 1955, pp. 451–454.
[23] J. W. Buehler, J. C. Kleinman, C. J. Hogue, L. T. Strauss, J. C. Smith. Birth weight-specific infant mortality, United States, 1960 and 1980. Public Health Reports, March–April 1987;102:151–161.
[24] Jeff Anderson, Gary Smith. A great company can be a great investment. Financial Analysts Journal, 2006;62:86–93.
[25] Anita Aurora, Lauren Capp, Gary Smith. The real dogs of the Dow. Journal of Wealth Management, 2008;10:64–72.
[26] Amos Tversky, Daniel Kahneman. On the psychology of prediction. Psychological Review, 1973;80:237–251.
[27] John A. Johnson. Sharing some ideas. Cape Cod Times, July 12, 1984.

Chapter 10 Multiple Regression

[1] Eugene F. Fama, Kenneth R. French. Common risk factors in the returns on bonds and stocks, Journal of Financial Economics, 1993;33:3–53.
[2] Alex Head, Gary Smith, Julia Wilson. Would a stock by any other ticker smell as sweet? Quarterly Review of Economics and Finance, 2009;49:551–561.
[3] Lester B. Lave, Eugene P. Seskin. Does air pollution shorten lives? In: John W. Pratt, editor, Statistical and mathematical aspects of pollution problems. New York: Marcel Dekker, 1974, pp. 223–247.
[4] James S. Coleman, Ernest Campbell, Carol Hobson, James McPartland, Alexander Mood, Frederick Weinfield, Robert York. Equality of educational opportunity. Washington, DC: U.S. Department of Health, Education, and Welfare, Office of Education, 1966.

[5] Margaret Hwang Smith, Gary Smith. Bubble, bubble, where's the housing bubble? Brookings Papers on Economic Activity, 2006;1:1–50.
[6] Margaret Hwang Smith, Gary Smith. Bubble, bubble, where's the housing bubble? Brookings Papers on Economic Activity, 2006;1:1–50.
[7] Franklin M. Fisher, John J. McGowan, Joen E. Greenwood. Folded, spindled and mutilated: economic analysis and *U.S. v. IBM*. Cambridge, MA: MIT Press, 1983.

Chapter 11 Modeling (Optional)

[1] Teddy Schall, Gary Smith. Career trajectories in baseball. Chance, 2000;13:35–38.
[2] Teddy Schall, Gary Smith. Career trajectories in baseball. Chance, 2000;13:35–38.

Index

Page numbers in *italics* indicate figures and tables.

I

J

K

L

M

N

O

P